*EXERCISES IN GAME THEORY*

*Volume 1: Basic Concepts*

Copyright © 2024 Giacomo Bonanno, all rights reserved

Kindle Direct Publishing

ISBN-13: 979-8326296146
ISBN-10: 8326296146

Giacomo Bonanno is Distinguished Professor of Economics at the University of California, Davis
`http://faculty.econ.ucdavis.edu/faculty/bonanno/`

# Preface

This book contains 180 exercises in Game Theory and is structured in such a way as to match the chapters and sections of *Game Theory, Volume 1: Basic Concepts* published in 2018, which can be found at:
`https://www.amazon.com/Game-Theory-1-Basic-Concepts/dp/1983604631`

I expect that there will be some typos and (hopefully minor) mistakes. If you come across any typos or mistakes, I would be grateful if you could inform me: I can be reached at gfbonanno@ucdavis.edu

I would like to thank Mathias Legrand for making the LaTeX template used for this book available for free (the template was downloaded from `http://www.latextemplates.com/template/the-legrand-orange-book`)

v.1 06-24

# Contents

# 1. Introduction

## 1.1 Introduction

This book is a companion to my 2018 textbook *Game Theory, Volume 1: Basic Concepts*, which can be found at:

https://www.amazon.com/Game-Theory-1-Basic-Concepts/dp/1983604631

(see also my web-site at

https://faculty.econ.ucdavis.edu/faculty/bonanno/Books.html).

It contains an additional set of 180 fully solved exercises.

The structure of this book mirrors, chapter by chapter and, within each chapter, section by section, that of *Game Theory, Volume 1: Basic Concepts*.

**Chapter 2** contains 48 exercises on ordinal games in strategic form and covers the notions of game-frame and game, strict and weak dominance, some auctions (in particular, the second-price auction), the pivotal mechanism, the procedures of iterated deletion of strictly / weakly dominated strategies, Nash equilibrium and games with infinite strategy sets.

**Chapter 3** contains 31 exercises on dynamic games with perfect information and covers the notions of backward induction and strategy as well as a special class of two-player, perfect-information games.

**Chapter 4** contains 23 exercises on general dynamic games. It covers the notions of imperfect information, strategy, subgame, subgame-perfect equilibrium and games with Chance moves.

**Chapter 5** contains 27 exercises on Expected Utility Theory.

**Chapter 6** contains 27 exercises on strategic-form games with cardinal (i.e. von Neumann-Morgenstern) payoffs and covers the notions of mixed strategy, mixed-strategy Nash equilibrium and the cardinal version of the procedure of iterated deletion of strictly dominated strategies.

**Chapter 7** contains 24 exercises on general extensive-form games with cardinal payoffs.

In Chapters 3, 4, 6 and 7 we have added a section at the end of the chapter containing more difficult and challenging exercises.

None of the concepts that are covered in this book are defined or explained again here. At the beginning of each section of each chapter, the reader is referred to the relevant definitions in the corresponding chapters of the 2018 textbook *Game Theory, Volume 1: Basic Concepts*.

# I

# Games with Ordinal Payoffs

# 2. Ordinal Games in Strategic Form

## 2.1 Game frames and games

The exercises in this section deal with the notions of *game-frame in strategic form* (Definition 2.1.1, Volume 1, Chapter 2, Section 2.1) and *game in strategic form with ordinal payoffs* (Definition 2.1.3, Volume 1, Chapter 2, Section 2.1).

Exercise 2.1 Consider the following situation. There are two players; each player gives a written instruction to the referee; the instruction can be either "give me \$100" or "give \$500 to the other player". The referee collects the instructions and carries them out.

**(a)** Represent this situation as a game-frame in strategic form where each outcome describes how much money Player 1 gets and how much money Player 2 gets.

From now on, use utility functions for the players that take on values in the set $\{0,1,2,3\}$.

**(b)** Write the strategic-form game based on the game-frame of Part (a), for the case where each player is selfish and greedy (that is, only cares about how much money she herself gets and prefers more money to less).

**(c)** Write the strategic-form game based on the game-frame of Part (a), for the case where each player is completely altruistic (that is, only cares about how much money the other player gets and considers more money for the other player to be better than less money for the other player).

**(d)** Write the strategic-form game based on the game-frame of Part (a), for the case where Player 1 is selfish and greedy and Player 2 is completely altruistic.

■

**Solution to Exercise 2.1.**

**(a)**

| | | Player 2: To me | | Player 2: To other | |
|---|---|---|---|---|---|
| Player 1 | To me | $100 | $100 | $600 | $0 |
| | To other | $0 | $600 | $500 | $500 |

**(b)**

| | | Player 2: To me | | Player 2: To other | |
|---|---|---|---|---|---|
| Player 1 | To me | 1 | 1 | 3 | 0 |
| | To other | 0 | 3 | 2 | 2 |

**(c)**

| | | Player 2: To me | | Player 2: To other | |
|---|---|---|---|---|---|
| Player 1 | To me | 1 | 1 | 0 | 3 |
| | To other | 3 | 0 | 2 | 2 |

**(d)**

| | | Player 2: To me | | Player 2: To other | |
|---|---|---|---|---|---|
| Player 1 | To me | 1 | 1 | 3 | 3 |
| | To other | 0 | 0 | 2 | 2 |

Exercise 2.2 Consider the following situation. There are three players. A referee gives \$5 to each player. Each player then chooses between keeping the \$5 or returning the \$5 to the referee. If two or more players return their \$5 to the referee, then the referee keeps the money that is returned to him and gives an additional \$10 to every player. If only one player returns her \$5 to the referee, then the referee gives the \$5 back to that player. If nobody returns her \$5 to the referee then the referee takes no further action. Thus each player may end up with \$5 or \$10 or \$15.

**(a)** Represent this situation as a game-frame in strategic form where each outcome describes the amounts of money that the three players end up with.

**(b)** For the case where each player is selfish (that is, only cares about how much money she herself gets) and greedy (prefers more money to less), represent the corresponding game using as utilities the numbers 0, 1 and 2.

**Solution to Exercise 2.2.**

**(a)** Let 'K' stand for 'keep' and 'R' for 'return'.

| Player 1 \ Player 2 | $K$ | $R$ |
|---|---|---|
| $K$ | \$5 \$5 \$5 | \$5 \$5 \$5 |
| $R$ | \$5 \$5 \$5 | \$10 \$10 \$15 |

Player 3: $K$

| Player 1 \ Player 2 | $K$ | $R$ |
|---|---|---|
| $K$ | \$5 \$5 \$5 | \$15 \$10 \$10 |
| $R$ | \$10 \$15 \$10 | \$10 \$10 \$10 |

Player 3: $R$

**(b)**

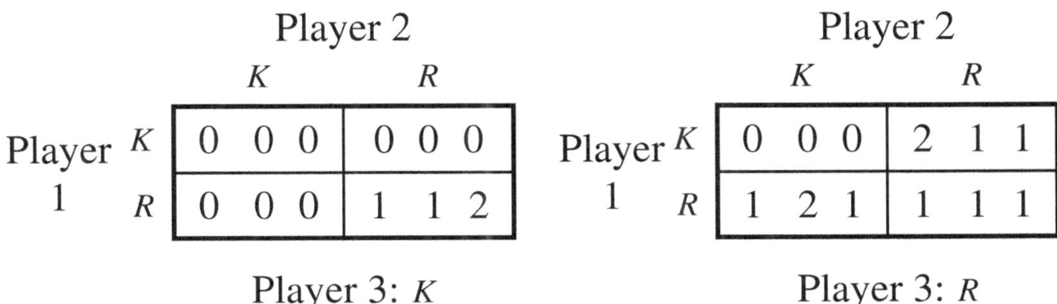

| Player 1 \ Player 2 | $K$ | $R$ |
|---|---|---|
| $K$ | 0 0 0 | 0 0 0 |
| $R$ | 0 0 0 | 1 1 2 |

Player 3: $K$

| Player 1 \ Player 2 | $K$ | $R$ |
|---|---|---|
| $K$ | 0 0 0 | 2 1 1 |
| $R$ | 1 2 1 | 1 1 1 |

Player 3: $R$

Exercise 2.3 Three people must drive from A to B at the same time. Each of them must choose a route. Two routes are available, one via X and one via Y, as shown in Figure 2.1. For each stretch of road, Figure 2.1 shows the amount of time it takes *per car* to travel it, as a function of the number of cars on the road. For example, if there is only one car traveling from A to B via Y then the total time taken is $20+4=24$ minutes (20 on the A-Y stretch and 4 on the Y-B stretch); if all three cars take the A to B route through X then each car takes a total amount of time equal to $10+21=31$ minutes.

**(a)** Represent this situation as a strategic-form game-frame by defining an outcome as the amount of time spent on the road by each driver.

**(b)** Assume that each driver only cares about how much time she spends on the road and her objective is to spend the least amount of time. Construct the strategic-form game based on the game-frame of Part (a) by using utility functions that take on values in the set $\{0,1,2,3,4,5\}$.

■

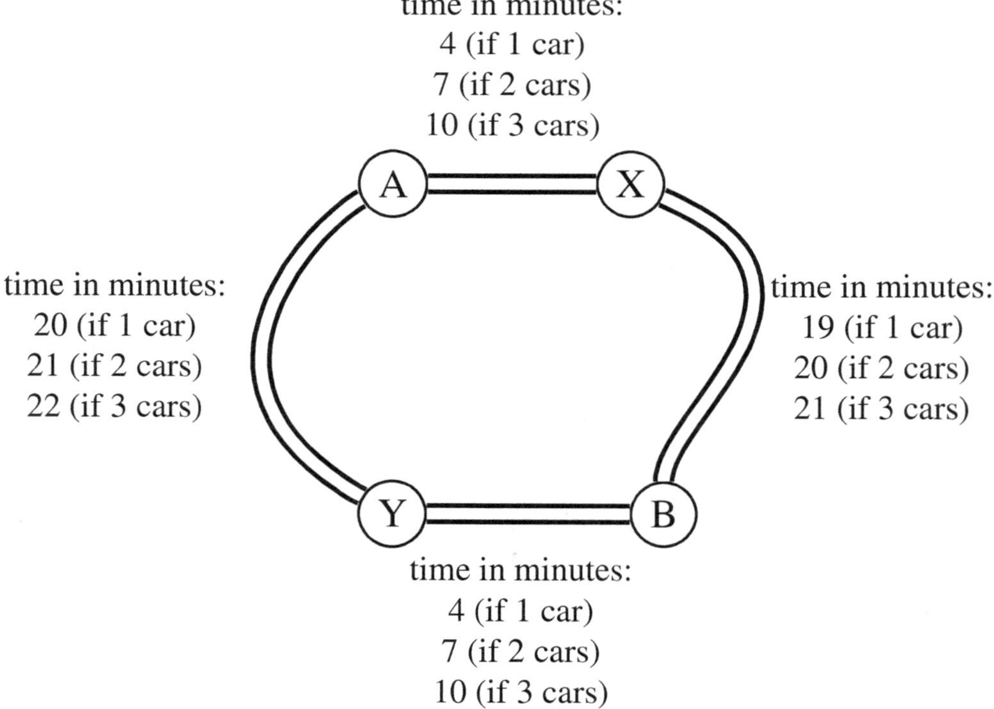


Figure 2.1: The situation described in Exercise 2.3.

**Solution to Exercise 2.3.**

**(a)** The game-frame is as follows:

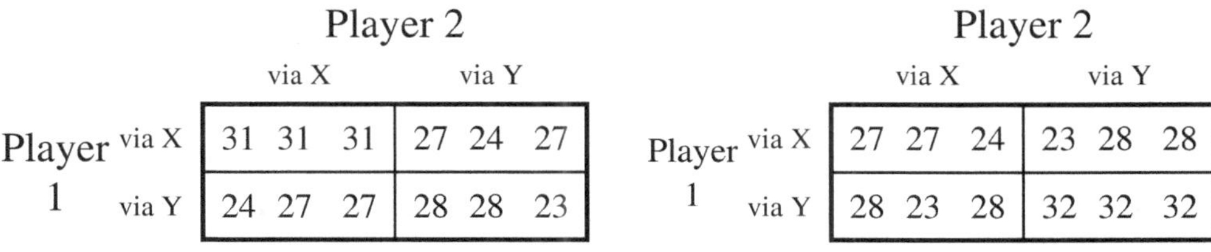

Player 3: via X

| Player 1 \ Player 2 | via X | via Y |
|---|---|---|
| via X | 31 31 31 | 27 24 27 |
| via Y | 24 27 27 | 28 28 23 |

Player 3: via Y

| Player 1 \ Player 2 | via X | via Y |
|---|---|---|
| via X | 27 27 24 | 23 28 28 |
| via Y | 28 23 28 | 32 32 32 |

**(b)** For each player we take the following utility function:

| outcome: | 23 | 24 | 27 | 28 | 31 | 32 |
|---|---|---|---|---|---|---|
| utility: | 5 | 4 | 3 | 2 | 1 | 0 |

Thus, the game is as follows:

Player 3: via X

| Player 1 \ Player 2 | via X | via Y |
|---|---|---|
| via X | 1 1 1 | 3 4 3 |
| via Y | 4 3 3 | 2 2 5 |

Player 3: via Y

| Player 1 \ Player 2 | via X | via Y |
|---|---|---|
| via X | 3 3 4 | 5 2 2 |
| via Y | 2 5 2 | 0 0 0 |

## 2.2 Strict and weak dominance

The exercises in this section deal with the notions of *strict dominance* and *weak dominance* (Definition 2.2.1, Volume 1, Chapter 2, Section 2.2) and with the notion of *dominant strategy* (Definitions 2.2.2 and 2.2.3, Volume 1, Chapter 2, Section 2.2).

Exercise 2.4 We will consider first a two-player version of a game and then a three-player version.

**(A)** There were only two students in a Game Theory class. They thought their professor was a nice guy but when they got their final exam they found on the first page a box with the following instructions:

> "You can check this box if you like. If you check your box and the other student does not check hers, then I will give you an extra 10 points. Conversely, if you do not check your box and the other student checks hers, then I will give her an extra 10 points. If both of you check your respective boxes then I will take away 10 points from each of you. If neither of you puts a check in your respective box then I neither add nor subtract any points."

We are only focusing here on the grade assigned in the final exam (not the class grade). The professor grades "on the curve" as follows: anybody with a score less than or equal to the mean gets a B and anybody with a score greater than the mean gets an A. The mean is calculated on the adjusted score, that is, the score after the extra 10 points are added or subtracted (depending on whether or not the students check their boxes, as explained above).
The two students are Ann and Brynn. Denote by $a$ the score that Ann got in the midterm exam and $b$ the score that Brynn got in the midterm exam. Assume that both Ann and Brynn came to the final exam expecting that they each would get the same score as in the midterm (that is, before they found out about the checking-the-box business). Furthermore, assume that each student only cares about what grade she gets (e.g. getting an A when the other student gets an A is as good as getting an A when the other gets a B).

**(a)** For each of the following cases represent the above situation as a game: $(i)\ a = b$, $(ii)\ a - b > 10$, $(iii)\ b - a > 10$, $(iv)\ 0 < a - b < 10$, $(v)\ 0 < b - a < 10$.

**(b)** For each of the five cases above, determine whether each student has a dominant strategy. Specify whether it is a weakly or strictly dominant strategy.

**(B)** The following year the same class has three students. The professor plays the same trick in the final: he puts a box with the following instructions.

> "You can check this box if you like. If you check your box and the other students do not, then I will give you an extra 10 points. If two or more of you put a check then I will subtract 10 points from the score of all those and only those who put a check. If all of you leave your boxes unchecked, then then I will neither add nor subtract any points."

The grading is as before: anybody with a score greater than the mean gets an A and anybody with a score less than or equal to the mean gets a B. Assume that the three students are equally good and went to the final expecting to get the same score (again, this was before they found out about the box business). Each student only cares about her grade. The students are Alice, Barbara and Carla.

**(a)** Represent this situation as a game.

**(b)** Do the students have a (weakly/strictly) dominant strategy?

**Solution to Exercise 2.4.**

**(A)** The two-player game.

**(a)** The five cases are as follows. For Case $(i)$ we first show the outcomes and then the payoffs, taking a utility function that assigns value 0 to a grade of B and value 1 to a grade of A. For the other cases we just write the payoffs.

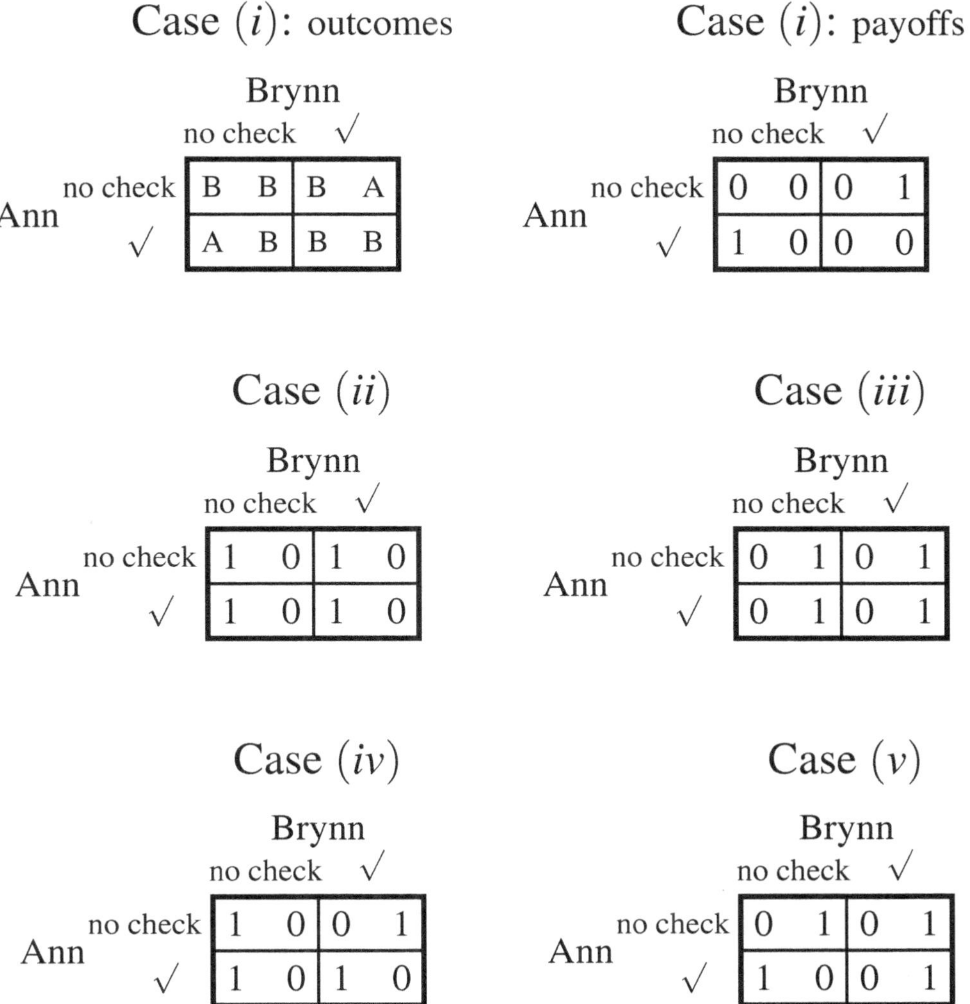

**Case $(i)$: outcomes**

| Ann \ Brynn | no check | √ |
|---|---|---|
| no check | B B | B A |
| √ | A B | B B |

**Case $(i)$: payoffs**

| Ann \ Brynn | no check | √ |
|---|---|---|
| no check | 0 0 | 0 1 |
| √ | 1 0 | 0 0 |

**Case $(ii)$**

| Ann \ Brynn | no check | √ |
|---|---|---|
| no check | 1 0 | 1 0 |
| √ | 1 0 | 1 0 |

**Case $(iii)$**

| Ann \ Brynn | no check | √ |
|---|---|---|
| no check | 0 1 | 0 1 |
| √ | 0 1 | 0 1 |

**Case $(iv)$**

| Ann \ Brynn | no check | √ |
|---|---|---|
| no check | 1 0 | 0 1 |
| √ | 1 0 | 1 0 |

**Case $(v)$**

| Ann \ Brynn | no check | √ |
|---|---|---|
| no check | 0 1 | 0 1 |
| √ | 1 0 | 0 1 |

**(b)** In cases $(i)$, $(iv)$ and $(v)$ checking the box is a weakly dominant strategy for each student. In cases $(ii)$ and $(iii)$ the two strategies are equivalent and the outcome is the same no matter what the students do.

**(B)** The three-player game.

**(a)** The game is as follows. As before, for each player, we use a utility function that assigns value 0 to a grade of B and value 1 to a grade of A.

Carla: no check

| Alice \ Barbara | no check | √ |
|---|---|---|
| no check | 0 0 0 | 0 1 0 |
| √ | 1 0 0 | 0 0 1 |

Carla: √

| Alice \ Barbara | no check | √ |
|---|---|---|
| no check | 0 0 1 | 1 0 0 |
| √ | 0 1 0 | 0 0 0 |

**(b)** Nobody has a dominant strategy (no check is best if the other two choose √, while √ is best if the other two choose no check).

Exercise 2.5

**(a)** For the game of Part (b) of Exercise 2.1 determine, for each player, whether one strategy dominates the other and state whether it is strict or weak dominance.

**(b)** For the game of Part (c) of Exercise 2.1 determine, for each player, whether one strategy dominates the other and state whether it is strict or weak dominance.

**(c)** For the game of Part (d) of Exercise 2.1 determine, for each player, whether one strategy dominates the other and state whether it is strict or weak dominance.

**(d)** Do the three games of Parts (b), (c) and (d) of Exercise 2.1 have a dominant-strategy profile?

■

**Solution to Exercise 2.5.**

**(a)** In the the game of Part (b) of Exercise 2.1, for each player "To me" strictly dominates "To other".

**(b)** In the the game of Part (c) of Exercise 2.1, for each player "To other" strictly dominates "To me".

**(c)** In the the game of Part (d) of Exercise 2.1, for Player 1 "To me" strictly dominates "To other", while for Player 2 "To other" strictly dominates "To me".

**(d)** For the game of Part (b), (To me, To me) is a strict dominant-strategy profile. For the game of Part (c), (To other, To other) is a strict dominant-strategy profile. For the game of Part (d), (To me, To other) is a strict dominant-strategy profile.

Exercise 2.6 For the game of Part (b) of Exercise 2.2 determine, for each player, whether one strategy dominates the other and state whether it is strict or weak dominance.

■

**Solution to Exercise 2.6.**

For each player it is neither the case that $K$ dominates $R$ (since $R$ is better than $K$ if one of the other two players chooses $K$ and the other chooses $R$), nor the case that $R$ dominates $K$ (since $K$ is better than $R$ if the other two players choose $R$).

Exercise 2.7 There is only one vacant bench on a bus and two passengers. Each passenger has to decide whether to stand or sit down. Suppose first that each passenger has the following preferences: he prefers to be the only one sitting; the next best outcome is one where they both sit (uncomfortably, since it is a narrow bench); the third best outcome is one where they both stand and the worst outcome is one where he stands and the other passenger is seated.

**(a)** Use a utility function that takes on values in the set $\{0,1,2,3\}$ to represent this situation as a strategic-form game.

**(b)** Is there a dominant-strategy profile? If so, is the corresponding outcome Pareto efficient, in the sense that every other outcome is considered to be worse by at least one of the players?

There is a saying that "if we were all better people the world would be a better place". Let's see if this is true in this example. Suppose now that each passenger is altruistic, in the sense that he only cares about the comfort of the other passenger (it is more comfortable to sit than to stand and it is better to sit alone than have to share the narrow bench) and he is polite, in the sense that he views as worst outcome one where he sits and the other stands.

**(c)** Use a utility function that takes on values in the set $\{0,1,2,3\}$ to represent this new situation as a strategic-form game.

**(d)** Is there a dominant-strategy profile? If so, is the corresponding outcome Pareto efficient?

**Solution to Exercise 2.7.**

**(a)** The game is as follows:

| 1 \ 2 | *sit* | *stand* |
|---|---|---|
| *sit* | 2 2 | 3 0 |
| *stand* | 0 3 | 1 1 |

**(b)** For each player *sit* strictly dominates *stand*. Thus, $(sit, sit)$ is a strict dominant-strategy profile. The corresponding outcome is Pareto efficient.

**(c)** The game is as follows:

| 1 \ 2 | *sit* | *stand* |
|---|---|---|
| *sit* | 2 2 | 0 3 |
| *stand* | 3 0 | 1 1 |

**(d)** For each player *stand* strictly dominates *sit*. Thus, $(stand, stand)$ is a strict dominant-strategy profile. The corresponding outcome is *not* Pareto efficient: the outcome where they both sit is preferred by both players.

Exercise 2.8 Consider the game shown in Figure 2.2 (where - as usual - the payoffs as in the following order, from left to right: Player 1, Player 2, Player 3):

**(a)** For what values of $z$ does Player 3 have a strictly dominant strategy?

**(b)** For what values of $z$ does Player 3 have a weakly, but not strictly, dominant strategy?

**(c)** Are there values of $y$ for which Player 2 has a strictly dominant strategy?

**(d)** Are there values of $y$ for which Player 2 has a weakly, but not strictly, dominant strategy?

**(e)** Are there values of $x$ for which Player 1 has a strictly dominant strategy?

**(f)** Are there values of $x$ for which Player 1 has a weakly, but not strictly, dominant strategy?

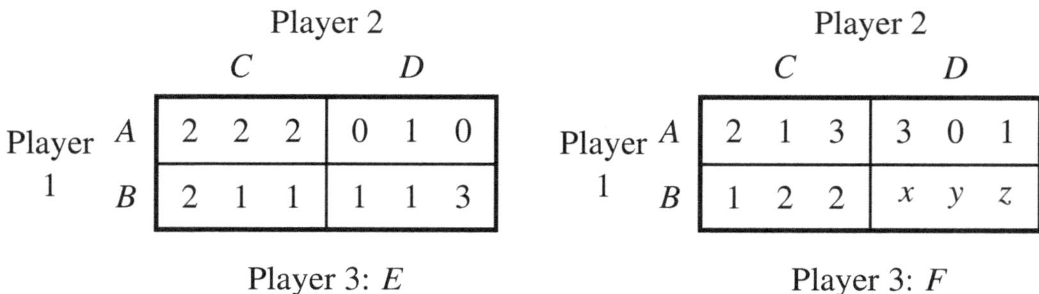

Player 3: $E$

| Player 1 \ Player 2 | $C$ | $D$ |
|---|---|---|
| $A$ | 2 2 2 | 0 1 0 |
| $B$ | 2 1 1 | 1 1 3 |

Player 3: $F$

| Player 1 \ Player 2 | $C$ | $D$ |
|---|---|---|
| $A$ | 2 1 3 | 3 0 1 |
| $B$ | 1 2 2 | $x$ $y$ $z$ |

Figure 2.2: Te game for Exercise 2.8.

**Solution to Exercise 2.8.**

**(a)** For $z > 3$.
**(b)** For $z = 3$.
**(c)** No, because if Player 1 plays $B$ and Player 3 plays $E$, then Player 2 gets the same payoff no matter whether she plays $C$ or $D$.
**(d)** Yes: for $y = 2$ (in which case $C$ weakly dominates $D$).
**(e)** No, because $A$ is better than $B$ against $(C,F)$ and $B$ is better than $A$ against $(D,E)$.
**(f)** No, for the same reason.

Exercise 2.9 You are facing a difficult decision. You and your friends Ben and Chris have been invited to a party, but during that time one of your favorite movies will be shown on TV. You prefer going to the party if you would have a good time there, but you prefer watching the movie rather than having a bad time at the party. More precisely, having a good time at the party gives you a utility of 3, watching the movie gives you a utility of 2, whereas having a bad time at the party gives you a utility of 0. Experience has taught you that you only have a good time at a party if both Ben and Chris are also there. The problem, however, is that you are not sure whether both will come to the party, since they also like the movie. In fact, the decision problems for Ben and Chris are similar to yours: watching the movie gives them both a utility of 2, having a good time at the party gives them a utility of 3, whereas having a bad time at the party gives them a utility of 0. However, Ben and Chris had a fight yesterday and, for that reason, would like to avoid each other. More precisely, Ben will only have a good time at the party if you will be there but not Chris, and Chris will only have a good time if you will be there but not Ben.

**(a)** Represent this situation as a strategic-form game.
**(b)** Does any of the players have a strategy which is weakly or strictly dominated?

Suppose now that you and Ben learn that Chris is for sure going to the party (he was forced to accept the invitation by his girlfriend). This is common knowledge between you and Ben.

**(c)** Do you now have a strategy which is weakly or strictly dominated?
**(d)** Does Ben now have a strategy which is weakly or strictly dominated?

■

**Solution to Exercise 2.9.**

**(a)** The game is as follows:

Chris: *Home*

| You \ Ben | *Home* | *Party* |
|---|---|---|
| *Home* | 2 2 2 | 2 0 2 |
| *Party* | 0 2 2 | 0 3 2 |

Chris: *Party*

| You \ Ben | *Home* | *Party* |
|---|---|---|
| *Home* | 2 2 0 | 2 0 0 |
| *Party* | 0 2 3 | 3 0 0 |

**(b)** None of the players has a (strictly or weakly) dominated strategy. For example, for You *Home* is better if both Ben and Chris also choose *Home*, but *Party* is better if both of them choose *Party*.

**(c)** We are now restricting attention to the table on the right, corresponding to Chris's decision to go to the party. No, you still don't have a (strictly or weakly) dominated strategy: *Home* is better for you if Ben also chooses *Home*, while *Party* is better for you if Ben also chooses *Party*.

**(d)** Yes, for Ben *Party* is now strictly dominated by *Home*.

Exercise 2.10 Consider the following two-player game. Each player chooses an integer from the set $\{0,1,2,\ldots,49,50\}$. The payoffs are sums of money and each player cares only about how much money she herself gets and prefers more money to less. Let $n_1$ be the number chosen by Player 1 and $n_2$ the number chosen by Player 2. Player 1 gets $\$|n_1 - n_2|$ (where $|x|$ is the absolute value of $x$, that is, $|x| = x$ if $x \geq 0$ and $|x| = -x$ if $x < 0$; for example, $|2| = |-2| = 2$). Player 2, on the other hand, gets $\$(150 - n_1 + n_2)$.

**(a)** Does Player 1 have a dominant strategy? If Yes, then name the strategy and state whether it is weak or strict dominance, if No then explain why not.

**(b)** Does Player 2 have a dominant strategy? If Yes, then name the strategy and state whether it is weak or strict dominance, if No then explain why not.

**Solution to Exercise 2.10.**

**(a)** No, Player 1 does not have a dominant strategy. If Player 2 chooses $n_2 = 50$ then player 1 gets the largest payoff by choosing $n_1 = 0$, while if Player 2 chooses $n_2 = 0$ then Player 1 gets the largest payoff by choosing $n_1 = 50$. On the other hand, any $0 < n_1 < 50$ is worse than either $n_1 - 1$ or $n_1 + 1$ against any $n_2$.

**(b)** Yes, $n_2 = 50$ is a strictly dominant strategy for Player 2.

**Exercise 2.11** Consider the strategic-form game shown in Figure 2.3, where $p$ is a real number strictly between 0 and 1: $0 < p < 1$. As usual, in each cell the number on the left is Player 1's payoff and the other number is Player 2's payoff.

**(a)** Are there values of $p$ for which Player 1 has a weakly dominant strategy?
**(b)** Are there values of $p$ for which Player 2 has a weakly dominant strategy?

Player 2

| | | $E$ | | $F$ | |
|---|---|---|---|---|---|
| | $A$ | $1+p$ | $2$ | $p$ | $4$ |
| | $B$ | $4$ | $1$ | $2-2p$ | $0$ |
| Player 1 | $C$ | $4-2p$ | $1+p$ | $2-p$ | $4p$ |
| | $D$ | $1+3p$ | $2-p$ | $0$ | $4-4p$ |

Figure 2.3: The game for Exercise 2.11.

**Solution to Exercise 2.11.**

**(a)** No. Since against $E$ the unique best reply is $B$, the only candidate for a dominant strategy is $B$. However, against $F$, $B$ is strictly worse than $C$, since $p > 0$ (and thus $2-2p < 2-p$).
**(b)** No, because $F$ is strictly better than $E$ against $A$, but $E$ is strictly better than $F$ against $B$.

**Exercise 2.12** Two players can make a total profit of up to \$100. They are in different rooms and cannot communicate. Each player has to write a multiple of 10 not greater than 100 (thus either 10 or 20 or 30 ... or 100) on a piece of paper and hand it to the referee. The referee then does the following:

- if the two amounts written down add up to 100 or less, she gives each player a sum of money equal to the amount he wrote (for example, if Player 1 writes 20 and Player 2 writes 60, then Player 1 gets \$20 and Player 2 gets \$60],
- if the two amounts written down add up to more than 100, the referee gives nothing to the players.

Each player is selfish and greedy, that is, only cares about how much money he gets and prefers getting more money to getting less money.

**(a)** Consider Player 1. Is the strategy of writing 10 strictly dominated by another strategy?
**(b)** Does Player 1 have any strictly dominated strategies?
**(c)** Consider Player 1. Is the strategy of writing 100 weakly dominated by another strategy?
**(d)** List all of Player 1's weakly dominated strategies.

**Solution to Exercise 2.12.**

**(a)** 10 is not strictly dominated for Player 1 (if Player 2 writes 90, then writing 10 gives Player 1 a payoff of \$10, while every other strategy gives him \$0).

**(b)** Player 1 does not have strictly dominated strategies.

**(c)** Yes, 100 is weakly dominated (for example by 10): writing 100 gives Player 1 \$0 in every possible case, while writing, say 10, would give him \$10 in some cases and \$0 in others.

**(d)** The only weakly dominated strategy is 100.

Exercise 2.13 Each of two players is given two one-dollar bills and an unmarked envelope. Each player then puts her voluntary contribution in her envelope, which can be either nothing or \$1 or \$2, seals the envelope and puts it in a box. A referee takes the two envelopes from the box, opens them and counts the total amount of money contributed by the two players, call it $x$. The referee increases this amount by 50% and distributes the resulting sum, call it $y$, in equal amounts to the two players. Thus, for example, if one player contributes \$1 and the other contributes \$2 then the total contribution is $x = 3$, the referee increases it by 50% so that it becomes $y = 4.50$ and gives \$2.25 to each player, so that the player who contributed \$1 ends up with $\$(1+2.25) = \$3.25$, while the player who contributed \$2 ends up with \$2.25.

**(a)** Represent this situation as a strategic-form game where the payoff of each player is the total amount of money she ends up with (initial \$2 minus the voluntary contribution plus the final distribution).

**(b)** Do the players have a (strictly/weakly) dominant strategy?

Now generalize the above situation to the case where there are $n \geq 2$ players and each player is given $m \geq 2$ dollars. Each player decides how much to contribute (any integer amount from the set $\{0, 1, 2, \dots m\}$), the total contribution $x$ is increased by $100\frac{n-1}{n}\%$ to $y = x + \frac{n-1}{n}x$ and the amount $y$ is divided equally among the $n$ players. For example, if $n = 4$, $m = 6$, Players 1 and 2 contribute \$2 each, Player 3 contributes nothing and Player 4 contributes \$4 then $x = 8$ and it is increased by 75% to $y = 8 + \frac{3}{4}8 = 14$ so that each player gets \$3.5. Thus the payoff of Players 1 and 2 is $4 + 3.5 = 7.5$, the payoff of Player 3 is $6 + 3.5 = 9.5$ and the payoff of Player 4 is $2 + 3.5 = 5.5$.

**(c)** Prove that each player has a strictly dominant strategy.

**Solution to Exercise 2.13.**

**(a)** The game is as follows:

| | | Player 2 | | |
|---|---|---|---|---|
| | | \$0 | \$1 | \$2 |
| Player 1 | \$0 | 2.00 2.00 | 2.75 1.75 | 3.50 1.50 |
| | \$1 | 1.75 2.75 | 2.50 2.50 | 3.25 2.25 |
| | \$2 | 1.50 3.50 | 2.25 3.25 | 3.00 3.00 |

**(b)** For each player contributing nothing is a strictly dominant strategy.

**(c)** Let $s_i$ denote the contribution of player $i$. Then the payoff function of player $i$ is

$$\pi_i(s_1, s_2, \ldots, s_n) = m - s_i + \frac{1}{n}\left(\frac{2n-1}{n}\right)(s_1 + s_2 + \ldots + s_n)$$

We shall prove that, for Player 1, $s_1 = 0$ is a strictly dominant strategy (the proof for any other player is similar). Fix an arbitrary strategy profile $(s_1, s_2, \ldots, s_n)$ with $s_1 > 0$. We want to show that $\pi_1(0, s_2, \ldots, s_n) > \pi_1(s_1, s_2, \ldots, s_n)$. Let $S = s_2 + s_3 + \ldots s_n$ denote the total contribution of the players other than 1. Then

$$\pi_1(0, s_2, \ldots, s_n) = m + \frac{2n-1}{n^2} S \quad \text{and}$$

$$\pi_1(s_1, s_2, \ldots, s_n) = m - s_1 + \frac{2n-1}{n^2}(s_1 + S)$$

Thus

$$\pi_1(0, s_2, \ldots, s_n) - \pi_1(s_1, s_2, \ldots, s_n) = \frac{n^2 - 2n + 1}{n^2} s_1 = \left(\frac{n-1}{n}\right)^2 s_1 > 0.$$

[Intuition for the case where $n = 2$: if a player increases her contribution by \$1, the pot increases by \$1.50 and each gets \$0.75; thus the contributor loses \$0.25 while the other player gains \$0.75. Since each player is selfish and greedy, she only takes into account the change in her own wealth, which is negative.]

Exercise 2.14 A worker has a reservation wage (or outside option) of $r \geq 0$. An employer values the worker at $v \geq 0$. The worker submits a wage demand $d \geq 0$ in a sealed envelope, while the employer submits a wage offer $o \geq 0$, also in a sealed envelope. A mediator opens the two envelopes and determines the outcome as follows:

- If $o \geq d$ the worker is hired at a wage of $o$, the employer's payoff is $v - o$ and the worker's payoff is $o$.
- If $o < d$, no employment takes place, the employer's payoff is zero and the worker's payoff is $r$.

**(a)** Does the worker have a dominant strategy? If your answer is 'No', explain why not. If your answer is 'Yes', state what the strategy is and prove that it is a dominant strategy.

**(b)** Show that, if $v > 0$, for the employer setting $o = v$ is not a dominant strategy.

**(c)** Assume that $r > 0$. Determine whether for the employer setting $o = r$ is a dominant strategy.

■

**Solution to Exercise 2.14.**

**(a)** For the worker choosing $d = r$ is a dominant strategy. Proof. We have to show that, no matter what value of $o$ the employer chooses, the worker cannot do better with some $d \neq r$ relative to $d = r$. Fix an arbitrary $o$. Two cases are possible: (1) $o < r$ and (2) $o \geq r$.

In case (1), with $d = r$ the worker is not hired and gets a payoff of $r$. The same would be true for any other $d$ with $d > o$. If, on the other hand, the worker were to choose a $d \leq o$ then he would be employed and his payoff would be $o$. Since, by

hypothesis, $o < r$, he would be worse off.
In case (2), with $d = r$ the outcome is employment with a payoff of $o$. Any other $d$ with $d \leq o$ would yield the same outcome. On the other hand, if the worker were to choose any $d > o$, then he would not be employed and his payoff would be $r$. Since, by hypothesis, $o \geq r$, he would not be better off.

**(b)** Suppose that $v > 0$ and consider the case where the worker has chosen a $d$ such that $d < v$. Then $o = v$ gives the employer a payoff of zero, while $o = d$ would give her a payoff of $v - d > 0$.

**(c)** If $r > 0$ then for the employer setting $o = r$ is not a dominant strategy: consider the case where the worker has chosen a $d$ such that $d < r$; then $o = r$ gives the employer a payoff of $v - r$, whereas any $o$ such that $d \leq o < r$ would give the employer a payoff of $v - o > v - r$.

## 2.3 Second-price auction

The exercises in this section deal with the second-price auction (Volume 1, Chapter 2, Section 2.3) and variations on it. The expression *selfish and greedy* refers to the preferences specified in Volume 1, Chapter 2, Section 2.3 which are as follows. If a player values the object that is being auctioned at $\$v$ then her utility is 0 if she does not win the auction and it is $v - p$ if she wins the auction and has to pay $\$p$ for the object.

Exercise 2.15 Two players play a second-price auction where the allowed bids are \$25, \$35 and \$45. Player 1 values the object at \$60, while Player 2 values the object at \$50. In case of ties the winner is Player 1. Both players are selfish and greedy (as defined above).

**(a)** Construct the corresponding strategic-form game.
**(b)** Does Player 1 have any strategies that are dominated? For each such strategy, specify what strategy dominates it and whether it is weak or strict dominance.
**(c)** Does Player 2 have any strategies that are dominated? For each such strategy, specify what strategy dominates it and whether it is weak or strict dominance.
**(d)** Is there a (weakly or strictly) dominant-strategy profile?

**Solution to Exercise 2.15.**

**(a)** The game is as follows (for example, if Player 1 bids \$35 and Player 2 bids \$25, then Player 1 wins and pays \$25, namely Player 2's bid, so that his utility is $60 - 25 = 35$).

Player 2

| Player 1 | bid \$25 | bid \$35 | bid \$45 |
|---|---|---|---|
| bid \$25 | 35, 0 | 0, 25 | 0, 25 |
| bid \$35 | 35, 0 | 25, 0 | 0, 15 |
| bid \$45 | 35, 0 | 25, 0 | 15, 0 |

**(b)** For Player 1 bidding \$25 is weakly dominated by bidding \$35 and also by bidding

$45. Furthermore, bidding $35 is weakly dominated by bidding $45. Thus bidding $45 is a weakly dominant strategy.

**(c)** For Player 2 bidding $25 is weakly dominated by bidding $35 and also by bidding $45. Furthermore, bidding $35 is weakly dominated by bidding $45. Thus bidding $45 is a weakly dominant strategy.

**(d)** Yes, the weakly dominant-strategy profile is (*bid* $45, *bid* $45).

Exercise 2.16 Three players play an auction where the allowed bids are $25, $35 and $45. Player 1 values the object at $36, Player 2 values the object at $32 and Player 3 at $38. In case of ties the winner is Player 1, if he was one of the players submitting the highest bid, otherwise the winner is Player 2 (thus for example, if Players 1 and 3 bid $35 and Player 2 bids $25, then the winner is Player 1, while if Player 1 bids $25 and Players 2 and 3 bid $35 then the winner is Player 2). Each player is selfish and greedy, in the sense explained at the beginning of this section.

Consider first the case where the rules of the auction such are that **the winner pays the lowest of the three bids** (the losers pay nothing and get nothing).

**(a)** Construct the corresponding strategic-form game.

**(b)** Does Player 1 have any strategies that are dominated? If so, specify whether it is weak or strict dominance.

**(c)** Does Player 1 have a dominant strategy?

**(d)** Does Player 2 have any strategies that are dominated? If so, specify whether it is weak or strict dominance.

**(e)** Does Player 2 have a dominant strategy?

**(f)** Does Player 3 have any strategies that are dominated? If so, specify whether it is weak or strict dominance.

**(g)** Does Player 3 have a dominant strategy?

**(h)** Is there a dominant strategy profile?

Consider now the case where the auction is a **second-price auction**.

**(i)** Construct the corresponding strategic-form game.

**(j)** Does Player 1 have any strategies that are dominated? If so, specify whether it is weak or strict dominance.

**(k)** Does Player 1 have a dominant strategy?

**(l)** Does Player 2 have any strategies that are dominated? If so, specify whether it is weak or strict dominance.

**(m)** Does Player 2 have a dominant strategy?

**(n)** Does Player 3 have any strategies that are dominated? If so, specify whether it is weak or strict dominance.

**(o)** Does Player 3 have a dominant strategy?

**(p)** Is there a dominant strategy profile?

■

**Solution to Exercise 2.16.**

**(a)** The game is as follows (for example, if the bids are $(35, 35, 25)$ then the winner is Player 1 and she pays \$25, so that her utility is $36 - 25 = 11$).

| Player 1 \ Player 2 | \$25 | \$35 | \$45 |
|---|---|---|---|
| \$25 | 11 0 0 | 0 7 0 | 0 7 0 |
| \$35 | 11 0 0 | 11 0 0 | 0 7 0 |
| \$45 | 11 0 0 | 11 0 0 | 11 0 0 |

Player 3: \$25

| Player 1 \ Player 2 | \$25 | \$35 | \$45 |
|---|---|---|---|
| \$25 | 0 0 13 | 0 7 0 | 0 7 0 |
| \$35 | 11 0 0 | 1 0 0 | 0 −3 0 |
| \$45 | 11 0 0 | 1 0 0 | 1 0 0 |

Player 3: \$35

| Player 1 \ Player 2 | \$25 | \$35 | \$45 |
|---|---|---|---|
| \$25 | 0 0 13 | 0 0 13 | 0 7 0 |
| \$35 | 0 0 13 | 0 0 3 | 0 −3 0 |
| \$45 | 11 0 0 | 1 0 0 | −9 0 0 |

Player 3: \$45

**(b)** For Player 1, \$25 is weakly dominated by \$35.
**(c)** Player 1 does not have a dominant strategy.
**(d)** For Player 2, \$25 is weakly dominated by \$35.
**(e)** Player 2 does not have a dominant strategy.
**(f)** For Player 3, \$25 is weakly dominated by \$35 and also by \$45; furthermore, \$35 is weakly dominated by \$45.
**(g)** For Player 3, \$45 is weakly dominant strategy.
**(h)** There is no dominant-strategy profile, because Players 1 and 2 do not have a dominant strategy.
**(i)** The game is as follows:

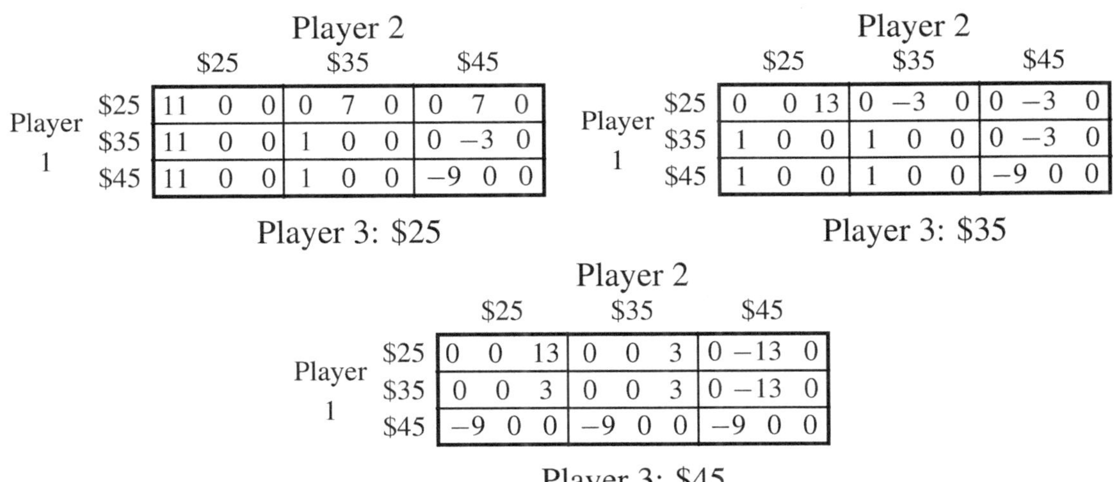

| Player 1 \ Player 2 | \$25 | \$35 | \$45 |
|---|---|---|---|
| \$25 | 11 0 0 | 0 7 0 | 0 7 0 |
| \$35 | 11 0 0 | 1 0 0 | 0 −3 0 |
| \$45 | 11 0 0 | 1 0 0 | −9 0 0 |

Player 3: \$25

| Player 1 \ Player 2 | \$25 | \$35 | \$45 |
|---|---|---|---|
| \$25 | 0 0 13 | 0 −3 0 | 0 −3 0 |
| \$35 | 1 0 0 | 1 0 0 | 0 −3 0 |
| \$45 | 1 0 0 | 1 0 0 | −9 0 0 |

Player 3: \$35

| Player 1 \ Player 2 | \$25 | \$35 | \$45 |
|---|---|---|---|
| \$25 | 0 0 13 | 0 0 3 | 0 −13 0 |
| \$35 | 0 0 3 | 0 0 3 | 0 −13 0 |
| \$45 | −9 0 0 | −9 0 0 | −9 0 0 |

Player 3: \$45

**(j)** For Player 1, \$25 and \$45 are weakly dominated by \$35.

**(k)** For Player 1, \$35 is a weakly dominant strategy.
**(l)** For Player 2, \$45 is weakly dominated by \$35.
**(m)** Player 2 does not have a dominant strategy.
**(n)** For Player 3, \$25 is weakly dominated by \$35 and also by \$45; furthermore, \$35 is weakly dominated by \$45.
**(o)** For Player 3, \$45 is a weakly dominant strategy.
**(p)** There is no dominant-strategy profile because Player 2 does not have a dominant strategy.

Exercise 2.17 A travel agent is auctioning two one-week vacations. Prize $A$ is a vacation in Hawaii and Prize $B$ is a vacation in Texas. There are three bidders. All bidders consider prize $A$ more desirable than prize $B$. Let $a_i$ be the monetary value of prize $A$ to bidder $i \in \{1,2,3\}$ (that is, winning prize $A$ is considered by bidder $i$ to be the same as being given $\$a_i$ ) and let $b_i$ be the monetary value of prize $B$ to bidder $i$. Thus $a_i > b_i$ for all $i \in \{1,2,3\}$. Having read about the good properties of Vickrey's second-price auction, the travel agent has decided to organize the following auction.

- It is a sealed-bid auction in which each bidder submits a non-negative bid $p_i$.
- The highest bidder wins prize $A$ and pays the second-highest bid.
- The second-highest bidder wins prize $B$ and pays the lowest bid.
- In case of ties, the *higher-numbered* bidder prevails over the bidder(s) who have submitted the same bid as she did. For example, if $p_1 = 10$, $p_2 = 12$ and $p_3 = 10$ then Bidder 2 wins prize $A$ and pays \$10, Bidder 3 wins prize $B$ and pays 10 and Bidder 1 pays nothing and wins nothing.

Note: The second-highest bid is the higher of the two bids remaining after removing the bid of the highest bidder. Examples: If the bids are 15, 15 and 8 then the highest bid is 15, the second highest is also 15 and the lowest is 8; if all the players bid 23 then the highest, second highest and lowest bid are all the same and equal to 23.
Assume that all bidders are selfish and greedy. Assume also that there are only two possible bids: 25 and 45 and that $a_1 = 90, b_1 = 70, a_2 = 80, b_2 = 40, a_3 = 70$ and $b_3 = 30$.

**(a)** Construct the strategic-form game corresponding to this auction.
**(b)** Does Player 1 have a dominant strategy? If Yes, specify whether it is strictly or weakly dominant. If No, explain why not.
**(c)** Does Player 2 have a dominant strategy? If Yes, specify whether it is strictly or weakly dominant. If No, explain why not.
**(d)** Does Player 3 have a dominant strategy? If Yes, specify whether it is strictly or weakly dominant. If No, explain why not.
**(e)** Is there a dominant-strategy profile?

**Solution to Exercise 2.17.**

**(a)** The game is as follows (for example, if everybody bids \$25, then Player 3 wins prize $A$ and pays \$25, so that her utility is $70 - 25 = 45$, while Player 2 wins prize $B$ and pays \$25, so that his utility is $40 - 25 = 15$).

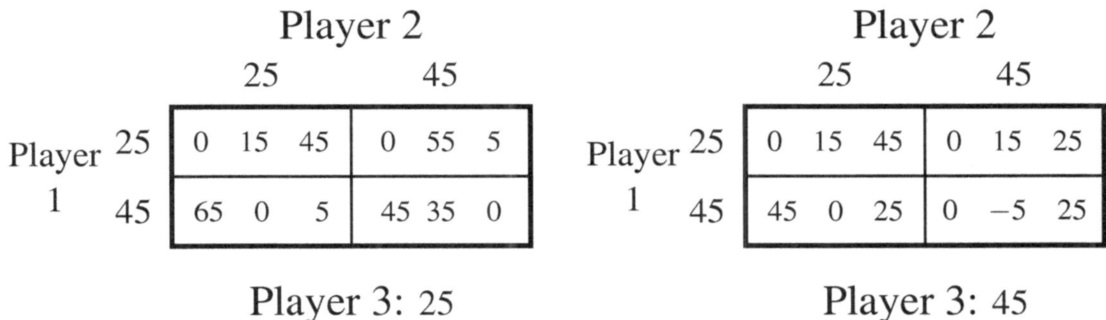

Player 3: 25

| Player 1 \ Player 2 | 25 | 45 |
|---|---|---|
| 25 | 0 15 45 | 0 55 5 |
| 45 | 65 0 5 | 45 35 0 |

Player 3: 45

| Player 1 \ Player 2 | 25 | 45 |
|---|---|---|
| 25 | 0 15 45 | 0 15 25 |
| 45 | 45 0 25 | 0 −5 25 |

Figure 2.4: The game for Exercise 2.17 (a).

**(b)** For Player 1 bidding 45 is a weakly dominant strategy.
**(c)** Player 2 does not have a dominant strategy (if $p_1 = 45$ and $p_3 = 25$ then for Player 2 bidding 45 is strictly better than biding 25, however if $p_1 = p_3 = 45$ then for Player 2 bidding 25 is strictly better than biding 45).
**(d)** For Player 3 bidding 45 is a weakly dominant strategy.
**(e)** Since there is one player who does not have a dominant strategy, there is no dominant strategy profile.

Exercise 2.18 Two objects are auctioned to 3 bidders: Amy, Bob and Claudia. Each bidder assigns zero value to each individual object but values the pair at \$100. Each player is selfish and greedy, in the sense that he/she only cares about his/her own net benefit. The bidding takes place as follows. There are two rounds. In each round each player can bid either \$0 or \$10 or \$20. At the end of the first round the following happens:

1. if they all bid 0 the game ends with nobody getting anything (there is no second round);
2. if at least one person bids a positive amount, then object 1 is assigned to the highest bidder (in case of ties the object is given to the first player in alphabetical order among those with the highest bid: thus to Amy if she is one of the highest bidders and to Brian otherwise). In this case, where somebody has won object 1, the auction proceeds to a second round of bidding.

In the second round what happened in the first round is common knowledge among the three players. At the end of the second round the following happens:

1. if they all bid 0 in the second round, then the game ends with nobody getting the second object;
2. if at least one person bids a positive amount in the second round, then object 2 is assigned to the highest bidder (in case of ties the object is given to the first player in alphabetical order among those with the highest second-round bid: thus to Amy if she is one of the highest bidders and to Brian otherwise).

In each round the winner pays his/her bid (thus this is not a second-price auction) and the others do not pay anything.

**(a)** Does any of the players have a dominant strategy in the second round?
**(b)** Is there a dominant-strategy profile in the second round?

**Solution to Exercise 2.18.**

**(a)** For every player, in round 2, it is a dominant strategy to bid 0 if he/she did not win object 1 in round 1; on the other hand, if he/she won object 1 in round 1, then there is no dominant strategy (for example, if Bob won object 1 in round 1 then for him bidding \$10 is worse than bidding \$20 if Amy bids \$10).

**(b)** No, because the player who won object 1 in round 1 does not have a dominant strategy.

## 2.4 The pivotal mechanism

The exercises in this section deal with the pivotal mechanism (Volume 1, Section 2.4).

Exercise 2.19 The pivotal mechanism is used to decide whether a new park should be built. There are five individuals. According to the proposed project, the cost of the park, which is \$120, would be allocated as follows:

| individual | 1 | 2 | 3 | 4 | 5 |
|---|---|---|---|---|---|
| cost | $c_1 = \$25$ | $c_2 = \$20$ | $c_3 = \$30$ | $c_4 = \$20$ | $c_5 = \$25$ |

For every individual $i \in \{1,..,5\}$, let $v_i$ be the perceived gross benefit (if positive; perceived gross loss, if negative) from having the park built. The $v_i$'s are as follows:

| individual | 1 | 2 | 3 | 4 | 5 |
|---|---|---|---|---|---|
| gross benefit | $v_1 = \$50$ | $v_2 = \$15$ | $v_3 = \$80$ | $v_4 = \$-15$ | $v_5 = \$-15$ |

The net benefit to individual $i$ is thus $v_i - c_i$.
Individual $i \in \{1,..,5\}$ has the following utility function (where $e_i$ is $i$'s initial wealth, which is assumed to be large enough to cover the assigned cost of the project as well as any potential taxes, and $t_i$ is the tax imposed on individual $i$ by the pivotal mechanism):

$$U_i = \begin{cases} e_i - t_i & \text{if the project is not carried out} \\ e_i + v_i - c_i - t_i & \text{if the project is carried out} \end{cases}$$

**(a)** What is the Pareto-efficient decision: to build the park or not?

Assume that the pivotal mechanism is used, so that each individual $i$ is asked to state a number $w_i$ which is going to be interpreted as the gross benefit to individual $i$ from carrying out the project. The are no restrictions on the number $w_i$: it can be positive, negative or zero. Suppose that the individuals make the following announcements:

| individual | 1 | 2 | 3 | 4 | 5 |
|---|---|---|---|---|---|
| stated benefit | $w_1 = \$90$ | $w_2 = \$10$ | $w_3 = \$65$ | $w_4 = \$-40$ | $w_5 = \$0$ |

**(b)** Given the above announcements, will the park be built?

**(c)** For each individual determine whether she is pivotal or not and what tax (if any) she has to pay.

**(d)** As you know, in the pivotal mechanism each individual has a dominant strategy. If all the individuals play their dominant strategies, will the park be built?

**(e)** Show that if every other individual reports her true benefit, then it is best for individual 4 to also report her true benefit.

**Solution to Exercise 2.19.**

**(a)** Let $net_i = v_i - c_i$. Then, since $\sum_{i=1}^{5} net_i = -5 < 0$ the Pareto efficient decision is not to build the park.

**(b)** Let $n_i = w_i - c_i$. Then, since $\sum_{i=1}^{5} n_i = 5 > 0$, based on the announcements the park will be built.

**(c)** Individual 1 is pivotal and has to pay a tax of \$60.
Individual 2 is not pivotal and thus does not have to pay a tax.
Individual 3 is pivotal and has to pay a tax of \$30.
Individuals 4 and 5 are not pivotal and thus do not have to pay a tax.

**(d)** For each individual, the dominant strategy is to report her true benefit. Thus if all the individuals do so the park will not be built.

**(e)** When all the individuals report their true benefits, individual 4 is pivotal and has to pay a tax of \$30, so that her utility is $e_4 - 30$. Any other report of individual 4 that, in conjunction with the truthful reports of the other individuals, has the effect that the park is not built will give individual 4 the same utility, namely $e_4 - 30$ (since the tax she has to pay depends only the reports of the other individuals). Any other report of individual 4 that, in conjunction with the truthful reports of the other individuals, has the effect that the park *is* built will give individual 4 a utility of $e_4 + v_4 - c_4 - 30 = e_4 - 15 - 20 - 30 = e_4 - 65 < e_4 - 30$ (again, the tax she has to pay depends only the reports of the other individuals).

Exercise 2.20 There are three individuals. A public project is being submitted to them and the pivotal mechanism is used. The cost of the project is $C = 9$. Below is the relevant data (all the amounts are in dollars):

| individual | 1 | 2 | 3 |
|---|---|---|---|
| initial wealth $e_i$ | 40 | 50 | 30 |
| gross benefit $v_i$ | 30 | −25 | 5 |
| assigned cost $c_i$ | 6 | 0 | 3 |

**(a)** Suppose that all the individuals report their true values. For each individual determine whether he is pivotal and calculate the tax that he has to pay.

**(b)** Show that, if the other two individuals report their true values,
- individual 1 gets lower utility by falsely reporting $w_1 = 0$ rather than his true value $v_2 = 30$,
- there is no other lie from which individual 1 would gain.
- Does this imply that for individual 1 telling the truth is a dominant strategy?

**(c)** Show that if the individuals lie and report $w_1 = 5, w_2 = -3, w_3 = 5$, then the decision made is Pareto inefficient. Show that there is an alternative situation that they all strictly prefer (construct the situation explicitly).

**Solution to Exercise 2.20.**

**(a)** If they all report their true values, the sum of the gross benefits is equal to 10, which is greater than $C$ and thus the project is carried out. Individual 1 is pivotal and has to pay a tax equal to $|-20-3| = 23$. Individual 2 is not pivotal and thus does not have to pay a tax. Individual 3 is pivotal and has to pay a tax equal to $|5-6| = 1$.

**(b)**

- If the reports are $w_1 = 0, w_2 = v_2 = -25, w_3 = v_3 = 5$ then the sum of the reported gross benefits is $-20 < C$ so that the project is not carried out. Individual 1 is not pivotal and thus $t_1 = 0$, so that his utility is $e_1 = 40$. If individual 1 switches to reporting his true value, namely $v_1 = 30$, then he becomes pivotal and $t_1 = 23$ so that his utility is $e_1 + v_1 - c_1 - t_1 = 40 + 30 - 6 - 23 = 41$, thus he is better off.
- Next we show that there is no lie that is advantageous for individual 1. The lies can be of two types. Type 1 lie: any $w_1$ such that $w_1 + v_2 + v_3 > 9$. Then the project would be carried out, individual 1 would be pivotal and would have to pay a tax of \$23 (recall that the tax depends only on the reports of the other two individuals). This case gives the same outcome as reporting the truth, with a corresponding utility of 41 for individual 1. Type 2 lie: any $w_1$ such that $w_1 + v_2 + v_3 \leq 9$. Then the project would not be carried out, individual 1 would not be pivotal and would not have to pay a tax. This case gives the same outcome as reporting $w_1 = 0$, with a corresponding utility of 40. Thus a lie gives either the same payoff or a worse payoff than the truth, when the other two individuals report their true values.
- The above does not imply that telling the truth is a dominant strategy for individual 1, because in principle individual 1 could gain by lying if the other two players also lie. However. we know from from Theorem 2.4.1, Volume 1, that this is not the case, that is, that telling the truth is indeed a dominant strategy.

**(c)** If the individuals lie and report $w_1 = 5, w_2 = -3, w_3 = 5$, then the decision is not to carry out the project (because the sum of the reported gross benefits is $7 < C$) and thus the outcome is Pareto inefficient (because, based on the true gross benefits, the Pareto efficient decision is to carry out the project). With these reported values, nobody is pivotal and thus the utility of each individual $i$ is $e_i$. To obtain a Pareto improvement, individuals 1 and 3 could could bribe individual 2 to convince her to have the project carried out by setting up a fund to which 1 contributes \$23.75 and individual 3 contributes \$1.50; the entire amount of this fund (namely, \$25.25) would then be given to individual 2. Then individual 1's utility would be $40 + 30 - 23.75 - 6 = 40.25 > 40 = e_1$, individual 2's utility would be $50 - 25 + 25.75 - 0 = 50.25 > 50 = e_2$ and individual 3's utility would be $30 + 5 - 1.50 - 3 = 30.50 > 30 = e_3$, so that everybody would be better off.

Exercise 2.21 Your friend Ann works for a company whose CEO is considering building an indoor swimming pool for the employees. The CEO is knowledgeable in game theory and has decided to use the pivotal mechanism. Ann's wealth is \$2,000 and she has to state how much she would value having a swimming pool. The cost of building the swimming pool is \$2,000 and it would be shared equally among the employees. There are four employees (Ann is one of them). Ann explained to you that she thinks her co-workers are not very keen on the pool. Bob can't swim and would find an indoor pool annoying. Ann thinks that his valuation (gross benefit) is $\$-500$ (that is, he considers having a pool as equivalent to having his wealth reduced by \$500). Charlie likes the idea of a swimming pool, but he is not so enthusiastic about it. His valuation (gross benefit) - Ann believes - is \$400. Donna shares Ann's enthusiasm and, according to what Ann believes, has a valuation (gross benefit) of \$800. Ann's own valuation (gross benefit) of the pool is \$1,500.

**(a)** What is Ann's utility going to be if she reports her true value of \$1,500 if her conjectures about her co-workers' valuations are correct and they indeed report their true valuations? item What would Ann's utility be if she lied and reported a gross benefit of \$500 if her conjectures about her co-workers' valuations are correct and they indeed report their true valuations?

**(b)** Ann tells you that, since she is very keen on having the pool built and she is not sure that her conjectures about her co-workers' valuations are correct, she intends to lie and report a value (gross benefit) of \$2,100. To convince her that it is not a good idea to do so, describe a situation to her (i.e. specify possible reported values of Bob, Charlie and Donna) where she is worse off reporting \$2,100 relative to reporting her true \$1,500.

**(c)** If Ann's conjectures about her co-workers' valuations are correct and they indeed report their true valuations, what would Ann's utility be if she reported \$2,100 instead of the true \$1,500?

**Solution to Exercise 2.21.**

**(a)** The sum of the reported values is $1,500-500+400+800=2,200>2,000$ so the decision is to build the pool. Ann is pivotal because $-500+400+800=700<500+500+500=1,500$. Her tax is $|700-1,500|=|-800|=800$. Thus her utility is $2,000+1,500-500-800=2,200$.

**(b)** The sum of the reported values would be $500-500+400+800=1,200<2,000$ so the decision would be not to build the pool. She would not be pivotal (same calculation as above). Thus her utility would be her initial wealth: 2,000.

**(c)** If each of the others reported a valuation of zero then if Ann reports 1,500 the decision is not to build the pool and her utility is 2,000 (she will not be pivotal), while if she reported 2,100 the decision would be to build the pool, she would be pivotal and would have to pay a tax of 1,500 and her utility would be $2,000+1,500-1,500-500=1,500$. Thus she would be worse off if she lied.

**(d)** The same as in the case where she reports the true value.

Exercise 2.22 A society consists of four individuals. They have agreed to use the pivotal mechanism to decide whether or not to carry out a project. The cost of the project is zero. Let $v_i$ be the true benefit from the project for individual $i$. Suppose that $v_1 = 2, v_2 = -8, v_3 = 5$ and $v_4 = 10$. They all have an initial of \$20.

**(a)** If they all report their true values, will the project be carried out? For every individual $i$ determine whether $i$ is pivotal or not and, if she is, what tax she has to pay.

**(b)** As you determined in Part (a), individual 4 has to pay a tax of \$1. She is upset about it and complains to her fellow citizens, who reply by saying "the tax you have to pay is a fair tax; in fact it equals the externality you impose on us; if you were not part of this society, we would not carry out the project and we would end up with a total utility of 60. Because of you, the project is carried out and our total utility is 59. Thus your action (your vote) imposes on us a negative externality of 1, which is exactly equal to the tax you have to pay." Individual 4 thinks for a while and then asks "how would you decide whether or not to carry out the project if I were not a member of this society?". They reply "we would use the pivotal mechanism, of course!". Then individual 4 says "if this is the case, then the externality that ought to be attributed to me is not $-1$ (a negative externality of 1) but +6 (a positive externality of 6)! So I should be paid, rather than be taxed! Explain how individual 4 calculated this number.

■

**Solution to Exercise 2.22.**

**(a)** If they report their true values, the sum is 9, which is greater than 0 (the cost of the project). Thus the project is carried out. Individuals 1, 2 and 3 are not pivotal and thus pay no taxes. Individual 4 is pivotal and pays a tax equal to $|2-8+5| = |-1| = 1$. The sum of the utilities of individuals 1, 2 and 3 (to be used in Part (b) is $(20+2)+(20-8)+(20+5) = 59$.

**(b)** Individual 4's is reasoning as follows. If we want to compute the externality that individual 4 imposes on the rest of society, we need to compute:

- the total utility of individuals 1, 2 and 3 when the pivotal mechanism is applied to the society consisting of individuals 1, 2, 3 and 4, (we computed this in Part (a): it is 59),
- the total utility of individuals 1, 2 and 3, call it $x$, when the pivotal mechanism is applied to the sub-society consisting of them only,
- the difference between 59 and $x$; this number should be construed at the true externality that individual 4 imposes on the other three individuals.

If individual 4 were not a member of society and the pivotal mechanism were applied to the sub-society consisting of individuals 1, 2 and 3, then the project would not be carried out, because $v_1 + v_2 + v_3 = -1 < 0$. Individuals 1 and 3 would not be pivotal and thus would pay no taxes. Individual 2 would be pivotal and would have to pay a tax equal to \$7. Hence the sum of utilities of individuals 1, 2 and 3 would be: $20+(20-7)+20) = 53$, which is less than 59, the latter being the sum of their utilities in the larger society that includes individual 4 also. Hence individual 4 can reasonably argue that her presence involves an increase in utility for the other three individuals equal to $59-53=6$, thus a positive externality.[1]

## 2.5 Iterated deletion procedures

The exercises in this section deal with the procedure of *iterated deletion of strictly dominated strategies* (IDSDS) and the procedure of *iterated deletion of weakly dominated strategies* (IDWDS; Volume 1, Chapter 2, Section 2.5).

### 2.5.1 IDSDS

**Exercise 2.23** Apply the IDSDS (iterated deletion of strictly dominated strategies) procedure to the game of Figure 2.5.

| | | Player 2 | | | |
|---|---|---|---|---|---|
| | | *E* | *F* | *G* | *H* |
| Player 1 | *A* | 0 0 | 2 1 | 1 2 | 0 3 |
| | *B* | 4 1 | 1 2 | 1 3 | 3 4 |
| | *C* | 1 0 | 3 1 | 2 0 | 1 1 |
| | *D* | 3 3 | 0 4 | 0 1 | 2 2 |

Figure 2.5: The game for Exercise 2.23

.

**Solution to Exercise 2.23.** For Player 1, *A* is strictly dominated by *C* and *D* is strictly dominated by *B*. For Player 2, *E* is strictly dominated by *F* and *G* is strictly dominated by *H*. After deleting *A*, *D*, *E* and *G* we are left with the following game:

[1]When individual 4 is added to the sub-society, the project is carried out. Individual 2 will have a utility loss of 8, because of the project, but will no longer have to pay a tax of 7; hence she will experience a net utility loss of 1. On the other hand, individuals 1 and 3 will experience an increase in utility equal to $2+5=7$. So the net externality that individual 4 imposes on individuals 1 to 3 is a positive one: $7-1=6$. Hence if transfers are to reflect externalities, individual 4 should receive a subsidy of 6 rather than have to pay a tax of 1!

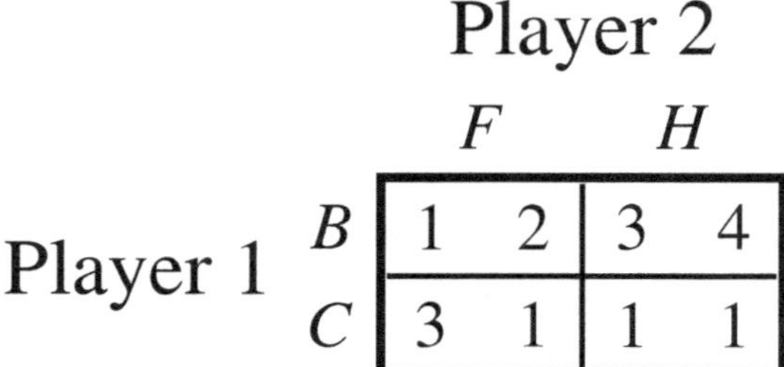

| Player 1 \ Player 2 | F | H |
|---|---|---|
| B | 1 2 | 3 4 |
| C | 3 1 | 1 1 |

In this game there are no strictly dominated strategies. Thus the output of the IDSDS procedure is the set of strategy profiles $\{(B,F), (B,H), (C,F), (C,H)\}$.

### 2.5.2 IDWDS

**Exercise 2.24** Apply the IDWDS to the game of Figure 2.6. ■

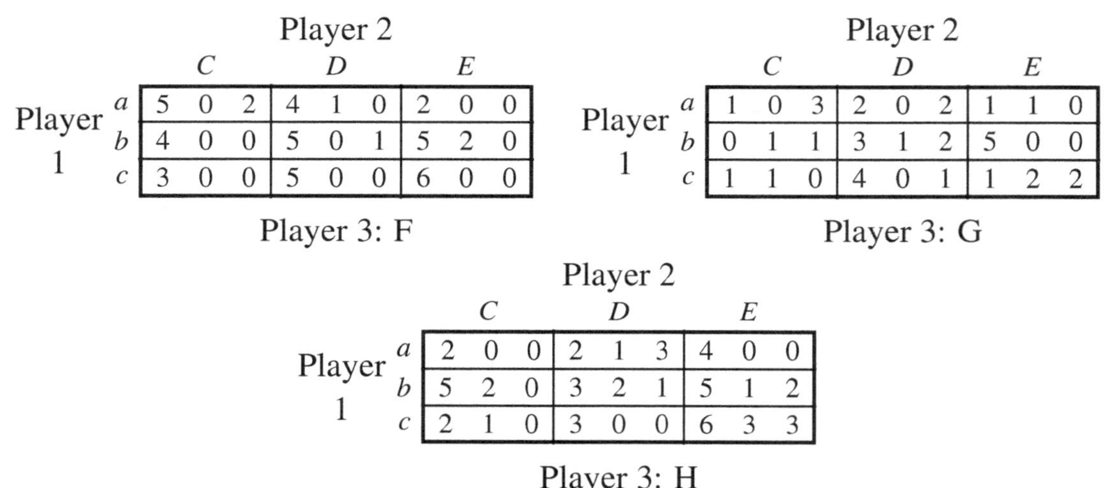

Player 3: F

| Player 1 \ Player 2 | C | D | E |
|---|---|---|---|
| a | 5 0 2 | 4 1 0 | 2 0 0 |
| b | 4 0 0 | 5 0 1 | 5 2 0 |
| c | 3 0 0 | 5 0 0 | 6 0 0 |

Player 3: G

| Player 1 \ Player 2 | C | D | E |
|---|---|---|---|
| a | 1 0 3 | 2 0 2 | 1 1 0 |
| b | 0 1 1 | 3 1 2 | 5 0 0 |
| c | 1 1 0 | 4 0 1 | 1 2 2 |

Player 3: H

| Player 1 \ Player 2 | C | D | E |
|---|---|---|---|
| a | 2 0 0 | 2 1 3 | 4 0 0 |
| b | 5 2 0 | 3 2 1 | 5 1 2 |
| c | 2 1 0 | 3 0 0 | 6 3 3 |

Figure 2.6: The game of Exercise 2.6

.

**Solution to Exercise 2.24.**
Only Player 3 has a weakly dominated strategy, namely $F$ (it is weakly dominated by $G$). Eliminating $F$ we are left with the following game:

Player 2

| Player 1 | *C* | | | *D* | | | *E* | | |
|---|---|---|---|---|---|---|---|---|---|
| *a* | 1 | 0 | 3 | 2 | 0 | 2 | 1 | 1 | 0 |
| *b* | 0 | 1 | 1 | 3 | 1 | 2 | 5 | 0 | 0 |
| *c* | 1 | 1 | 0 | 4 | 0 | 1 | 1 | 2 | 2 |

Player 3: G

Player 2

| Player 1 | *C* | | | *D* | | | *E* | | |
|---|---|---|---|---|---|---|---|---|---|
| *a* | 2 | 0 | 0 | 2 | 1 | 3 | 4 | 0 | 0 |
| *b* | 5 | 2 | 0 | 3 | 2 | 1 | 5 | 1 | 2 |
| *c* | 2 | 1 | 0 | 3 | 0 | 0 | 6 | 3 | 3 |

Player 3: H

In this game only Player 1 has a weakly dominated strategy, namely *a* (it is weakly dominated by *c*). After deleting *a*, we are left with a game where only Player 2 has a weakly dominated strategy, namely *D* (it is weakly dominated by *C*). After deleting *D* we are left with the following game:

Player 2

| Player 1 | *C* | | | *E* | | |
|---|---|---|---|---|---|---|
| *b* | 0 | 1 | 1 | 5 | 0 | 0 |
| *c* | 1 | 1 | 0 | 1 | 2 | 2 |

Player 3: *G*

Player 2

| Player 1 | *C* | | | *E* | | |
|---|---|---|---|---|---|---|
| *b* | 5 | 2 | 0 | 5 | 1 | 2 |
| *c* | 2 | 1 | 0 | 6 | 3 | 3 |

Player 3: *H*

In this game no player has a dominated strategy. Thus the iterative deletion procedure stops here.

Exercise 2.25

**(a)** Apply the IDWDS procedure to the game of Part (a) of Exercise 2.16.

**(b)** Apply the IDWDS procedure to the game of Part (b) of Exercise 2.16. ■

**Solution to Exercise 2.25.**

**(a)** In the first round delete all of the following: \$25 for Player 1, \$25 for Player 2 and both \$25 and \$35 for Player 3. In the second round delete \$45 for Player 2. In the third, and final, round delete \$35 for Player 1. Thus the output of the IDWDS algorithm is the strategy profile $(\$45, \$35, \$45)$.

**(b)** In the first, and only, round delete all of the following: \$25 and \$45 for Player 1, \$45 for Player 2 and both \$25 and \$35 for Player 3. Thus the output of the IDWDS algorithm is the set of strategy profile $\{(\$35, \$25, \$45), (\$35, \$35, \$45)\}$.

Exercise 2.26 Consider again the two-stage bidding game of Exercise 2.18. Recall that, in round 2, for every player it is a dominant strategy to bid 0 if he/she did not win object 1 in round 1, while, if he/she won object 1 in round 1, then there is no dominant strategy.

**(a)** What is the output of the IDWDS procedure applied to round 2?

**(b)** Assuming that everybody predicts the outcome of the second round to be the one determined in Part (a), construct the strategic form for the stage-1 bidding.

**(c)** What is the output of the IDWDS procedure applied to the game of Part (b)? ■

**Solution to Exercise 2.26.**

**(a)** The output of the IDWDS in the second round is as follows: the winner of the first round bids \$10 and the other two bid \$0.

**(b)** The game is as follows:

| | | Bob | | |
|---|---|---|---|---|
| | | \$0 | \$10 | \$20 |
| | \$0 | 0 0 0 | 0 80 0 | 0 70 0 |
| Amy | \$10 | 80 0 0 | 80 0 0 | 0 70 0 |
| | \$20 | 70 0 0 | 70 0 0 | 70 0 0 |

Claudia: \$0

| | | Bob | | |
|---|---|---|---|---|
| | | \$0 | \$10 | \$20 |
| | \$0 | 0 0 80 | 0 80 0 | 0 70 0 |
| Amy | \$10 | 80 0 0 | 80 0 0 | 0 70 0 |
| | \$20 | 70 0 0 | 70 0 0 | 70 0 0 |

Claudia: \$10

| | | Bob | | |
|---|---|---|---|---|
| | | \$0 | \$10 | \$20 |
| | \$0 | 0 0 70 | 0 0 70 | 0 70 0 |
| Amy | \$10 | 0 0 70 | 0 0 70 | 0 70 0 |
| | \$20 | 70 0 0 | 70 0 0 | 70 0 0 |

Claudia \$20

**(c)** For each player, \$0 is weakly dominated by \$10 (and also by \$20); on the other hand, none of the other two strategies is weakly dominated. Thus in the first step of the deletion procedure we eliminate \$0 for every player. In the resulting game, Amy does not have a dominated strategy, while for Bob \$20 weakly dominates \$10 and the same is true for Claudia. Thus in the second step of the procedure we delete \$10 for both Bob and Claudia. In the resulting game, for Amy \$10 is strictly dominated by \$20. Thus the outcome of the IDWDS is the strategy profile (\$20,\$20,\$20).

Exercise 2.27 Note: this exercise looks very similar to Exercise 2.18 (and Exercise 2.26) but the strategic aspects are quite different.
Two complementary objects (e.g. a left shoe and a right shoe) are auctioned to three bidders: Ann, Bob and Carla. Each bidder assigns zero value to each individual object but values the pair at \$100. The bidding takes place as follows. There are two rounds. In each round each player can bid either \$0 or \$10 or \$20. At the end of the first round the following happens:

1. if one player bid more than the others, he/she gets the first object,
2. if the highest bid was submitted by at least two players then Ann gets the object, **even if she was not one of the highest bidders.**

At the end of the first round, whatever the outcome, **everybody, including those who lost the first round, has to pay his/her first-round bid**.
Then we proceed to the next stage where the same rules apply. [Thus, at the end of the second round the following happens: (1) if one player bid more than the others, he/she gets the second object and everybody (including those who lost the second round) has to pay his/her second-round bid; (2) if the highest bid was submitted by at least two players then Ann gets the second object (even if she was not one of the highest bidders) and everybody has to pay his/her second-round bid (even those who lost)]
Note that, in the second round, the bids that are used to determine the outcome of the second round, i.e. who gets the second object, are not the cumulative bids (first-round bid + second-round bid) but just the second-round bids.
Each player is selfish and greedy, in the sense that he/she only cares about his/her own net benefit.

**(a)** (a.1) Write the strategic form of the second-round game in the case where in the first round Ann bid \$0, Bob bid \$10 and Carla bid \$10. Note: write the payoffs of the entire game, not only of the second stage.
(a.2) In this game, does any of the players have a dominant strategy?
(a.3) What is the output of the IDSDS procedure in this game?
(a.5) What is the output of the IDWDS procedure in this game?
**(b)** In general, does any of the players have a dominant strategy in the second round?
**(c)** What is the output of the IDWDS procedure in the second round?
**(d)** Write the strategic form for the first-round game assuming that everybody predicts the outcome of the second round to be the one given in Part (c)
**(e)** What is the output of the IDWDS procedure for the game of Part (d)?

■

**Solution to Exercise 2.27.**

**(a)** (a.1) Given the rules, since there were two highest bids, the first object went to Ann (even though she bid less than the others). Thus the strategic form of this second-round game is as follows:

Carla: $0

| Ann \ Bob | $0 | $10 | $20 |
|---|---|---|---|
| $0 | 100 −10 −10 | 0 −20 −10 | 0 −30 −10 |
| $10 | 90 −10 −10 | 90 −20 −10 | −10 −30 −10 |
| $20 | 80 −10 −10 | 80 −20 −10 | 80 −30 −10 |

Carla: $10

| Ann \ Bob | $0 | $10 | $20 |
|---|---|---|---|
| $0 | 0 −10 −20 | 100 −20 −20 | 0 −30 −20 |
| $10 | 90 −10 −20 | 90 −20 −20 | −10 −30 −20 |
| $20 | 80 −10 −20 | 80 −20 −20 | 80 −30 −20 |

Carla: $20

| Ann \ Bob | $0 | $10 | $20 |
|---|---|---|---|
| $0 | 0 −10 −30 | 0 −20 −30 | 100 −30 −30 |
| $10 | −10 −10 −30 | −10 −20 −30 | 90 −30 −30 |
| $20 | 80 −10 −30 | 80 −20 −30 | 80 −30 −30 |

(a.2) For both Bob and Carla, bidding $0 is a strictly dominant strategy. Ann does not have a dominant strategy.

(a.3) The output of the IDSDS procedure is the strtegy profile ($0,$0,$0).

(a.4) The IDWDS procedure yields the same output as the IDSDS procedure.

**(b)** For every player it is a dominant choice to bid $0 in round 2 if he/she did not win object 1 in round 1. There is no dominant choice for any player if he/she won the first round (e.g. for Ann bidding $0 is worse than bidding $20 if only one other person bids $20).

**(c)** The output of the IDWDS procedure in the second round is as follows: (1) if Amy won the first round, everybody bids $0; (2) if Bob won the first round, Bob bids $10 and the others bid $0; (3) if Carla won the first round, Carla bids $10 and the others bid $0.

**(d)** Let us first figure out who wins the first object:

Carla: $0

| Ann \ Bob | $0 | $10 | $20 |
|---|---|---|---|
| $0 | Ann | Bob | Bob |
| $10 | Ann | Ann | Bob |
| $20 | Ann | Ann | Ann |

Carla: $10

| Ann \ Bob | $0 | $10 | $20 |
|---|---|---|---|
| $0 | Carla | Ann | Bob |
| $10 | Ann | Ann | Bob |
| $20 | Ann | Ann | Ann |

Carla: $20

| Ann \ Bob | $0 | $10 | $20 |
|---|---|---|---|
| $0 | Carla | Carla | Ann |
| $10 | Carla | Carla | Ann |
| $20 | Ann | Ann | Ann |

Thus the payoffs are as follows:

Carla: \$0

| | | Bob \$0 | Bob \$10 | Bob \$20 |
|---|---|---|---|---|
| Ann | \$0 | 100 0 0 | 0 80 0 | 0 70 0 |
| | \$10 | 90 0 0 | 90 −10 0 | −10 70 0 |
| | \$20 | 80 0 0 | 80 −10 0 | 80 −20 0 |

Carla: \$10

| | | Bob \$0 | Bob \$10 | Bob \$20 |
|---|---|---|---|---|
| Ann | \$0 | 0 0 90 | 100 −10 −10 | 0 70 −10 |
| | \$10 | 90 0 −10 | 90 −10 −10 | −10 70 −10 |
| | \$20 | 80 0 −10 | 80 −10 −10 | 80 −20 −10 |

Carla: \$20

| | | Bob \$0 | Bob \$10 | Bob \$20 |
|---|---|---|---|---|
| Ann | \$0 | 0 0 70 | 0 −10 70 | 100 −20 −20 |
| | \$10 | −10 0 70 | −10 −10 70 | 90 −20 −20 |
| | \$20 | 80 0 −20 | 80 −10 −20 | 80 −20 −20 |

**(e)** In the game of Part (d) no player has dominated strategies. Thus the IDWDS procedure leaves the game unchanged.

Exercise 2.28 Consider the game of Exercise 2.12. What is the output of the IDWDS?

**Solution to Exercise 2.28.** In the first round delete 100 for both players. In the resulting smaller game, neither player has any weakly dominated strategies. Thus the procedure stops here and the output of the IDWDS procedure consists of all the strategy profiles $(x,y)$ such that $10 \le x \le 90$ and $10 \le y \le 90$.

Exercise 2.29 Father, mother and daughter have to decide where to spend their vacation. There are three possibilities: mountain (M), lake (L) and sea (S). The father prefer M to L and L to S. The mother prefers S to M and M to L. The daughter prefers L to S and S to M. They decide that each should write either M or S or L on a piece of paper and give it to their neighbor. The neighbor will choose the alternative that gets at least two votes (i.e. the one that is chosen by at least two people). If no alternative gets more than 1 vote, then no choice is made and they spend their vacation at home (which each of them considers to be equivalent to the worst of the other three alternatives).

**(a)** Represent this situation as a strategic-form game.

**(b)** Does any of the players have a dominant strategy?

**(c)** What is the output of the IDWDS procedure?

Suppose that each player voted for his/her most preferred alternative so that they would have to stay at home. In order to avoid this outcome, mother and daughter suggest that they go through a second round of voting, but this time they suggest that the father's choice be given 1½ votes, while the mother's choice gets one vote and the daughter's choice also gets one vote. Thus, the three of them are again to write a choice on a piece of paper and give it to the neighbor. The neighbor will choose the alternative that gets the largest number of votes with the understanding that the father's choice carries 1½ votes, so that a tie is now no longer possible and thus the outcome of them staying at home is no longer possible.

**(d)** Represent this new situation as a strategic-form game.
**(e)** Does any of the players have a dominant strategy?
**(f)** What is the output of the IDWDS procedure?
**(g)** Using the IDWDS procedure to predict the outcome, can we conclude that the extra half vote given to the father favors him?

**Solution to Exercise 2.29.**

**(a)** We can use the following utility functions to represent the players' preferences:

| | Father | Mother | Daughter |
|---|---|---|---|
| $L$ | 1 | 0 | 2 |
| $M$ | 2 | 1 | 0 |
| $S$ | 0 | 2 | 1 |
| home | 0 | 0 | 0 |

With these utility functions the strategic-form game is as follows:

Daughter: L

| Father \ Mother | L | M | S |
|---|---|---|---|
| L | 1 0 2 | 1 0 2 | 1 0 2 |
| M | 1 0 2 | 2 1 0 | 0 0 0 |
| S | 1 0 2 | 0 0 0 | 0 2 1 |

Daughter: M

| Father \ Mother | L | M | S |
|---|---|---|---|
| L | 1 0 2 | 2 1 0 | 0 0 0 |
| M | 2 1 0 | 2 1 0 | 2 1 0 |
| S | 0 0 0 | 2 1 0 | 0 2 1 |

Daughter: S

| Father \ Mother | L | M | S |
|---|---|---|---|
| L | 1 0 2 | 0 0 0 | 0 2 1 |
| M | 0 0 0 | 2 1 0 | 0 2 1 |
| S | 0 2 1 | 0 2 1 | 0 2 1 |

**(b)** None of the players has a dominant strategy.

**(c)** S is a dominated strategy for the Father (weakly dominated by M and also by L), L for the Mother (weakly dominated by S and also by M) and M for the daughter (weakly dominated by L and also by S). Thus in the first round we delete S for the Father, L for the Mother and M for the Daughter and we are left with the following smaller game:

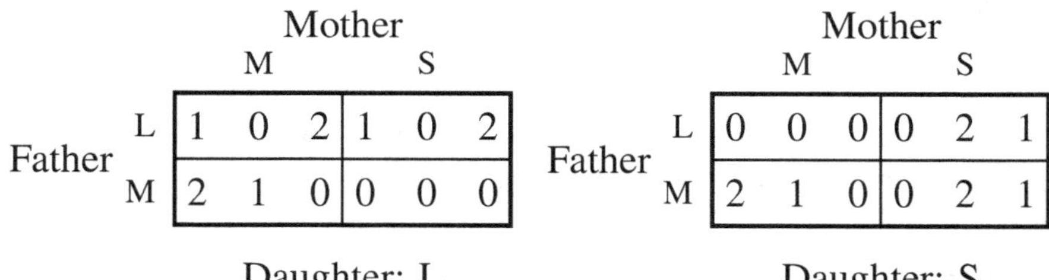

Daughter: L

| Father \ Mother | M | S |
|---|---|---|
| L | 1 0 2 | 1 0 2 |
| M | 2 1 0 | 0 0 0 |

Daughter: S

| Father \ Mother | M | S |
|---|---|---|
| L | 0 0 0 | 0 2 1 |
| M | 2 1 0 | 0 2 1 |

In this reduced game no player has a dominated strategy. Thus the IDWDS procedure stops here.

**(d)** Using the same utility functions as in Part (a), the game is now as follows:

Daughter: L

| Father \ Mother | L | M | S |
|---|---|---|---|
| L | 1 0 2 | 1 0 2 | 1 0 2 |
| M | 1 0 2 | 2 1 0 | 2 1 0 |
| S | 1 0 2 | 0 2 1 | 0 2 1 |

Daughter: M

| Father \ Mother | L | M | S |
|---|---|---|---|
| L | 1 0 2 | 2 1 0 | 1 0 2 |
| M | 2 1 0 | 2 1 0 | 2 1 0 |
| S | 0 2 1 | 2 1 0 | 0 2 1 |

Daughter: S

| Father \ Mother | L | M | S |
|---|---|---|---|
| L | 1 0 2 | 1 0 2 | 0 2 1 |
| M | 2 1 0 | 2 1 0 | 0 2 1 |
| S | 0 2 1 | 0 2 1 | 0 2 1 |

**(e)** The Father now has M as a dominant strategy. No other player has a dominant strategy.

**(f)** In the first round, delete L and S for the Father.
In the resulting game, S becomes a dominant strategy for the Mother and, for the Daughter, L weakly dominates M. Thus, in the second round delete L and M for the Mother and M for the Daughter.
In the resulting game, S becomes a dominant strategy for the Daughter. Thus, in the third round delete L for the Daughter. Hence the output of the IDWDS is the strategy profile (M,S,S).

**(g)** According to the output of the IDWDS, the outcome is that they go to the Sea, which - from the point of view of the Father - is the worst of the three alternatives. Thus, the extra half vote given to the Father hurts him, rather than favor him (it makes his vote predictable).

## 2.6 Nash equilibrium

The exercises in this section deal with the notion of *Nash equilibrium* (Definition 2.6.2, Volume 1, Section 2.6).

Exercise 2.30 Find all the Nash equilibria of the game of Part (b) of Exercise 2.2 ■

**Solution to Exercise 2.30.**
There are four Nash equilibria: $(K,K,K)$, $(K,R,R)$, $(R,K,R)$ and $(R,R,K)$.

Exercise 2.31 Find all the Nash equilibria of the game Exercise 2.3. ■

**Solution to Exercise 2.31.**
There are three Nash equilibria: (via X, via X, via Y), (via X, via Y, via X), (via Y, via X,

via X). Thus, at a Nash equilibrium two drivers take the route via X and one driver takes the route via Y.

Exercise 2.32 There are ten Davis residents who commute to Berkeley every day. For simplicity, let us assume that they are the only car owners in that area and that car-pooling is not possible (so that there is only one person per car). Consider the simultaneous game they play on any given day. Each commuter decides whether to travel by train ($T$) or by car ($C$). Taking the train avoids congestion and each commuter who travels by train gets a utility (or payoff) of 2. Traveling by car, on the other hand, involves congestion: if there are $n \in \{1, 2, ..., 10\}$ cars on the road, the utility (or payoff) to each driver is equal to $10 - n$.

**(a)** Find all the Nash equilibria of this game.

**(b)** Define an outcome to be *socially optimal* if it maximizes the sum of the utilities. Do any of the Nash equilibria give rise to a socially optimal outcome?

**(c)** Define an outcome $z$ to be *Pareto efficient* if there is no other outcome $z'$ such that (1) at least one player prefers $z'$ to $z$ and (2) every other player considers $z'$ to be at least as good as $z$ (equivalently, $z$ is Pareto efficient if, for every other outcome $z'$, at least one player prefers $z$ to $z'$). Do any of the Nash equilibria give rise to a Pareto efficient outcome?

In what follows, assume that the road from Davis to Berkeley is now privatized and the owner charges a toll of \$$p$ per car, where $p \in \{0, 1, 2, ..., 8\}$. A toll of \$$p$ reduces the payoff of a driver by an amount exactly equal to $p$.

**(d)** Find the toll that maximizes the road-owner's revenue (defined as the product $pm$, where $m$ is the number of commuters who decide to drive instead of taking the train) if she expects the **largest** number of drivers compatible with Nash equilibrium.

**(e)** Find the toll that maximizes the road-owner's revenue (defined as the product $pm$, where $m$ is the number of commuters who decide to drive) if she expects the **smallest** number of drivers compatible with Nash equilibrium.

**Solution to Exercise 2.32.**

**(a)** The Nash equilibria are all those strategy profiles where

- either exactly 8 commuters decide to drive (and everybody gets a payoff of 2; there are 45 such equilibria, since there are 45 different ways of selecting two individuals from a set of ten), or
- exactly 7 commuters decide to drive (the drivers get a payoff of 3 and the others a payoff of 2; there are 120 such equilibria).

In the first type of equilibria, a driver who unilaterally deviated would get the same utility (namely 2), while a train-rider who unilaterally deviated would see his utility decrease from 2 to 1. In the second type of equilibria, a driver who unilaterally deviated would see his utility decrease from 3 to 2, while a train-rider who unilaterally deviated would get the same utility (namely 2).

**(b)** Let $n$ be the number of drivers (so that the number of train-riders is $10 - n$). The sum of the payoffs is $f(n) = 2(10 - n) + (10 - n)n = 20 + 8n - n^2$, which is maximized when $n = 4$. Thus none of the Nash equilibria give rise to a socially optimal outcome, since they all involve more than 4 drivers on the road (either 8 or 7).

**(c)** The Nash equilibria with 8 drivers are not Pareto efficient (move one driver to the train and the utilities of the remaining drivers increase, while nobody's utility decreases). The Nash equilibria with 7 drivers *are* Pareto efficient (no reallocation of commuters to a different means of transportation makes somebody better off without making anybody worse off).

**(d)** Given $p$, the owner expects the number of drivers $n$ to be such that $10 - n - p = 2$, that is, $n = 8 - p$. Hence the revenue is $R = p(8 - p)$, which is maximized when $p = 4$ (in which case $R = 16$).

**(e)** Given $p$, the owner expects the number of drivers $n$ to be such that $10 - n - p = 3$, that is, $n = 7 - p$. Hence the revenue is $R = p(7 - p)$, which is maximized when either $p = 3$ or $p = 4$ (in either case $R = 12$).

Exercise 2.33 Consider the three-player game of Figure 2.2 in Section 2.

**(a)** What strategy profile is a Nash equilibrium, whatever the values of $x, y$ and $z$?

**(b)** For what values of $x, y$ and $z$ are there Nash equilibria where Player 1 plays $B$?

**(c)** For what values of $x, y$ and $z$ does the game have a unique Nash equilibrium?

**Solution to Exercise 2.33.**

**(a)** $(A, C, F)$ is always a Nash equilibrium.

**(b)** The only strategy profiles which could be Nash equilibria, besides $(A, C, F)$, are $(B, D, E)$, which is a Nash equilibrium if and only if $z \leq 3$, and $(B, D, F)$, which is a Nash equilibrium if and only if $x \geq 3$ and $y \geq 2$ and $z \geq 3$.

**(c)** We need to eliminate both $(B, D, E)$ and $(B, D, F)$. To eliminate $(B, D, E)$ we need $z > 3$ and to eliminate $(B, D, F)$ we need either $x < 3$ or $y < 2$. Thus the answer is: $z > 3$ and either $x < 3$ or $y < 2$.

Exercise 2.34 Find the Nash equilibria of the game of Part (a) of Exercise 2.9.

**Solution to Exercise 2.34.**
There is only one Nash equilibrium, namely $(Home, Home, Home)$.

Exercise 2.35 Find the Nash equilibria of the game of Exercise 2.10.

**Solution to Exercise 2.35.**
There is only one Nash equilibrium given by $(n_1 = 0, n_2 = 50)$.

Exercise 2.36 Consider the game of Exercise 2.10. For every strategy profile determine whether there are values of $p$ strictly between 0 and 1 for which that strategy profile is a Nash equilibrium. If a strategy profile can be a Nash equilibrium, find all the values of $p$ for which it is a Nash equilibrium.

**Solution to Exercise 2.36.**

- Since, for every $p$ strictly between 0 and 1, $1 + p < 4$, $(A, E)$ is not a Nash equilibrium (Player 1 can get a higher payoff with $B$).
- Since, for every $p$ strictly between 0 and 1, $4 - 2p < 4$, $(C, E)$ is not a Nash equilibrium (Player 1 can get a higher payoff with $B$).

- Since, for every $p$ strictly between 0 and 1, $1+3p < 4$, $(D,E)$ is not a Nash equilibrium (Player 1 can get a higher payoff with $B$).
- $(B,E)$ is a Nash equilibrium for every value of $p$ between 0 and 1.
- Since $0 < p$, $(D,F)$ is not a Nash equilibrium (Player 1 can get a higher payoff with $A$).
- $(B,F)$ is not a Nash equilibrium (Player 2 can get a higher payoff with $E$).
- Since, for every $p$ strictly between 0 and 1, $p < 2-p$, $(A,F)$ is not a Nash equilibrium (Player 1 can get a higher payoff with $C$).
- In order for $(C,F)$ to be a Nash equilibrium we need the following inequalities to be satisfied.
  For Player 1: $2-p \geq p$ (this is true for every $p$ strictly between 0 and 1), $2-p \geq 2-2p$ (this is true for every $p$ strictly between 0 and 1) and $2-p \geq 0$ (this is true for every $p$ strictly between 0 and 1).
  For Player 2: $4p \geq 1+p$, which is true if and only if $p \geq \frac{1}{3}$.
  Thus $(C,F)$ is a Nash equilibrium for every $p \geq \frac{1}{3}$.

Exercise 2.37 A referee auctions \$1,000 to $n$ players ($n \geq 2$). Each player independently submits an envelope containing her bid in cash; any amount of cash can be put in the envelope and submitting an empty envelope is allowed; bids have to be integers (e.g. submitting \$12 is allowed, but submitting 11.50 is not). If one player's bid is larger than all the other bids, then she wins the \$1,000. If two or more players bid the highest amount, the referee gives nothing to the players, that is, nobody wins. **The submitted envelopes are not returned to the players, that is, no player ever recovers her bid.** Assume that all the players are selfish and greedy, that is, each player cares only about her own net monetary gain. Explain why this game has no Nash equilibria. ■

**Solution to Exercise 2.37.**
Consider an arbitrary strategy profile $(x_1, x_2, \ldots, x_n)$. We want to show that it cannot be a Nash equilibrium. Let player $i$ be the highest bidder or one of the highest bidders, that is, suppose that $x_i \geq x_j$, for every $j \in \{1,2,\ldots,n\}$.

1. if $x_i = 0$ then nobody wins and any player could have done better by increasing her bid from \$0 to \$1.
2. If $x_i > 0$ and, for every other $j \in \{1,2,\ldots,n\}$, $x_j = 0$ then player $i$ is the winner and gets $\$(1,000 - x_i)$. If $x_i > 1$ then player $i$ could have increased her payoff by reducing her bid to \$1. If $x_i = 1$ the any other player could have increased her payoff by increasing her bid from \$0 to \$2.
3. If $x_i > 0$ and, for some $j \in \{1,2,\ldots,n\}$, $x_j > 0$, then player $j$ could have increased her payoff by reducing her bid to \$0.

Exercise 2.38 In a second price auction (where the winner pays not her bid but the second highest bid), for every player bidding her true value is a dominant strategy. Consider the following modification of this auction. For simplicity we consider only two players. The player who submits the highest bid wins (in case of a tie, Player 1 wins). The winner pays not her bid but the **average** of her bid and the other player's bid. The true value of the object to player $i \in \{1,2\}$ is $v_i > 0$. Assume the usual preferences: a player's payoff is zero, if she does not win the object, and the difference between her true value and what she has to pay, if she does win the object. Bids can be any non-negative numbers.

**(a)** Write the payoff function of Player 1 and the payoff function of Player 2.

**(b)** Suppose that $v_1 = 80$. Is bidding \$80 a dominant strategy for Player 1? Whatever your answer, prove your claim. [Note: your proof has to be general enough to cover every possible value of $v_2$.]

**(c)** Suppose that $v_2 = 60$. Is bidding \$60 a dominant strategy for Player 2? Whatever your answer, prove your claim. [Note: your proof has to be general enough to cover every possible value of $v_1$.]

**(d)** Suppose that $0 < v_2 < v_1$. Find at least one Nash equilibrium of this game.

**Solution to Exercise 2.38.**

**(a)** The payoff functions are as follows:

$$\text{Payoff function of Player 1: } \begin{cases} v_1 - \left(\dfrac{b_1+b_2}{2}\right) & \text{if } b_1 \geq b_2 \\ 0 & \text{if } b_1 < b_2 \end{cases}$$

$$\text{Payoff function of Player 2: } \begin{cases} v_2 - \left(\dfrac{b_1+b_2}{2}\right) & \text{if } b_1 < b_2 \\ 0 & \text{if } b_1 \geq b_2 \end{cases}$$

**(b)** No, for Player 1 it is not a dominant strategy to bid \$80 (her true value). To prove this it is sufficient to show that there is at least one case (that is, one possible bid of Player 2) where Player 1 does better with a bid different from \$80. Suppose that Player 2's bid is \$10. Then, if Player 1 bids \$80, the average of the bids is 45 and Player 1's payoff is $80 - 45 = 35$. If, instead, she were to bid, say \$20, she would still win, the average would be 15 and her payoff would be $80 - 15 = 65 > 35$. Note that this argument is completely independent of the value of $v_2$.

**(c)** No, for Player 2 it is not a dominant strategy to bid \$60. Similar argument to Part (a): suppose that Player 1's bid is \$40. Then, if Player 2 bids \$60, the average of the bids is 50 and Player 2's payoff is $60 - 50 = 10$. If, instead, she were to bid, say \$50, she would still win, the average would be 45 and her payoff would be $60 - 45 = 15 > 10$. Note, again, that this argument is independent of the value of $v_1$.

**(d)** For every $b$ such that $v_2 \leq b \leq v_1$, the strategy profile where both players bid \$$b$ is a Nash equilibrium. Proof: given the tie-breaking rule, the object goes to Player 1 for a payment of \$$b$; hence Player 1's payoff is $v_1 - b \geq 0$; if Player 1 were to reduce her bid then she would not win the object and her payoff would be 0; if she bid more than \$$b$ she would still win the object but would have to pay more than \$$b$ (the average of the bids would be higher than $b$), thus her payoff would be less than $v_1 - b$; as for Player 2, her payoff is 0; if she reduced her bid, she would still have a payoff of 0; if she increased her bid, she would win the object but would have to pay more than $v_2$, hence her payoff would be negative.

Exercise 2.39 Consider the game of Exercise 2.14 in Section 2. Find a Nash equilibrium of this game for the case where $v \geq r$ and for the case where $v < r$.

**Solution to Exercise 2.39.** If $v \geq r$, a Nash equilibrium is $d = o = r$, with payoffs of $r$ for the worker and $v - r$ for the employer. If $v < r$, a Nash equilibrium is $d = r$ and $o = 0$, with payoffs of $r$ for the worker and 0 for the employer.

Exercise 2.40 The company eToys is going out of business and seeks a buyer for a truckload of dolls in its warehouse. There are two retailers (Players 1 and 2) who are potentially interested. eToys holds an auction with the following rules. The retailers simultaneously and independently submit sealed bids and then eToys gives the merchandise to the bidder whose bid exceeds the bid of the other bidder (if there is such a bidder) and the winner must pay his own bid (this is called a *first-price auction*). If the retailers submit the same bids ($b_1 = b_2$), then eToys tosses a fair coin to pick the winner. It is common knowledge that the retailer who obtains the load of dolls can resell it for a total of \$15,000. If player $i \in \{1,2\}$ bids more than the other player, then he wins the auction and his payoff is $(15,000 - b_i)$, where $b_i$ denotes his bid; on the other hand, the losing retailer gets a payoff of 0. If the retailers submit the same bids ($b_1 = b_2$), then the payoff of each player $i$ is $\frac{1}{2}(15,000 - b_i)$ (his expected payoff). Find all the Nash equilibria of this game. Prove that they are Nash equilibria and the only Nash equilibria.

■

**Solution to Exercise 2.40.** There is only one Nash equilibrium: $b_1 = b_2 = 15,000$. This is indeed a Nash equilibrium: each player wins with probability $\frac{1}{2}$ and gets a payoff of zero ($\frac{1}{2}(15,000 - 15,000) = 0$); a player who unilaterally switches to a lower bid loses the auction and still gets a payoff of zero, while a player who unilaterally switches to a higher bid wins and gets a negative payoff. Next we prove that no other strategy pair is a Nash equilibrium.

- $(b_1, b_2)$ with $b_1 \geq b_2$ and $b_1 > 15,000$ is not a Nash equilibrium because Player 1 gets a negative payoff and can increase his payoff by reducing $b_1$ to 0. Similarly for $(b_1, b_2)$ with $b_2 \geq b_1$ and $b_2 > 15,000$.
- $(b_1, b_2)$ with $b_2 < b_1 \leq 15,000$ is not a Nash equilibrium, because Player 1 can increase his payoff by reducing $b_1$ slightly (while keeping it above $b_2$). Similarly for $(b_1, b_2)$ with $b_1 < b_2 \leq 15,000$.
- $(b_1, b_2)$ with $b_2 = b_1 < 15,000$ is not a Nash equilibrium, because Player 1's expected payoff is $\frac{1}{2}(15,000 - b_1)$ and he can increase his payoff to $15,000 - b_1'$ by choosing a $b_1'$ slightly greater than $b_1$: $15,000 - b_1' > \frac{1}{2}(15,000 - b_1)$ if and only if $b_1' < \frac{15,000 + b_1}{2}$ (which is less than 15,000 since $b_1 < 15,000$ by hypothesis). Thus any $b_1'$ such that $b_1 < b_1' < \frac{15,000 + b_1}{2}$ gives Player 1 a higher payoff.

Exercise 2.41 FoodQuiet (from now on, FQ) is a restaurant that has been operating for a number of years. It has developed a regular clientele and, until last year, its yearly profits have been \$$R$. Things have changed this year: the MegaSound discotheque (from now on, MS) opened just opposite FQ and changed the character of the neighborhood: the noise and traffic have made the experience of eating at FQ much less pleasant. As a consequence, FQ has suffered a loss in profits equal to \$$L$ (with $0 < L \leq R$). ). MS, on the other hand, has been quite successful, with a yearly profit of \$$P$. FQ threatened to sue MS, but in the end - in order to avoid expensive litigation - they agreed to submit to an arbitrator, the highly regarded retired judge Solomon. Since both $L$ and $P$ are private and non-verifiable information to the respective parties, the judge decides to have the two firms play the following game. FQ will report a non-negative number $\ell$, which will be interpreted by the judge as the profit loss suffered by FQ. MS will report a non-negative number $\Pi$ that the judge will interpret as the yearly profit of MS. The reports are made simultaneously and independently. Note that nothing prevents either party from making a false report (since the true values are private information). The judge then will make the following decision: if $\ell > \Pi$ then MS will have to shut down for a year; if $\Pi \geq \ell$ then MS will be allowed to remain open, but it will have to pay compensation to FQ in the amount of $\ell$. After one year the matter will be reconsidered. Thus the payoff functions are as follows:

$$\text{for FQ: } \begin{cases} R-L+\ell & \text{if } \Pi \geq \ell \\ R & \text{if } \ell > \Pi \end{cases} \qquad \text{for MS: } \begin{cases} P-\ell & \text{if } \Pi \geq \ell \\ 0 & \text{if } \ell > \Pi \end{cases}$$

**(a)** Is it a dominant strategy for FQ to make a truthful report, that is, to choose $\ell = L$?
**(b)** Is it a dominant strategy for MS to make a truthful report, that is, to choose $\Pi = P$?
**(c)** For the case where $P > L$ (recall that $P$ and $L$ denote the *true* amounts), find all the Nash equilibria.
**(d)** For the case where $P < L$, find all the Nash equilibria.
**(e)** Assuming that FQ and MS end up playing a Nash equilibrium, is the outcome Pareto efficient? [a]

[a] An outcome $x$ is *Pareto efficient* if there is no other outcome $y$ such that at least one player prefers $y$ to $x$ and the other player either prefers $y$ to $x$ or is indifferent between $y$ and $x$.

**Solution to Exercise 2.41.**

**(a)** No, for FQ choosing $\ell = L$ is not a dominant strategy. For example, if $\Pi > L$ then choosing $\ell = L$ yields FQ a payoff equal to $R$, while choosing an $\ell$ such that $L < \ell \leq \Pi$ yields a payoff equal to $R + \ell - L > R$.

**(b)** For MS choosing $\Pi = P$ is a weakly dominant strategy. Proof. Fix an arbitrary $\ell$. We must show that $\Pi = P$ gives at least as high a payoff against $\ell$ as any other $\Pi$. Three cases are possible.
Case 1: $\ell < P$. In this case $\Pi = P$ or any other $\Pi$ such that $\Pi \geq \ell$ yields MS a payoff of $P - \ell > 0$, while any $\Pi < \ell$ yields a payoff of 0.
Case 2: $\ell = P$. In this case MS's payoff is zero no matter what $\Pi$ it chooses.
Case 3: $\ell > P$. In this case $\Pi = P$ or any other $\Pi$ such that $\Pi < \ell$ yields a payoff of 0, while any $\Pi \geq \ell$ yields a payoff of $P - \ell < 0$.

**(c)** Suppose that $P > L$. If $(\Pi, \ell)$ is a Nash equilibrium with $\Pi \geq \ell$ then it must be that $\ell \leq P$ (otherwise MS could increase its payoff by reducing $\Pi$ below $\ell$) and it must be that $\ell \geq L$ (otherwise FQ would be better off by increasing $\ell$ above $\Pi$). Thus it must be $L \leq \ell \leq P$, which is possible, given our assumption. However, it cannot be that $\Pi > \ell$, because FQ would be getting a higher payoff by increasing $\ell$ to $\Pi$. Thus it must be $\Pi \leq \ell$, which implies that $\Pi = \ell$. Thus, the following are Nash equilibria:

$$\text{all the pairs } (\Pi, \ell) \text{ with } L \leq \ell \leq P \text{ and } \Pi = \ell.$$

Now consider pairs $(\Pi, \ell)$ with $\Pi < \ell$. Then it cannot be that $\ell < P$, because MS could increase its payoff by increasing $\Pi$ to $\ell$. Thus it must be $\ell \geq P$ (hence, by our supposition, $\ell > L$). Furthermore, it must be that $\Pi \leq L$ (otherwise FQ could increase its profits by reducing $\ell$ to, or below, $\Pi$). Thus

$$(\Pi, \ell) \text{ with } \Pi < \ell \text{ is a Nash equilibrium if and only if } \Pi \leq L \text{ and } \ell \geq P.$$

**(d)** Suppose that $P < L$. For the same reasons given above, an equilibrium with $\Pi \geq \ell$ requires $L \leq \ell \leq P$. However, this is not possible given that $P < L$. Hence,

$$\text{there is no Nash equilibrium } (\Pi, \ell) \text{ with } \Pi \geq \ell.$$

Thus we must restrict attention to pairs $(\Pi, \ell)$ with $\Pi < \ell$. As explained before, it must be that $\ell \geq P$ and $\Pi \leq L$. Thus,

$$(\Pi, \ell) \text{ with } \Pi < \ell \text{ is a Nash equilibrium if and only if } P \leq \ell \text{ and } \Pi \leq L.$$

**(e)** Pareto efficiency requires that the discotheque be shut down if $P < L$ and that it remain open if $P > L$.[2] When $P < L$ all the equilibria have $\Pi < \ell$ which leads to shut-down, hence a Pareto efficient outcome. When $P > L$, there are two types of equilibria: one where $\Pi = \ell$ and the discotheque remains open (a Pareto efficient outcome) and the other where $\Pi < \ell$ in which case the discotheque shuts down, yielding a Pareto inefficient outcome.

Exercise 2.42 Consider again the situation described in Exercise 2.3, but modified as follows: there are now four (rather than three) drivers and the amounts of time required for each stretch of the road are as indicated Figure 2.7. Continue to assume that each traveler only cares about how much time she spends on the road and her objective is to spend the least amount of time on the road.

**(a)** Verify that the only Nash equilibria are those strategy profiles where two drivers take the route via X and the other two take the route via Y (there are six such equilibria).

Now modify the situation illustrated in Figure 2.7 by adding a freeway joining X and Y. The times (in minutes) taken on this new freeway are as follows:

| | |
|---|---|
| if 1 car: | 7 |
| if 2 cars: | 8 |
| if 3 cars: | 9 |
| if 4 cars: | 10 |

Assume that a person who takes the A-X-Y-B route travels the A-X portion at the same time as someone who takes the A-X-B route, and the Y-B portion at the same time as someone who takes the A-Y-B route. In other words, the time it takes to travel each old segment is the same as it was before the X-Y freeway was added. Thus, in the new situation each driver has three options for travel from A to B: A-X-B, A-Y-B and A-X-Y-B.

**(b)** Verify that the strategy profiles that were Nash equilibria in the pre-freeway

[2]For example, suppose that $P < L$ and the discotheque remains open. Then FQ suffers a larger loss than MS gains in profit so that FQ could bribe MS to shut down in exchange for a payment equal to $P + \frac{1}{2}(L - P)$ and both parties would be better off.

situation are no longer Nash equilibria.

**(c)** Prove that the following are now Nash equilibria: one driver takes the A–X–B route, two drivers take the A–X–Y–B route, and one driver takes the A–Y–B route. (In fact, these are the only Nash equilibria.)

**(d)** Did the construction of the X-Y freeway make the travelers better off?

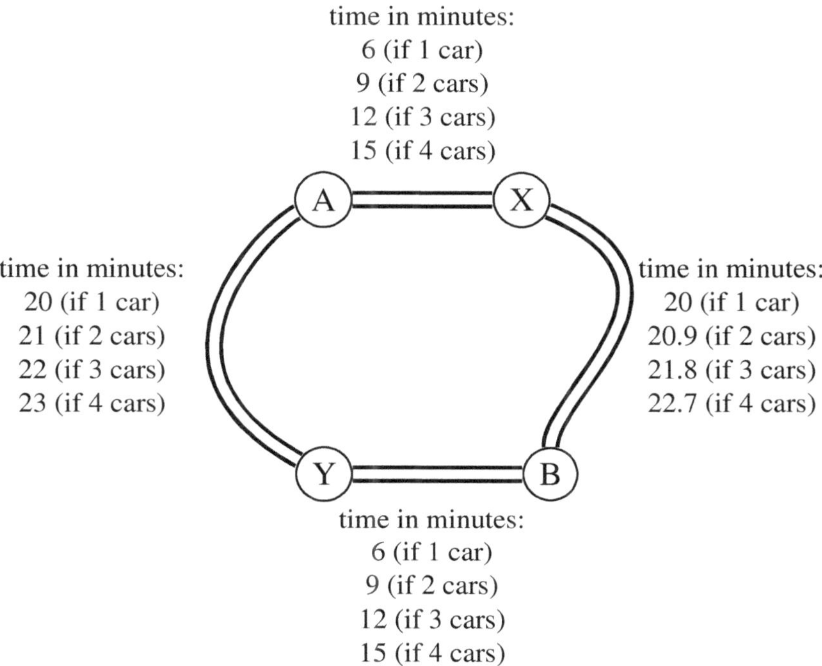


Figure 2.7: The situation described in Exercise 2.38.

**Solution to Exercise 2.42.**

**(a)** When two drivers take the route via X and the other two take the route via Y then the following is true:

- each driver who takes the route via X spends 29.9 minutes on the road and if she were to unilaterally switch to the route via Y her time on the road would increase to 34 minutes,
- each driver who takes the route via Y spends 30 minutes on the road and if she were to unilaterally switch to the route via X her time on the road would increase to 33.8 minutes.

Thus all of these are Nash equilibria. Simple calculations show that any situation where 3 or 4 drivers take the same route is not a Nash equilibrium, since each of these 3 or 4 drivers would spend less time on the road by switching to the alternative route.

**(b)** In the pre-freeway situation the only equilibria were the ones where two drivers took the route A–X–B and two the route A–Y–B, resulting in a total travel time for each person of either 29.9 or 30 minutes. However, if a person taking A–X–B switches to the freeway at X and then takes Y–B her total travel time becomes $9+7+12=28$

minutes. Thus the old equilibria are no longer equilibria.
Now we show that the following situation is a Nash equilibrium: one person takes the A–X–B route, two people take the A–X–Y–B route, and one person takes the A–Y–B route. In this situation, each driver's travel time is 32 minutes. We have to show that no person can change her route and decrease her travel time.

- If the driver taking A–X–B switches to A–X–Y–B, her travel time increases to 12 + 9 + 15 = 36 minutes; if she switches to A–Y–B her travel time increases to 21 + 15 = 36 minutes.
- If one of the drivers taking A–X–Y–B switches to A–X–B, her travel time increases to 12 + 20.9 = 32.9 minutes; if she switches to A–Y–B her travel time increases to 21 + 12 = 33 minutes.
- If the driver taking A–Y–B switches to A–X–B, her travel time increases to 15 + 20.9 = 35.9 minutes; if she switches to A–X–Y–B, her travel time increases to 15 + 9 + 12 = 36 minutes.

[There are no other Nash equilibria: for every other allocation of people to routes at least one driver can switch routes and reduce her travel time. For example, if one driver takes A–X–B, one driver takes A–X–Y–B, and two drivers take A–Y–B, then the travel time of those taking A–Y–B is 21 + 12 = 33 minutes; if one of them switches to A–X–B then her travel time falls to 12 + 20.9 = 32.9 minutes. Another example: if one driver takes A–Y–B, one driver takes A–X–Y–B, and two drivers take A–X–B, then the travel time of those taking A–X–B is 12 + 20.9 = 32.9 minutes; if one of them switches to A–X–Y–B then her travel time falls to 12 + 8 + 12 = 32 minutes.]

**(c)** The addition of the freeway makes *everybody worse off*: in the equilibrium with the freeway every driver's travel time increases from either 29.9 or 30 minutes to 32 minutes.

Exercise 2.43 Give an example of a three-player game where there is a Nash equilibrium $(s_1^*, s_2^*, s_3^*)$ such that, for every player $i = 1,2,3$, there is a strategy $\hat{s}_i \in S_i$ that weakly dominates $s_i^*$. ■

**Solution to Exercise 2.43.**
There are, of course, many such games. In the game shown below, (A,A,A) is a Nash equilibrium where, for each player, A is weakly dominated by B.

Player 3: A

| Player 1 \ Player 2 | A | B |
|---|---|---|
| A | 0 0 0 | 1 0 1 |
| B | 0 1 0 | 2 2 1 |

Player 3: B

| Player 1 \ Player 2 | A | B |
|---|---|---|
| L | 0 0 0 | 1 0 2 |
| M | 0 1 0 | 2 2 2 |

Exercise 2.44 A two-player game is symmetric if $S_1 = S_2$ and, for every $(x,y) \in S_1 \times S_2$, $\pi_1(x,y) = \pi_2(y,x)$. For each of the following claims either provide a proof (if you think the claim is correct) or provide a counterexample (if you think the claim is false).

**(a)** If $(x^*, y^*)$ is a Nash equilibrium then $x^* = y^*$, that is, at every Nash equilibrium the two players make the same choices.

**(b)** If the set of Nash equilibria is non-empty, then it must include one of the form $(x^*, y^*)$ with $x^* = y^*$, that is, perhaps some Nash equilibria involve different choices but at least one involves the same choice for both players.

■

**Solution to Exercise 2.44.**

Both claims are false. The following is a counterexample to both. $S_1 = S_2 = \{a,b\}$, $\pi_1(a,a) = \pi_2(a,a) = \pi_1(b,b) = \pi_2(b,b) = 0$ and $\pi_1(a,b) = \pi_2(b,a) = \pi_1(b,a) = \pi_2(a,b) = 1$. This game has two Nash equilibria: $(a,b)$ and $(b,a)$.

Exercise 2.45 Two travelers returning home from a vacation discover that their identical statuettes they bought have been smashed in transit. The airline wants to reimburse the travelers in an amount equal to the price they paid. However, this is private information to the travelers; in principle they could claim that they paid more than they actually did and the airline would have no way of finding out. The airline manager, having taken a course in game theory, sets up the following scheme to elicit the value of the article. The two travelers are put in separate rooms and asked to independently fill in a compensation claim between \$2 and \$100 (in increments of \$1; thus only integer amounts). The airline will then reimburse each traveler at the smallest of the two claims; in addition, if the claims differ, a reward of \$2 will be paid to the person making the smaller claim and a penalty of \$2 will be deducted from the reimbursement to the larger claimant (for example, if Player 1 claims \$78 and Player 2 claims \$55 then Player 1 gets $\$(55-2) = \$53$ while Player 2 gets $\$(55+2) = \$57$). For parts (a)-(e) assume that each player is selfish and greedy in the sense that each player only cares about how much money he gets and prefers more money to less.

**(a)** Sketch the strategic-form of this game. Since you probably do not have the time or inclination to fill in almost 10,000 entries, show the pattern by showing the payoffs when the travelers claim \$2, \$3, \$4, \$98, \$99 and \$100 (and every combination of these). [Exploit the symmetry of the problem.]

**(b)** Do the players have a dominant strategy?

**(c)** What do you get by applying the IDWDS (iterative deletion of weakly dominated strategies) procedure?

**(d)** Find all the Nash equilibria of this game. Prove that they are Nash equilibria and that there are no other Nash equilibria.

**(e)** Assuming that the true value of the damaged article is indeed between \$2 and \$100, does this game induce the travelers to reveal the true price they paid?

Given a strategy pair $(m,n)$ define the *regret* of Player 1, denoted by $R_1(m,n)$, as the difference between the maximum amount of money Player 1 could have got, given that Player 2 chose $n$, and the amount of money he actually gets at $(m,n)$.

**(f)** Find $R_1(m,n)$ for all $(m,n) \in \{2,...,100\} \times \{2,...,100\}$.

**(g)** What are the maximum possible regret and the minimum possible regret (over the entire set $\{2,...,100\} \times \{2,...,100\}$)?

**(h)** Defining Player 2's regret $R_2(m,n)$ similarly (as the difference between the maximum amount of money Player 2 could have got, given that Player 1 chose $m$, and the amount of money she actually gets at $(m,n)$), write a table similar to the one for Part (a) but with regrets instead of payoffs. As before, you only need to show the pattern by showing regrets when the travelers claim \$2, \$3, \$4, \$96, \$97, \$98, \$99 and \$100 (and every combination of these; please note that, relative to Part (a), we have added \$96 and \$97). [Exploit the symmetry of the problem.]

**(i)** Suppose that each player chooses in a MinMax way in the following sense: for each claim of his the player finds the maximum possible regret and then chooses a claim that minimizes the maximum regret. What strategy profiles are consistent with both players choosing in a MinMax way?

**(j)** Suppose that it is common knowledge between the players that they choose in a MinMax way (as explained above). Apply the following iterative deletion procedure: in each round eliminate all those claims that are weakly dominated (in the current table). What do you get by applying this iterative deletion procedure?

**Solution to Exercise 2.45.**

**(a)** The strategy sets are $S_1 = S_2 = \{\$2, \$3, \$4, \ldots, \$100\}$. We can take as utility function for Player 1 the function $U_1(\$x,\$y) = x$ and for Player 2 the function $U_2(\$x,\$y) = y$ so that the payoff functions are:

$$\pi_1(\$m,\$n) = \begin{cases} n-2 & \text{if } m > n \\ m & \text{if } m = n \\ m+2 & \text{if } m < n \end{cases} \quad \text{and} \quad \pi_2(\$m,\$n) = \begin{cases} n+2 & \text{if } m > n \\ m & \text{if } m = n \\ m-2 & \text{if } m < n \end{cases}$$

| | \$2 | \$3 | \$4 | ... | \$98 | \$99 | \$100 |
|---|---|---|---|---|---|---|---|
| \$2 | 2 2 | 4 0 | 4 0 | ... | 4 0 | 4 0 | 4 0 |
| \$3 | 0 4 | 3 3 | 5 1 | ... | 5 1 | 5 1 | 5 1 |
| \$4 | 0 4 | 1 5 | 4 4 | ... | 6 2 | 6 2 | 6 2 |
| ... | ... | ... | ... | ... | ... | ... | ... |
| \$98 | 0 4 | 1 5 | 2 6 | ... | 98 98 | 100 96 | 100 96 |
| \$99 | 0 4 | 1 5 | 2 6 | ... | 96 100 | 99 99 | 101 97 |
| \$100 | 0 4 | 1 5 | 2 6 | ... | 96 100 | 97 101 | 100 100 |

**(b)** Neither player has a dominant strategy.

**(c)** \$99 dominates \$100 for each player; deleting \$100 for both, we get a smaller game where \$98 dominates \$99 for each player, and so on. Thus the output of the IDWDS procedure is the strategy pair (\$2,\$2).

**(d)** There is only one Nash equilibrium, namely (\$2,\$2). Given that the other player is reporting \$2, reporting \$2 gives a payoff of 2, while reporting more than \$2 gives a payoff of 0. $(\$m,\$n)$ with $m = n > 2$ is not a Nash equilibrium because Player 1 can increase his payoff from $m$ to $(m+1)$ by switching his strategy to $(m-1)$. $(\$m,\$n)$ with $m > n \geq 2$ is not a Nash equilibrium because Player 1 can increase his payoff from $(n-2)$ to $n$ by switching his strategy to $n$. Finally, $(\$m,\$n)$ with $2 \leq m < n$ is not a Nash equilibrium because Player 2 can increase her payoff from $(m-2)$ to $m$ by switching her strategy to $m$.

**(e)** If we base the prediction of what the players will do either on the output of the IDWDS or on the notion of Nash equilibrium, the prediction is that they will both report \$2, no matter what the true value of the object. Thus the answer is: No.

**(f)** If Player 2 chooses to report \$$n$ with $n > 2$ the best reply of Player 1 (if he wants to get as much money as possible) is to choose $n-1$ (if $n = 2$ player 1's best reply is to choose 2 also). Let $B_1(n)$ denote the maximum amount of money that Player 1 can get when Player 2 chooses $n$. Then $B_1(n) = \begin{cases} 2 & \text{if } n = 2 \\ n+1 & \text{if } n > 2 \end{cases}$. Thus,

$$R_1(m,2) = \begin{cases} 0 & \text{if } m = 2 \\ 2 & \text{if } m > 2 \end{cases} \quad \text{and, for } n > 2, R_1(m,n) = \begin{cases} n-m-1 & \text{if } m \le n-1 \\ 1 & \text{if } m = n \\ 3 & \text{if } m > n \end{cases}.$$

**(g)** The minimum regret is 0 (e.g. when $m = n = 2$) and the maximum regret is 97 (when $m = 2$ and $n = 100$).

**(h)** The regret table is as follows:

| | \$2 | | \$3 | | \$4 | | ... | \$96 | | \$97 | | \$98 | | \$99 | | \$100 | |
|---|---|---|---|---|---|---|---|---|---|---|---|---|---|---|---|---|---|
| \$2 | 0 | 0 | 0 | 2 | 1 | 2 | ... | 93 | 2 | 94 | 2 | 95 | 2 | 96 | 2 | 97 | 2 |
| \$3 | 2 | 0 | 1 | 1 | 0 | 3 | ... | 92 | 3 | 93 | 3 | 94 | 3 | 95 | 3 | 96 | 3 |
| \$4 | 2 | 1 | 3 | 0 | 1 | 1 | ... | 91 | 3 | 92 | 3 | 93 | 3 | 94 | 3 | 95 | 3 |
| ... | ... | | ... | | ... | | ... | ... | | ... | | ... | | ... | | ... | |
| \$96 | 2 | 93 | 3 | 92 | 3 | 91 | ... | 1 | 1 | 0 | 3 | 1 | 3 | 2 | 3 | 3 | 3 |
| \$97 | 2 | 94 | 3 | 93 | 3 | 92 | ... | 3 | 0 | 1 | 1 | 0 | 3 | 1 | 3 | 2 | 3 |
| \$98 | 2 | 95 | 3 | 94 | 3 | 93 | ... | 3 | 1 | 3 | 0 | 1 | 1 | 0 | 3 | 1 | 3 |
| \$99 | 2 | 96 | 3 | 95 | 3 | 94 | ... | 3 | 2 | 3 | 1 | 3 | 0 | 1 | 1 | 0 | 3 |
| \$100 | 2 | 97 | 3 | 96 | 3 | 95 | ... | 3 | 3 | 3 | 2 | 3 | 1 | 3 | 0 | 1 | 1 |

**(i)** Let $M_1(m) = \max_n\{R_1(m,n)\}$. Then $M_1(m) = \max\{3, (99-m)\}$. Thus MinMax requires choosing from the set $\{96, 97, 98, 99, 100\}$.

**(j)** If the MinMax criterion is common knowledge, then it is common knowledge that each player will choose from the set $\{96, 97, 98, 99, 100\}$. By referring to the lower right portion of the matrix in Part (i), it can be seen that \$100 is weakly dominated by \$99. After deleting \$100 for both players, \$99 is weakly dominated by \$98, etc. Thus the iterative deletion procedure leads to each player choosing \$96 and getting \$96.

## 2.7 Games with infinite strategy sets

The exercises in this section deal with games with infinite strategy sets (Volume 1, Chapter 2, Section 2.7).

Exercise 2.46 Consider the following two-player game. The strategy sets are $S_1 = S_2 = [0,1]$ (the set of real numbers from 0 to 1). The payoff function $\pi_i(x,y)$ of player $i \in \{1,2\}$ is as follows, where $x$ denotes the strategy of Player 1 and $y$ the strategy of Player 2:

$$\pi_1(x,y) = \begin{cases} x & \text{if } y=0 \text{ and } x<\frac{1}{2} \\ 1-x & \text{if } y=0 \text{ and } x \geq \frac{1}{2} \\ 1-x & \text{if } y>0 \text{ and } \frac{y}{2} \leq x \\ x & \text{if } y>0 \text{ and } x<\frac{y}{2} \end{cases}$$

$$\pi_2(x,y) = \begin{cases} y & \text{if } x=0 \text{ and } y<\frac{1}{2} \\ 1-y & \text{if } x=0 \text{ and } y \geq \frac{1}{2} \\ 1-y & \text{if } x>0 \text{ and } \frac{x}{2} \leq y \\ y & \text{if } x>0 \text{ and } y<\frac{x}{2} \end{cases}$$

**(a)** Show that each player has at least one strictly dominated strategy.
**(b)** What is the output of the IDSDS (iterated deletion of strictly dominated strategies) procedure?
**(c)** Prove that this game has no Nash equilibria.

**Solution to Exercise 2.46.**

**(a)** Consider first Player 1. First of all, note that $\pi_1(x,y) \geq 0$ for all $(x,y)$. Choosing $x=0$ is strictly dominated by $x=\frac{1}{2}$. In fact, $x=0$ gives a payoff of 0 no matter what $y$ is, while $x=\frac{1}{2}$ gives a payoff of $\frac{1}{2}$ (for any $y$; note that, for every $y \in [0,1]$, $\frac{y}{2} \leq \frac{1}{2}$). Furthermore, any $x$ such that $\frac{1}{2} < x \leq 1$ is strictly dominated by $x=\frac{1}{2}$, because the largest possible value for $\frac{y}{2}$ is $\frac{1}{2}$ so a choice of $x \in (\frac{1}{2},1]$ gives a payoff of $1-x<\frac{1}{2}$ (for any $y$), while $x=\frac{1}{2}$ gives a payoff of $\frac{1}{2}$ (for any $y$). Since the game is symmetric, the same reasoning applies to Player 2.

**(b)** By what we established in Part (a), in the first round of deletion we can delete 0 as well as any number in the range $(\frac{1}{2},1]$ for both players, thus reducing the game to a smaller game where the strategy sets are $S_1 = S_2 = \left(0,\frac{1}{2}\right]$. In this smaller game, for Player 1 any $x \in (\frac{1}{4},\frac{1}{2}]$ gives a payoff of $1-x<\frac{3}{4}$ so that it is strictly dominated by $x=\frac{1}{4}$ which gives a payoff of $\frac{3}{4}$ (since the largest possible value for $\frac{y}{2}$ is now $\frac{1}{4}$). Again, by symmetry, the same is true for Player 2. Thus a second round of deletion shrinks each strategy set to $\left(0,\frac{1}{4}\right]$. Continuing this reasoning, a further round shrinks the sets to $\left(0,\frac{1}{8}\right]$, etc. That is, round $n$ of deletion shrinks the strategy sets to $\left(0,\frac{1}{2^n}\right]$. Taking the limit as $n \to \infty$ we get the empty set! In other words, every strategy pair is deleted at some round of the IDSDS procedure.

**(c)** Since 0 is a strictly dominated strategy, if $(x,y)$ is a Nash equilibrium it must be that both $x$ and $y$ are positive. Now, $(x,y)$ with $x \geq y > 0$ (so that $x > \frac{y}{2}$) is not a Nash equilibrium since Player 1 can increase his payoff by switching to $x = \frac{y}{2}$. Similarly, $(x, y)$ with $y \geq x > 0$ is not a Nash equilibrium because Player 2 can increase her payoff by switching to $y = \frac{x}{2}$. Thus there are no Nash equilibria.

Exercise 2.47 An industry consists of two firms that produce a homogeneous product. Each firm can produce any non-negative amount of output. Let $q_1$ denote the output of Firm 1 and $q_2$ the output of Firm 2. The inverse demand function is given by: $P = e^{-Q}$ where $Q = q_1 + q_2$ is total industry output.[a] Each firm's aim is to maximize its own revenue (the revenue of firm $i \in \{1,2\}$ is defined as $q_i P$).

**(a)** Represent this situation as a strategic-form game by specifying the strategy sets and the payoff functions.

**(b)** Suppose that Firm 1 expects Firm 2 to produce 2 units; what output should Firm 1 choose? What if Firm 1 expects Firm 2 to produce 3 units? And if it expects Firm 2 to produce 4 units?

**(c)** Does Firm 1 have a dominant strategy?

**(d)** Is there a dominant-strategy profile?

■

[a] Recall that $e$ is the irrational number 2.71828... and that $\frac{d}{dx}e^x = e^x$ and $e^{-x} = \frac{1}{e^x}$, which is positive for every non-negative $x$.

**Solution to Exercise 2.47.**

**(a)** The strategy sets are $S_1 = S_2 = [0, \infty)$. The payoff function of Firm 1 is $R_1(q_1, q_2) = q_1 e^{-(q_1+q_2)}$ and the payoff function of Firm 2 is $R_2(q_1, q_2) = q_2 e^{-(q_1+q_2)}$.

**(b)** Fix and arbitrary level of output for Firm 2, call it $\hat{q}_2$. If Firm 1 expects Firm 2 to produce $\hat{q}_2$ units, then the output level of Firm 1 that maximizes Firm 1's profit is given by the solution (with respect to $q_1$) to the equation $\frac{\partial}{\partial q_1} R_1(q_1, \hat{q}_2) = 0$, that is, the equation $(1 - q_1)e^{-(q_1+\hat{q}_2)} = 0$ which is $q_1 = 1$. Note that the solution is independent of $\hat{q}_2$. Thus the answer is: Firm 1 should produce 1 unit, no matter what the output of Firm 2 is.

**(c)** Yes, $q_1 = 1$ is a dominant strategy for Firm 1.

**(d)** The same reasoning applies to Firm 2 (the game is symmetric). Thus the dominant-strategy profile is $(q_1 = 1, q_2 = 1)$.

Exercise 2.48 Two players are involved in a dispute over an object. The value of the object to player $i \in \{1,2\}$ is $\$vi > 0$. Time is modeled as a continuous variable that starts at 0 and runs indefinitely. Each player chooses when to concede the object to the other player; if the first player to concede does so at time $t$, then the other player obtains the object at that time. If both players concede simultaneously, the object is split equally between them, so that player $i$ receives something worth $\$\frac{v_i}{2}$ to her. Time is valuable: until a concession is made by either player, each player loses \$1 per unit of time. The payoff of player $i$ is equal to the value of what she gets (0 if she gets nothing) minus the cost of waiting.

Formulate this situation as a strategic-form game and find all the pure-strategy Nash equilibria (prove that they are Nash equilibria and that they are the only Nash equilibria).

■

**Solution to Exercise 2.48.** The set of strategies of player $i$ is $S_i = [0,\infty)$ and the payoff functions are

$$\pi_1(t_1,t_2) = \begin{cases} -t_1 & \text{if } t_1 < t_2 \\ \frac{v_1}{2} - t_1 & \text{if } t_1 = t_2 \\ v_1 - t_1 & \text{if } t_1 > t_2 \end{cases} \quad \text{and} \quad \pi_2(t_1,t_2) = \begin{cases} v_2 - t_1 & \text{if } t_1 < t_2 \\ \frac{v_2}{2} - t_1 & \text{if } t_1 = t_2 \\ -t_2 & \text{if } t_1 > t_2 \end{cases}$$

Let $(t_1,t_2)$ be a strategy profile. If $t_1 = t_2$ then by conceding slightly later than $t_1$ Player 1 can obtain the object in its entirety instead of getting just half of it, so this is not a Nash equilibrium. If $0 < t_1 < t_2$ then Player 1 can increase her payoff to from $-t_1$ to zero by switching to $t_1 = 0$; thus this is not a Nash equilibrium either. The same reasoning applies to Player 2 if $0 < t_2 < t_1$. If $0 = t_1 < t_2$ then Player 1 can increase her payoff by deviating to a time slightly after $t_2$ unless $v_1 - t_2 \leq 0$. Similarly for Player 2. Thus, $(t_1,t_2)$ is a Nash equilibrium if and only if

$$\text{either} \quad 0 = t_1 < t_2 \text{ and } t_2 \geq v_1 \quad \text{or} \quad 0 = t_2 < t_1 \text{ and } t_1 \geq v_2.$$

# 3. Perfect-information Games

## 3.1 Trees, frames and games

The exercises in this section deal with the notions of *finite extensive form (or frame) with perfect information* and *finite extensive game with perfect information* (Definitions 3.1.2 and 3.1.3, Volume 1, Chapter 3, Section 3.1).

Exercise 3.1 Three salesmen (Drew, Jeff, and Luke) are deciding which of two clients (X or Y) to pursue. There are two sign-up sheets (one for each client), and a salesman makes his decision by writing his name on the sign-up sheet for his chosen client. At least one salesman must pursue each client, so once two have signed up for a given client, the third salesman is automatically assigned to the other client (no action is necessary on his part in this case). Drew is the most senior salesman, so he gets to make his decision first; Jeff sees what Drew chose and makes his choice next; finally Luke sees what Drew and Jeff chose and makes his decision (unless both Drew and Jeff signed up for the same client). For example, if Drew and Jeff sign up for Y then Luke has no choice and is assigned to X; if Drew signs up for X and Jeff for Y then Luke has a choice between signing up for X or signing up for Y.

**(a)** Draw an extensive-form frame with perfect information to represent this situation.

**(b)** Convert the frame of Part (a) into a game using the following information and utility values from the set $\{0,1,2\}$. The two clients are identical, so each salesman cares only about which other salesman (if any) he is partnered with. It is common knowledge among the three of them that they rank the possible outcomes as follows:

| | Drew | Jeff | Luke |
|---|---|---|---|
| best | working with Jeff | working with Luke | working with Drew |
| middle | working alone | working with Drew | working with Jeff |
| worst | working with Luke | working alone | working alone |

**Solution to Exercise 3.1.**

**(a)** The extensive-form frame is shown in Figure 3.1.

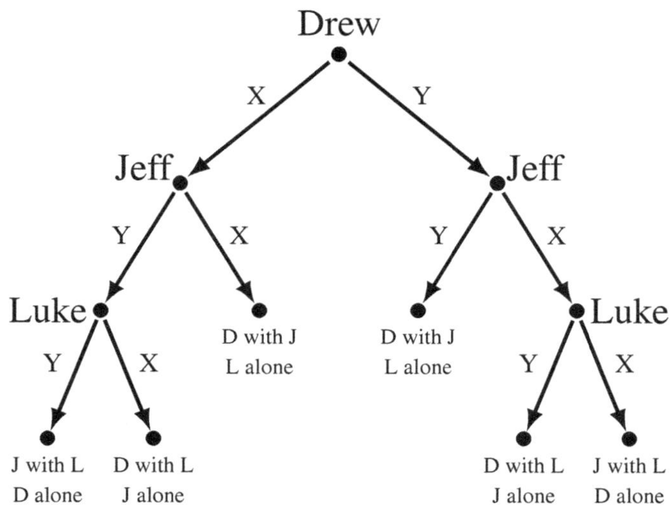


Figure 3.1: The extensive-form frame of Exercise 3.1.

**(b)** The extensive-form game is shown in Figure 3.2.

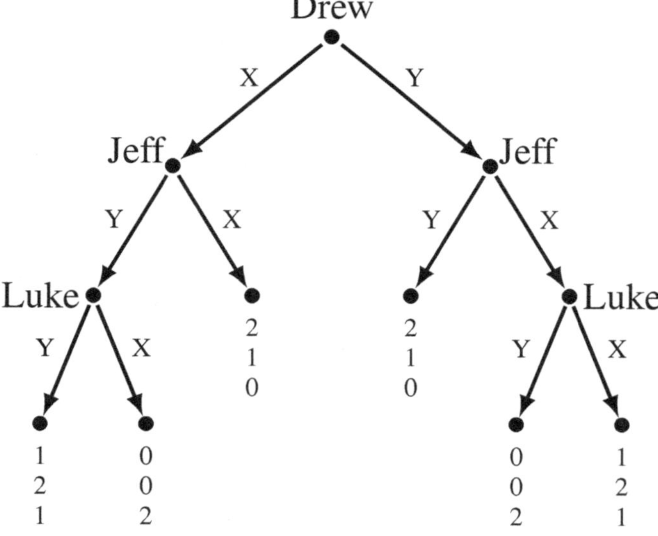


Figure 3.2: The extensive-form game based on the frame of Figure 3.1.

Exercise 3.2 There are two firms in an industry. They produce an identical good. Each firm can produce either 3 or 4 or 6 units of output. Let $q_1$ denote the output of Firm 1 and $q_2$ the output of Firm 2. The price of the product, denoted by $P$, is determined by total industry output $Q$ (which is defined as $Q = q_1 + q_2$) according to the following formula: $P = 24 - Q$. The cost of producing each unit of output is equal to \$12 and it is the same for both firms. Firm 1 chooses its output first, then Firm 2 observes the choice of Firm 1 and chooses its own output. Draw an extensive form game-frame with perfect information to represent this situation. Denote an outcome as quadruple $(Q, P, C_1, C_2)$ where $Q$ is industry output, $P$ the price, $C_1$ the total cost of Firm 1 and $C_2$ the total cost of Firm 2. ■

**Solution to Exercise 3.2.** The extensive-form frame is shown in Figure 3.3.

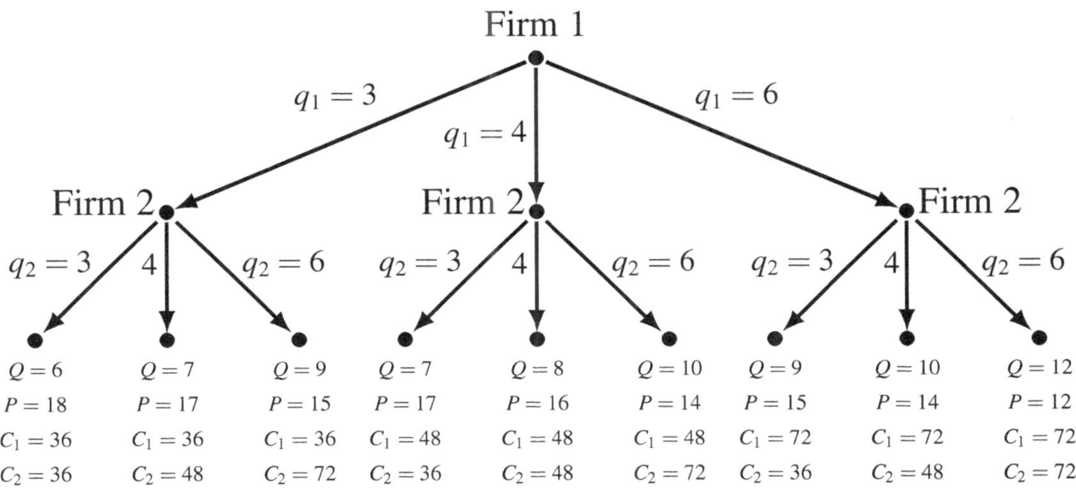


Figure 3.3: The extensive-form game-frame of Exercise 3.2.

Exercise 3.3 Consider again the extensive-form frame of Exercise 3.2 (Figure 3.3). Suppose that each firm is interested in, and only in, its own profit, which is defined as: revenue minus cost. A firm's revenue is equal to the product of the firm's output and the price; its cost is equal to the product of its unit cost (which is \$12) and its output. Thus the profit of Firm 1 is $\Pi_1(q_1, q_2) = q_1(24 - q_1 - q_2) - 12q_1$ and the profit of Firm 2 is $\Pi_2(q_1, q_2) = q_1(24 - q_1 - q_2) - 12q_2$. Convert the extensive-form frame of Exercise 3.1 into a game in extensive form by taking as utility of an outcome for firm $i$ the corresponding profit for firm $i$. ■

**Solution to Exercise 3.3.** The extensive-form game is shown in Figure 3.4.

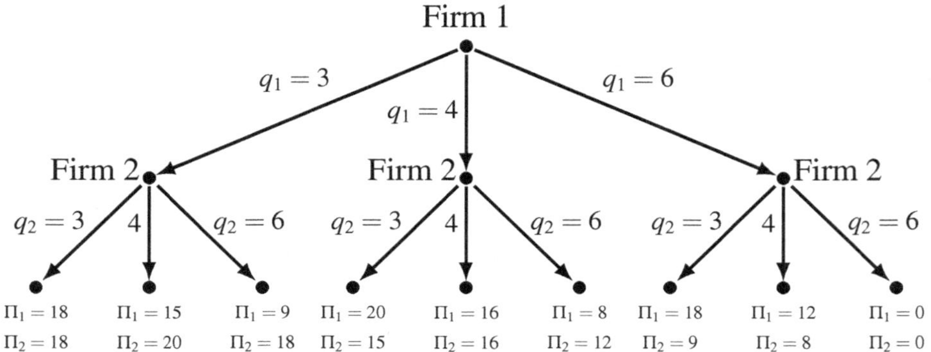


Figure 3.4: The extensive-form game of Exercise 3.3.

Exercise 3.4 There are two consumers potentially interested in a product. One is willing to pay up to \$10 for one unit of it and would buy at most one unit. Thus this consumer, call her "Low", will buy exactly one unit if and only if $p \leq 10$, where $p$ is the price of the product. The other consumer, call him "High", will buy one unit if and only if the price is not greater than \$12. The manufacturer produces the good at zero cost and sells it to a retailer at a price of \$$w$ per unit. The retailer has no other costs and sets the price $p$ to consumers. Assume that $w$ can only be either 5 or 9 and $p$ can only be either 9 or 12. Draw an extensive-form game-frame to represent this situation where the players are the manufacturer and the retailer and each outcome specifies the profit of each player (profit is defined as revenue minus cost, revenue is defined as the quantity sold times the price and cost as the quantity sold times the cost per unit). ■

**Solution to Exercise 3.4.** The extensive-form frame is shown in Figure 3.5, where, at each terminal node, the top number is the profit of the manufacturer and the bottom number is the profit of the retailer.

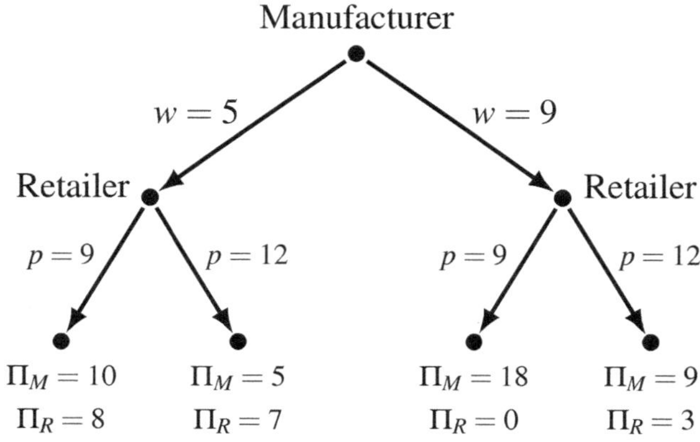


Figure 3.5: The extensive-form frame for Exercise 3.4.

Exercise 3.5 Players $A$ and $B$ have to decide how to divide \$100 between them. They agree to the following procedure. $A$ starts by making an offer of $\$x_1$ to $B$ where $x_1 \in \{25, 50, 75\}$. Player $B$ either accepts or rejects. If $B$ accepts then the game ends with $B$ receiving $\$x_1$ and $A$ receiving $\$(100 - x_1)$. If $B$ rejects the offer, then \$25 is given to a charity and it becomes $B$'s turn to make an offer of $\$x_2$ to $A$ where $x_2 \in \{25, 50\}$. Player $A$ can accept or reject. If $A$ rejects then the remaining \$75 is given to the charity and both $A$ and $B$ end up with nothing. If $A$ accepts then the game ends with $A$ receiving $\$x_2$ and $B$ receiving $\$(75 - x_2)$. Describe an outcome as a triple $\begin{matrix} a \\ b \\ c \end{matrix}$ where $a$ is the amount that $A$ receives, $b$ the amount that $B$ receives and $c$ is the amount that the charity receives. The only players are $A$ and $B$ (the charity is not a player). Draw an extensive-form game-frame with perfect information to represent this bargaining situation.

**Solution to Exercise 3.5.** The extensive-form frame is shown in Figure 3.6.

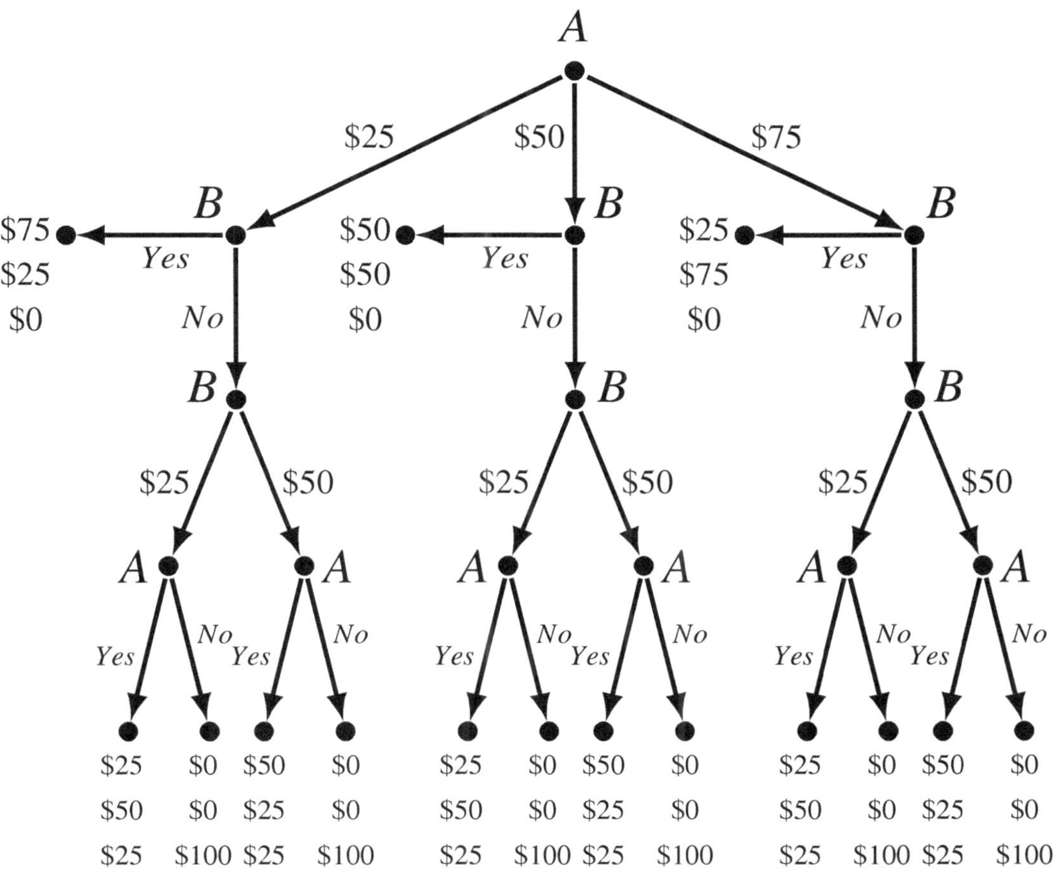


Figure 3.6: The extensive-form frame for Exercise 3.5.

## 3.2 Backward induction

The exercises in this section deal with the notion of *backward induction* (Definition 3.2.1, Volume 1, Chapter 3, Section 3.2).

Exercise 3.6 Ann and Barbara, who do not know each other, have been chosen randomly to play the following game. They are kept in separate rooms. A referee gives each of them \$10. Ann makes the first move, by deciding whether to keep her \$10 (in which case the game ends and each player gets to keep the \$10 given to her) or return her \$10 to the referee, who will then add \$30 and give the resulting sum of \$40 to Barbara. Barbara now decides whether to keep her \$50 (\$10 given to her initially and \$40 added now) – in which case the game ends and she goes home with \$50 and Ann goes home with nothing – or to send half of the money back to Ann, in which case the game ends with each player getting \$25.

**(a)** Draw an extensive-form game-frame to represent this situation.

**(b)** Do you agree with the following statement (taken from Dan Arieli, *The upside of irrationality*, Harper Collins, 2010, p. 125)?

> "The prediction of rational economics is very simple: no one would ever give back half of their \$50, and, since this behavior is so glaringly predictable from a rational economic perspective, no one would ever send over their \$10 in the first place."

**(c)** What assumptions need to be made in order to conclude that the backward induction solution is as described in the above quotation?

A team of Swiss researchers led by Ernst Fehr modified the game by adding a further choice by Ann in case she returns her \$10 and Barbara decides to keep her \$50: Ann can choose to pay, out of her own pocket, any amount from the set $\{\$0,\$10,\$20,\$25\}$ and if she chooses the amount $x$ then twice that amount is taken away from Barbara's \$50 and the game ends.

**(e)** Draw an extensive-form game-frame to represent the modified interaction.

**(f)** Suppose that it is common knowledge that each player cares only about her own wealth (she prefers more money to less and is indifferent between any two outcomes where her own wealth is the same). Express this hypothesis in terms of a ranking of the terminal nodes and find the backward induction solution.

**(g)** Suppose now that Barbara believes that (1) Ann essentially cares about her own wealth (and prefers more money to less) but (2) Ann derives enjoyment from punishing non-reciprocating behavior, in the sense that if she observes Barbara being selfish, then she strictly prefers paying more money to paying less money in order to punish Barbara; however, she prefers not to have to punish Barbara. Furthermore Ann believes that Barbara believes the above and Ann and Barbara in fact do prefer more money to less. Express the above preferences in terms of a ranking of the terminal nodes and find the backward induction solution.[a]

[a] It is interesting to note that in the Swiss experiment, the brain of the participant in the role of Ann was scanned by positron emission tomography (PET). The results showed increased activity in the striatum (an area of the brain associated with the experience or rewards) during the punishment phase. In other

words, according to the PET scan it looked as though the decision to punish was related to a feeling of pleasure. Furthermore, those who had a high level of striatum activation punished others to a greater degree by giving up more of their own money.

**Solution to Exercise 3.6.**

**(a)** The extensive-form frame is shown in Figure 3.7

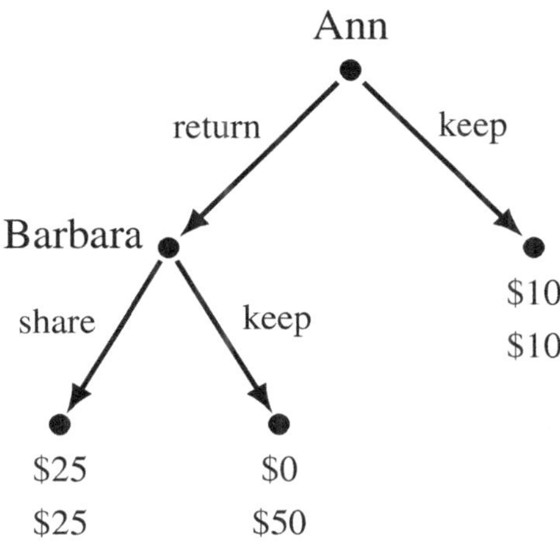


Figure 3.7: The extensive-form frame for Exercise 3.6.

**(b)** The claim is **not** correct: in game theory (and economics) rationality is defined in terms of choosing an action which is optimal given one's preferences and beliefs. It is quite possible that Barbara strictly prefers the outcome ($25,$25) to the outcome ($0,$50), even though she gets half the amount of money in the first outcome: such preferences reflect a notion of fairness or reciprocation or moral obligation. There is no assumption in game theory or economics that rules out such preferences. Thus it is quite possible that a rational Ann will decide to send her $10 and a rational Barbara will decide to reciprocate by sending $25 back to Ann.

**(c)** In order to conclude that Ann will keep her $10 we need to assume that Ann believes (possibly erroneously) that Barbara ranks ($0,$50) above ($25,$25) and that Ann herself ranks ($10,$10) above ($0,$50). In the backward-induction solution there is also the counterfactual choice of Barbara to keep her $50 if given the move. If we interpret this as a statement about what Barbara would actually do if Ann were to return her $10, then we need to add the assumption that Ann's beliefs about Barbara's preferences are in fact correct.

**(d)** The extensive-form frame is shown in Figure 3.8

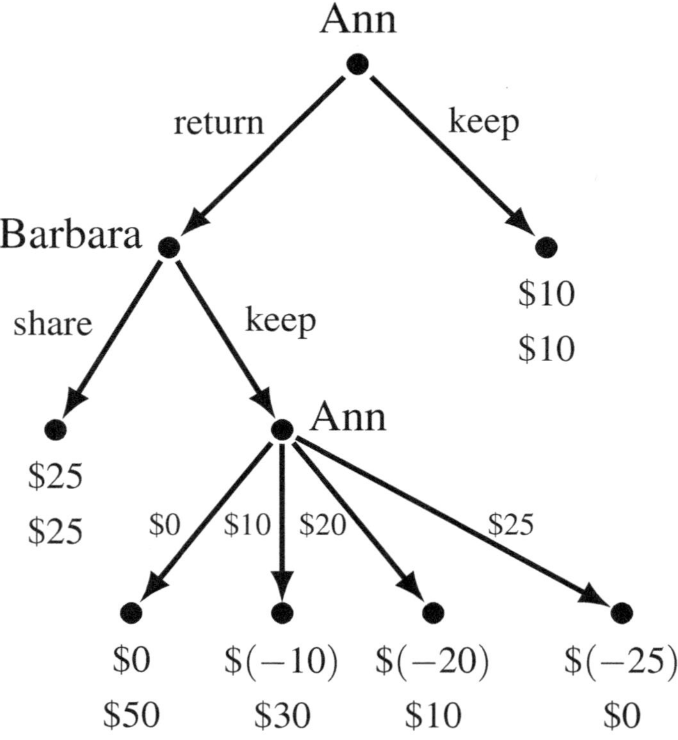


Figure 3.8: The case where Ann can punish Barbara.

**(e)** Ann's ranking is:

$$(\$25,\$25) \succ (\$10,\$10) \succ (\$0,\$50) \succ (\$-10,\$30) \succ (\$-20,\$10) \succ (\$-25,\$0)$$

Barbara's ranking is:

$$(\$0,\$50) \succ (\$-10,\$30) \succ (\$25,\$25) \succ (\$10,\$10) \sim (\$-20,\$10) \succ (\$-25,\$0)$$

The backward-induction solution is: Ann keeps her \$10 and, if she were at her second node, she would pay \$0 (thus not punishing Barbara) and Barbara would keep the entire amount (if given a chance to move).

**(f)** Ann's ranking is:

$$(\$25,\$25) \succ (\$10,\$10) \succ (\$-25,\$0) \succ (\$-20,\$10) \succ (\$-10,\$30) \succ (\$0,\$50)$$

Barbara's ranking is:

$$(\$0,\$50) \succ (\$-10,\$30) \succ (\$25,\$25) \succ (\$10,\$10) \sim (\$-20,\$10) \succ (\$-25,\$0)$$

The backward-induction solution is: Ann returns her \$10 and, if she were at her second node, she would pay \$25 and Barbara reciprocates by sharing the \$50.

Exercise 3.7 Consider the extensive-form game-frame with perfect information shown in Figure 3.9. The players have complete and transitive preferences over the set of outcomes $\{z_1, z_2, z_3, z_4\}$ and their preferences are such that the corresponding game has exactly three backward-induction solutions, namely $(a,(c,e))$, $(b,(c,e))$ and $(b,(c,f))$.

**(a)** What can you infer about the preferences of Player 1?

**(b)** What can you infer about the preferences of Player 2?

**(c)** Write all the complete and transitive preference relations of Player 1 that are compatible with the above information.

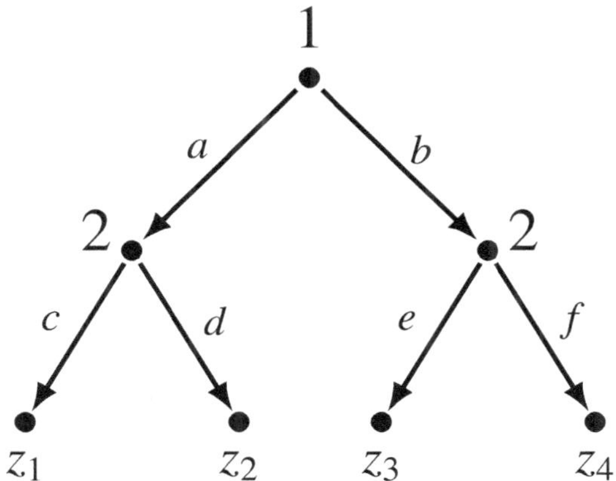


Figure 3.9: The game-frame for Exercise 3.7.

**Solution to Exercise 3.7.**

**(a)** Since $(a,(c,e))$ and $(b,(c,e))$ are both backward-induction solutions, it must be that Player 1 is indifferent between $z_1$ and $z_3$. Since $(b,(c,f))$ is a backward-induction solution but $(a,(c,f))$ is not, it must be that Player 1 prefers $z_4$ to $z_1$. Thus there are only two inferences that we can make about Player 1's preferences: (1) $z_1 \sim_1 z_3$ and (2) $z_4 \succ_1 z_1$ (and thus, by transitivity, also $z_4 \succ_1 z_3$).

**(b)** Since both $e$ and $f$ are part of a backward-induction solution, it must be that Player 2 is indifferent between $z_3$ and $z_4$. Since $c$ is part of a backward-induction solution but $d$ is not, it must be that Player 2 prefers $z_1$ to $z_2$. Thus there are only two inferences that we can make about Player 2's preferences: (1) $z_3 \sim_2 z_4$ and (2) $z_1 \succ_2 z_2$.

**(c)** There are five possibilities, depending on where $z_2$ appears in the ranking:

$$\begin{bmatrix} best & z_2 \\ & z_4 \\ worst & z_1, z_3 \end{bmatrix}, \quad \begin{bmatrix} best & z_2, z_4 \\ worst & z_1, z_3 \end{bmatrix}, \quad \begin{bmatrix} best & z_4 \\ & z_2 \\ worst & z_1, z_3 \end{bmatrix},$$

$$\begin{bmatrix} best & z_4 \\ worst & z_1, z_2, z_3 \end{bmatrix}, \quad \begin{bmatrix} best & z_4 \\ & z_1, z_3 \\ worst & z_2 \end{bmatrix}.$$

**Exercise 3.8** Consider the following situation. Three players have to pick a location for their next meeting from the set $\{a,b,c,d,e,f,g,h\}$. They proceed as follows. First Player 1 decides whether the location should be chosen from the set $\{a,b,c,d\}$ or from the set $\{e,f,g,h\}$ and informs Players 2 and 3 of his decision. Then Player 2 decides whether the location should be one of the first two or one of the last two (in alphabetical order) from the set that Player 1 chose (e.g. if Player 1 chose $\{a,b,c,d\}$ then Player 2 chooses between $\{a,b\}$ and $\{c,d\}$). Finally Player 3, having been informed of Player 2's decision, picks the location from the two-element set chosen by Player 2.

**(a)** Draw an extensive-form game-frame to represents this situation.

**(b)** Suppose that the Players' preferences are as follows, where $x \succ_i y$ means that player $i$ prefers $x$ to $y$:

- Player 1: $a \succ_1 c \succ_1 f \succ_1 h \succ_1 e \succ_1 b \succ_1 d \succ_1 g$
- Player 2: $a \succ_2 b \succ_2 d \succ_2 c \succ_2 g \succ_2 f \succ_2 h \succ_2 e$
- Player 3: $e \succ_3 f \succ_3 b \succ_3 d \succ_3 c \succ_3 a \succ_3 g \succ_3 h$.

Find the backward-induction solution of the corresponding game.

**Solution to Exercise 3.8.**

**(a)** The extensive-form frame is shown in Figure 3.10.

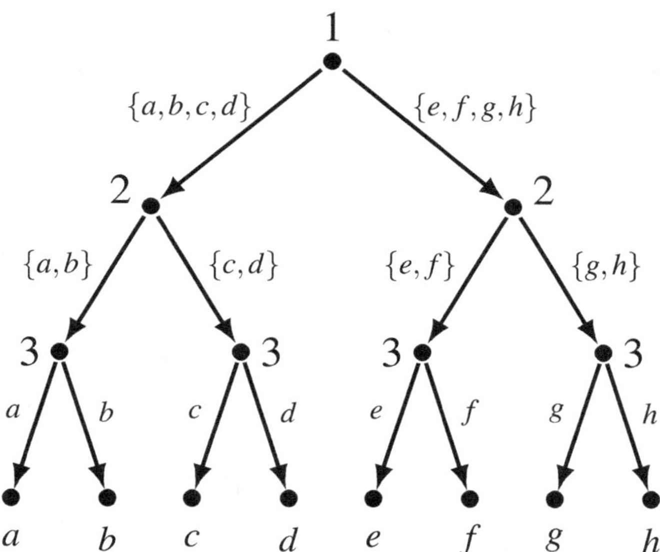


Figure 3.10: The the game-frame of Exercise 3.8.

**(b)** The backward-induction solution is as follows:

- Player 1: choose $\{a,b,c,d\}$.
- Player 2: if Player 1 chooses $\{a,b,c,d\}$ then pick $\{a,b\}$, while if Player 1 chooses $\{e,f,g,h\}$ then pick $\{g,h\}$.
- Player 3: if Player 2 chooses $\{a,b\}$ then pick $b$, if Player 2 chooses $\{c,d\}$ then pick $d$, if Player 2 chooses $\{e,f\}$ then pick $e$, if Player 2 chooses $\{g,h\}$ then pick $g$.

Exercise 3.9 Find the backward-induction solutions of the game of Figure 3.2. ■

**Solution to Exercise 3.9.** There are two backward-induction solutions. One solution is shown as double edges in Figure 3.11. The other solution is the same in terms of Jeff's and Luke's choices, but has Drew choosing Y instead of X. The outcome is the same in both solutions, namely Drew working with Jeff and Luke working alone (the only difference is which client they work with).

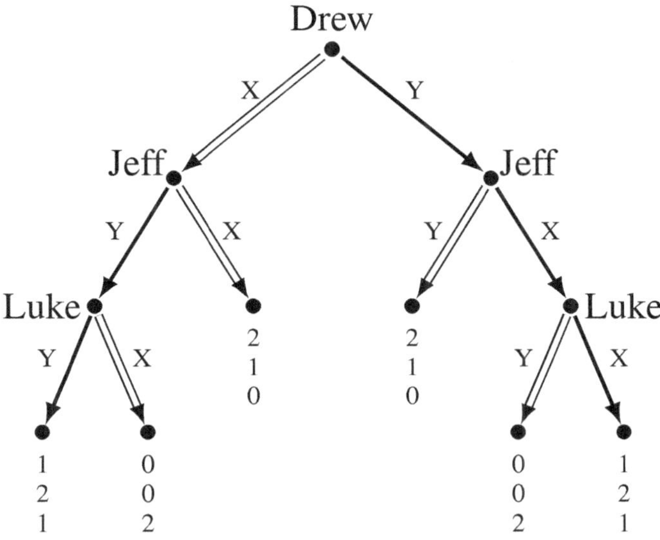


Figure 3.11: The the backward-induction solution of the game of Figure 3.2.

Exercise 3.10 Find the backward-induction solution of the extensive-form game of Exercise 3.2. Is the outcome Pareto efficient? ■

**Solution to Exercise 3.10.** The extensive-form game is shown as double edges in Figure 3.12. Thus Firm 1 will produce 6 units and Firm 2 will respond by producing 3 units and would produce 4 units if Firm 1 were to produce either 3 or 4 units. The outcome is not Pareto efficient because both firms would make higher profits if Firm 1 were to choose $q_1 = 4$ and Firm 2 $q_2 = 3$.

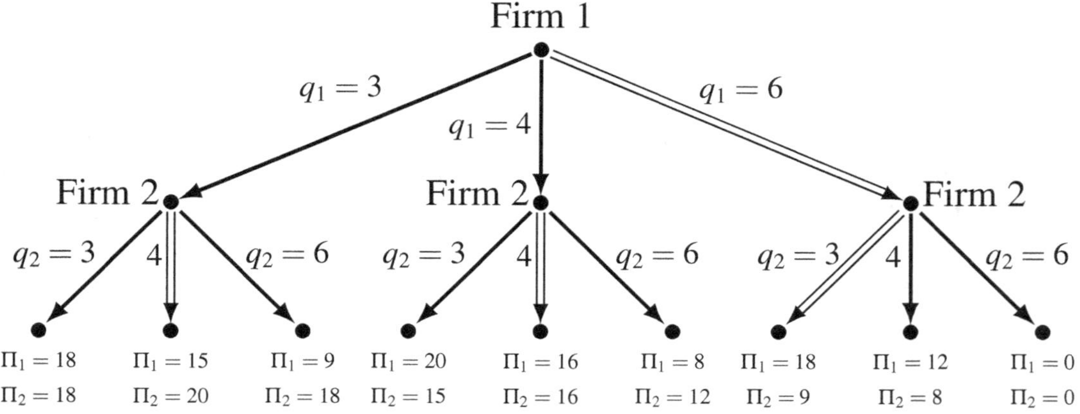


Figure 3.12: The the backward-induction solution of the game of Exercise 3.2.

Exercise 3.11 Consider again the extensive-form frame of Exercise 3.3 (Figure 3.4).

**(a)** Consider first the case where each firm is selfish and greedy, that is, it is only interested in its own profits and prefers higher profits to lower profits. Find the backward-induction solution.

**(b)** Consider now the case where the retailer is selfish and greedy while the manufacturer's objective is to make sure that the sum of the profits of the two firms is as large as possible. Find the backward-induction solution.

**(c)** Consider now the case where the manufacturer is selfish and greedy while the retailer's objective is to make sure that the sum of the profits of the two firms is as large as possible. Find the backward-induction solution(s).

■

**Solution to Exercise 3.11.**

**(a)** If we denote an outcome as a pair $(x,y)$ where $x$ is the manufacturer's profit and $y$ is the retailer's profit, then we can use the following utility functions: for the manufacturer $U_m(x,y) = x$ and for the retailer $U_r(x,y) = y$. Then the backward-induction solution is as follows:

- The retailer chooses $p = 9$ if the manufacturer charges $w = 5$ and chooses $p = 12$ if the manufacturer charges $w = 9$.
- The manufacturer chooses $w = 5$.

The outcome is that the manufacturer makes a profit of 10 and the retailer a profit of 8.

**(b)** If we denote an outcome as a pair $(x,y)$ where $x$ is the manufacturer's profit and $y$ is the retailer's profit, then we can use the following utility functions: for the manufacturer $U_m(x,y) = x + y$ and for the retailer $U_r(x,y) = y$. Then the backward-induction solution is the same as in Part (a).

**(c)** In this case we can use the following utility functions: for the manufacturer $U_m(x,y) = x$ and for the retailer $U_r(x,y) = x + y$. Then the backward-induction solution is:

- The retailer chooses $p = 9$ if the manufacturer charges $w = 5$ and also if the manufacturer charges $w = 9$.
- The manufacturer chooses $w = 9$.

The outcome is that manufacturer makes a profit of 18 and the sum of manufacturer's and retailer's profits is 18 (because the retailer's profit is 0).

Exercise 3.12 We will consider several games based on the extensive-form frame of Exercise 3.5 (Figure 3.6). In the following $z \succ_i z'$ means that player $i$ strictly prefers $z$ to $z'$ and $z \sim_i z'$ means that player $i$ is indifferent between $z$ and $z'$.

**GAME 1** (both selfish). $A$'s preferences are as follows. Let $z = (a,b,c)$ and $z' = (a',b',c')$. Then $z \succ_A z'$ if and only if $a > a'$; in every other case $z \sim_A z'$. $B$'s preferences are analogous: $z \succ_B z'$ if and only if $b > b'$; in every other case $z \sim_B z'$.

**(a)** Find the backward induction *outcomes* of this game (that is, specify the play generated by the backward-induction solutions; no need to specify the entire strategies).

**GAME 2** ($A$ selfish, $B$ selfish but also charity-concerned). $A$'s preferences are as in Part (a). $B$'s preferences are as follows. Let $z = (a,b,c)$ and $z' = (a',b',c')$. Then $z \succ_B z'$ if and only if either (1) $b > b'$ or (2) $b = b'$ and $c > c'$; in every other case $z \sim_B z'$.

**(b)** Find the backward induction *outcomes* of this game.

**GAME 3** ($A$ selfish, $B$ somewhat altruistic). $A$'s preferences are as in Part (a). $B$'s preferences are as follows. Let $z = (a,b,c)$ and $z' = (a',b',c')$. Then $z \succ_B z'$ if and only if either (1) $b > b'$ or (2) $b = b'$ and $c > c'$ or (3) $b = b'$ and $c = c'$ and $a > a'$; in every other case $z \sim_B z'$.

**(c)** Find the backward induction *outcomes* of this game.

**GAME 4** ($A$ selfish, $B$ selfish and spiteful). $A$'s preferences are as in Part (a). $B$'s preferences are as follows. Let $z = (a,b,c)$ and $z' = (a',b',c')$. Then $z \succ_B z'$ if and only if either (1) $b > b'$ or (2) $b = b'$ and $a < a'$; in every other case $z \sim_B z'$.

**(d)** Find the plays associated with the backward-induction *solutions* of this game. For one solution of your choice, explain in words the backward-induction strategy of player $A$ (the entire strategy).

===

**(e)** What type of player $B$ would you like as an opponent if you were player $A$? In other words, as player $A$ which of the above games would you want to play?

■

**Solution to Exercise 3.12.** First note that – in all the four games – player $A$, being selfish, will always prefer saying Yes to saying No to a counteroffer of player $B$. Thus the extensive-form frame of Figure 3.6 can be simplified as shown in Figure 3.13.

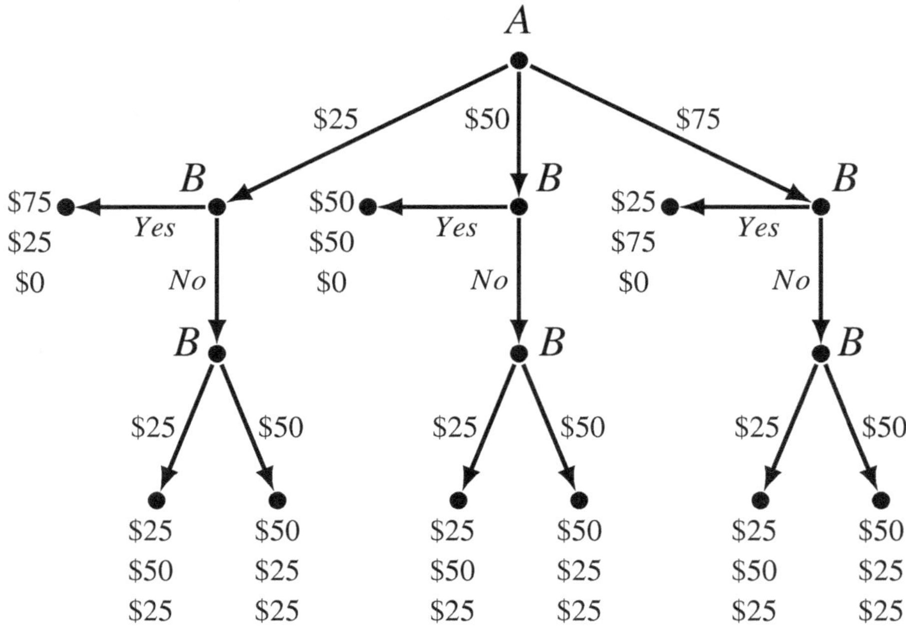


Figure 3.13: First reduction of the extensive form of Figure 3.4.

In the extensive form of Figure 3.13 at the lower decision nodes player *B* (after rejecting player *A*'s offers) gets more by counter-offering \$25 and thus in all four games he will prefer to counter-offer \$25 (in all cases he ranks an outcome where he gets more money higher than an outcome where he gets less money, irrespective of what player *A* and the charity get). Thus the extensive form can be further simplified as shown in Figure 3.14.

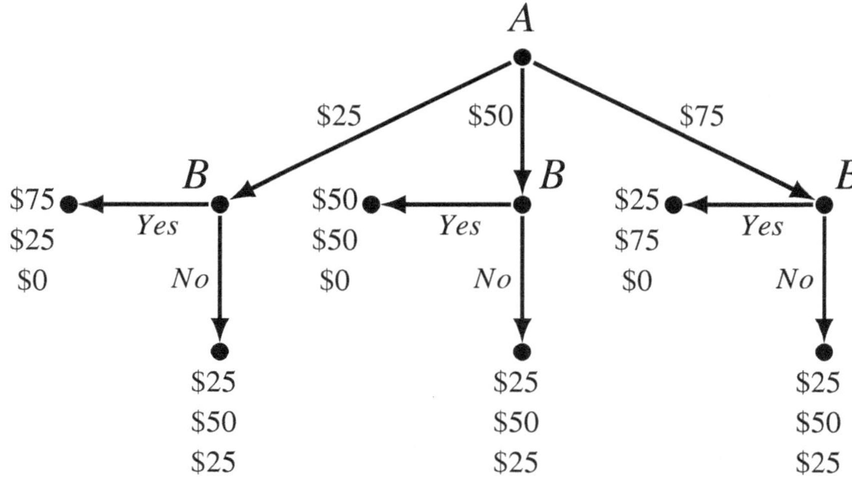


Figure 3.14: Further reduction of the extensive form of Figure 3.4.

Referring to the extensive form of Figure 3.14, in all four games, at the node on the left player *B* prefers No to Yes (he gets more) and at the node on the right he prefers Yes to No (again, he gets more). Thus the extensive form can be further simplified as shown in

Figure 3.15.

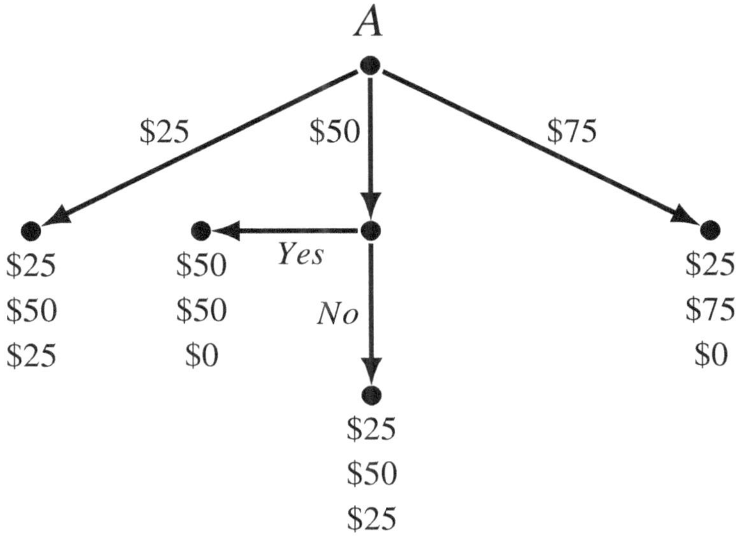


Figure 3.15: Final reduction of reduction of the extensive form of Figure 3.4.

Thus the issue is: what would player *B* do if offered $50?

**(a)** In Game 1, player *B* is indifferent between the outcomes ($50, $50, $0) and ($25, $50, $25). Thus we have four backward induction solutions, with the following associated plays (or actual sequences of moves):

1. *A* offers $50 and *B* says Yes.
2. *A* offers $25, *B* says No and counter-offers $25 and *A* says Yes.
3. *A* offers $50, *B* says No and counter-offers $25 and *A* says Yes.
4. *A* offers $75 and *B* says Yes.

**(b) and (c) and (d)** In Games 2, 3 and 4, player *B* prefers outcome ($25, $50, $25) to outcome ($50, $50, $0) although for different reasons). Thus, if offered $50 he will say No. Hence there are three backward induction solutions, with the following associated plays (or actual sequences of moves):

1. *A* offers $25, *B* says No and counter-offers $25 and *A* says Yes.
2. *A* offers $50, *B* says No and counter-offers $25 and *A* says Yes.
3. *A* offers $75 and *B* says Yes.

In the first solution, the backward induction strategy of *A* is "offer $25 and if I find myself at a later node where *B* has made a counteroffer, always accept".

**(e)** In Games 2,3 and 4 all the solutions involve player *A* getting $25, while in Game 1 there are solutions where *A* gets $25 and one solution where he gets $50. Thus if I were player *A*, I would want to have a selfish opponent (Game 1), as this is the only type of opponent for which any solution gives Player *A* more than $25, and at worst it gives the same payoff as any solution involving another type of Player *B*.

Exercise 3.13 Consider the following perfect-information game between the proprietor of a business, $P$, and two potential managers, $M1$ and $M2$. There are four dates $t \in \{0,1,2,3\}$. At date t = 0, $P$ selects one of the managers to run the business in period 1 (period 1 occurs between dates 1 and 2). At date $t = 2$, $P$ again selects one of the managers (possibly the same one that was selected for period 1 or possibly a different manager) to run the business in period 2 (period 2 occurs between dates 3 and 4). At date 1 (at the beginning of period 1) and at date 3 (at the beginning of period 2) the manager selected to run the business decides what fraction of the business income (for that period) she will expropriate for herself. Let $x_T \in \{q,r\}$ (with $0 < q < r < 1$) denote the fraction of the business income expropriated in period $T \in \{1,2\}$ by the manager selected for that period. Total business income in each period is equal to 1. The fraction of business income not expropriated by the manager in period $T$, i.e. $1 - x_T$, is consumed by $P$ and constitutes $P$'s payoff for that period. A manager's per-period payoff is given by $x_T$ if selected to run the firm in that period, and zero otherwise. There is no discounting of future payoffs, that is, a player's payoff is equal to the sum for her payoffs in the two periods and **payoffs are collected at date 4**.

**(a)** Draw the extensive-form game.

**(b)** What is the highest possible payoff that $P$ can obtain at a backward-induction solution? Provide a strategy profile that sustains such a payoff.

**(c)** What is the lowest possible payoff that $P$ can obtain at a backward-induction solution? Provide a strategy profile that sustains such a payoff.

■

**Solution to Exercise 3.13.**

**(a)** The extensive-form game-frame is shown in Figure 3.16.

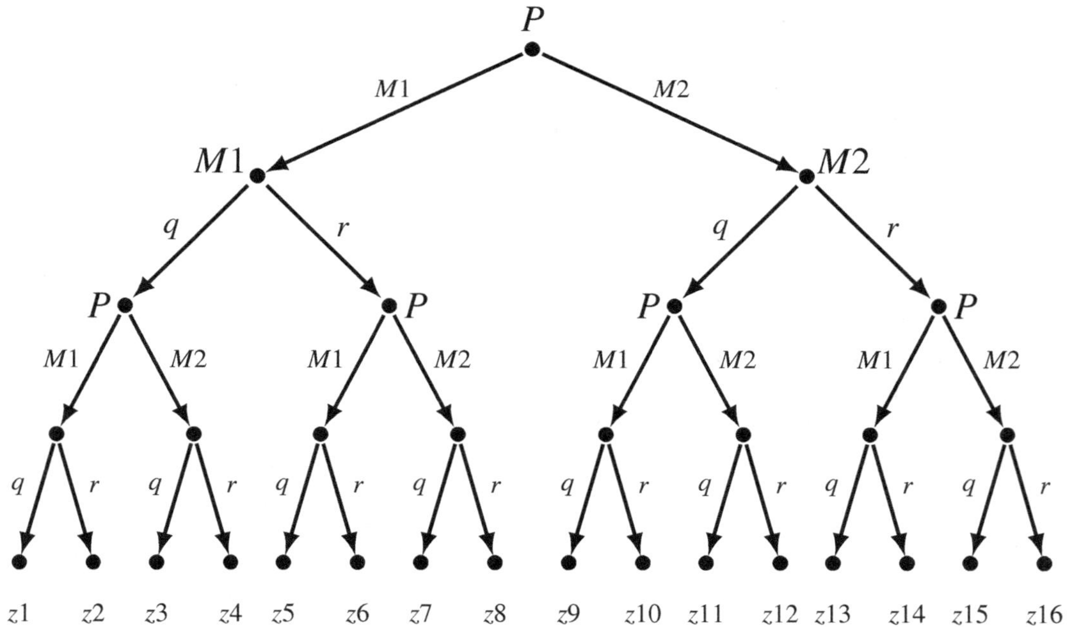


Figure 3.16: The extensive-game for Exercise 3.13.

The game is obtained by adding the following payoffs:

| node | $P's$ payoff | $M1's$ payoff | $M2's$ payoff |
|---|---|---|---|
| $z1$ | $2-2q$ | $2q$ | $0$ |
| $z2$ | $2-q-r$ | $q+r$ | $0$ |
| $z3$ | $2-2q$ | $q$ | $q$ |
| $z4$ | $2-q-r$ | $q$ | $r$ |
| $z5$ | $2-q-r$ | $q+r$ | $0$ |
| $z6$ | $2-2r$ | $2r$ | $0$ |
| $z7$ | $2-q-r$ | $r$ | $q$ |
| $z8$ | $2-2r$ | $r$ | $r$ |
| $z9$ | $2-2q$ | $q$ | $q$ |
| $z10$ | $2-q-r$ | $r$ | $q$ |
| $z11$ | $2-2q$ | $0$ | $2q$ |
| $z12$ | $2-q-r$ | $0$ | $q+r$ |
| $z13$ | $2-q-r$ | $q$ | $r$ |
| $z14$ | $2-2r$ | $r$ | $r$ |
| $z15$ | $2-q-r$ | $0$ | $q+r$ |
| $z16$ | $2-2r$ | $0$ | $2r$ |

At the beginning-of-the-second-period nodes (date 3), the manager who has been appointed for that period will choose $r$ ($r$ gives an extra payoff of $r-q$ relative to $q$). Thus the game can be reduced as shown in Figure 3.17.

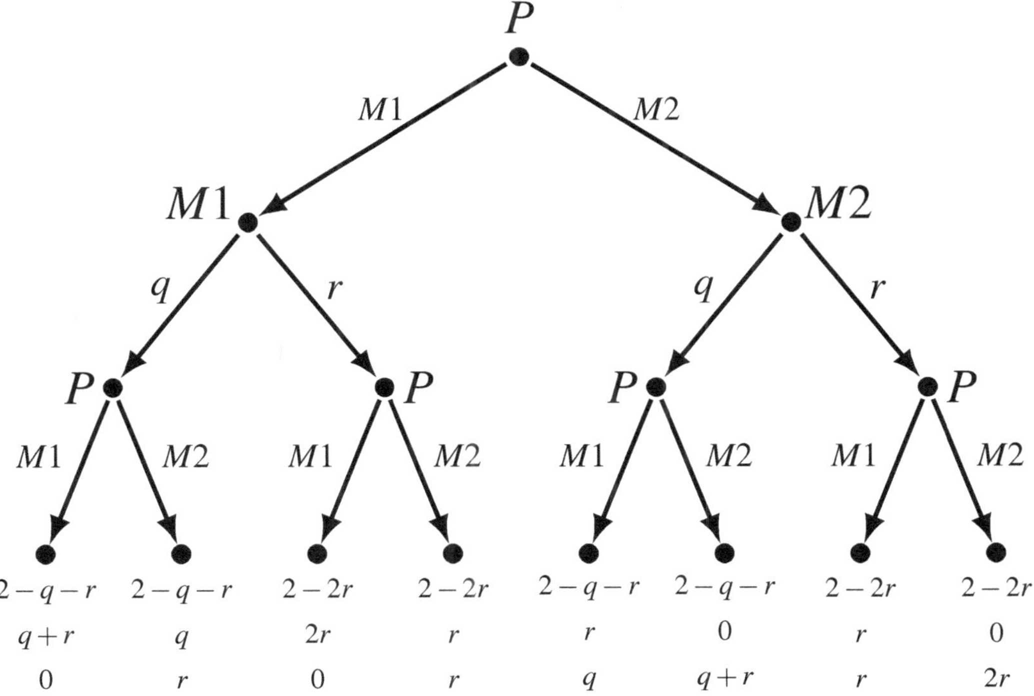


Figure 3.17: A reduction of the extensive-game of Figure 3.16.

**(b)** At date 2, $P$ is indifferent between appointing $M1$ or $M2$; thus there are many backward-induction solutions. The highest possible payoff for $P$ is $2-q-r$. It is achieved, for example, with the following backward-induction solution:

- P's strategy: appoint $M1$ in period 1 and then if $M1$ chooses $q$ appoint $M1$ again in period 2, otherwise appoint $M2$ in period 2.
- M1's strategy: choose $q$ in period 1 and then choose $r$ at any second-period node if appointed in period 2.
- M2's strategy: if appointed choose $r$ always.

There are two such backward-induction solutions: the other switches the roles of manager 1 and manager 2.

**(c)** The lowest possible payoff for P is $2-2r$. It is associated with the following backward-induction solution:

- P's strategy: appoint $M1$ in both periods.
- M1's strategy: choose $r$ always.
- M2's strategy: choose $r$ always.There are two such backward-induction solutions: the other switches the roles of manager 1 and manager 2.

Exercise 3.14 Player 1 and Player 2 bargain over \$300. They take turns making offers: first Player 1 makes a proposal that Player 2 can accept or reject; if Player 2 rejects then she makes a proposal that Player 1 can accept or reject and then the game ends. Each offer takes one period, and the players are impatient: they discount payoffs received in later periods by a factor of 0.8 per period. For example, $\$x$ received in period 1 is worth $\$x$ but $\$x$ received in period 2 is worth only $\$0.8x$, from the point of view of period 1, and $\$x$ received in period 3 is worth only $\$(0.8)^2x$, from the point of view of period 1. Each player is selfish and greedy (that is, cares only about how much money he himself gets and prefers more money to less). Below is a more detailed description of the game.

- At the beginning of the first period, Player 1 proposes to take a share $s_1 \in \{\frac{1}{4}, \frac{1}{3}\}$ of the \$300, leaving the share $(1-s_1)$ for Player 2.
- Player 2 either accepts the offer – in which case the game ends in period 1 with Player 1 receiving the share $s_1$ and Player 2 receiving the share $(1-s_1)$ – or rejects the offer, in which case play proceeds to the second period.
- At the beginning of the second period (if the play proceeds to the second period), Player 2 proposes to take a share $s_2 \in \{\frac{1}{4}, \frac{1}{3}\}$ of the \$300, leaving the share $(1-s_2)$ for Player 1.
- Player 1 either accepts the offer – in which case the game ends in period 2 with Player 1 receiving the share $(1-s_2)$ and Player 2 receiving the share $s_2$ – or rejects the offer, in which case play proceeds to the third period.
- In the third period (if the play proceeds to the third period), each player receives \$150.

**(a)** Represent this game as an extensive-form game with perfect information, associating with the terminal nodes the payoffs as viewed from the point of view of period 1.

**(b)** Find the backward induction solution.

**Solution to Exercise 3.14.** Let $U(\$x, i)$ be the utility of $\$x$ received in period $i \in \{1,2,3\}$. Then $U(\$x,1) = x$, $U(\$x,2) = 0.8x$ and $U(\$x,3) = (0.8)^2 x$. For example, $U(\$75,2) = 60$ and $U(\$150,3) = 96$.

**(a)** The extensive-form game is shown in Figure 3.18.

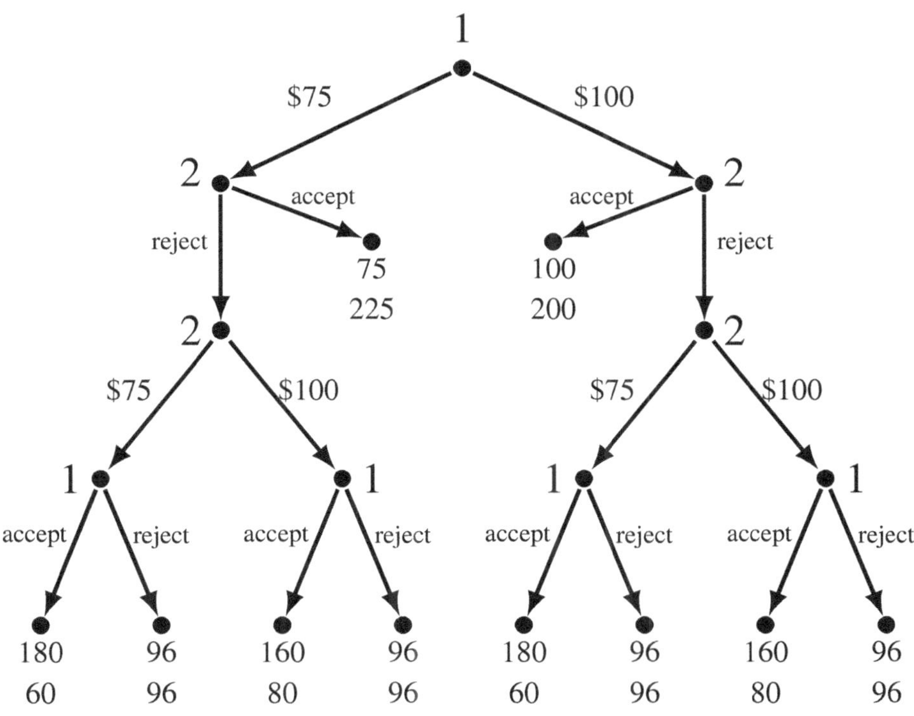


Figure 3.18: The extensive-form game of Exercise 3.14.

**(b)** The backward-induction solution is shown as double edges in Figure 3.19.

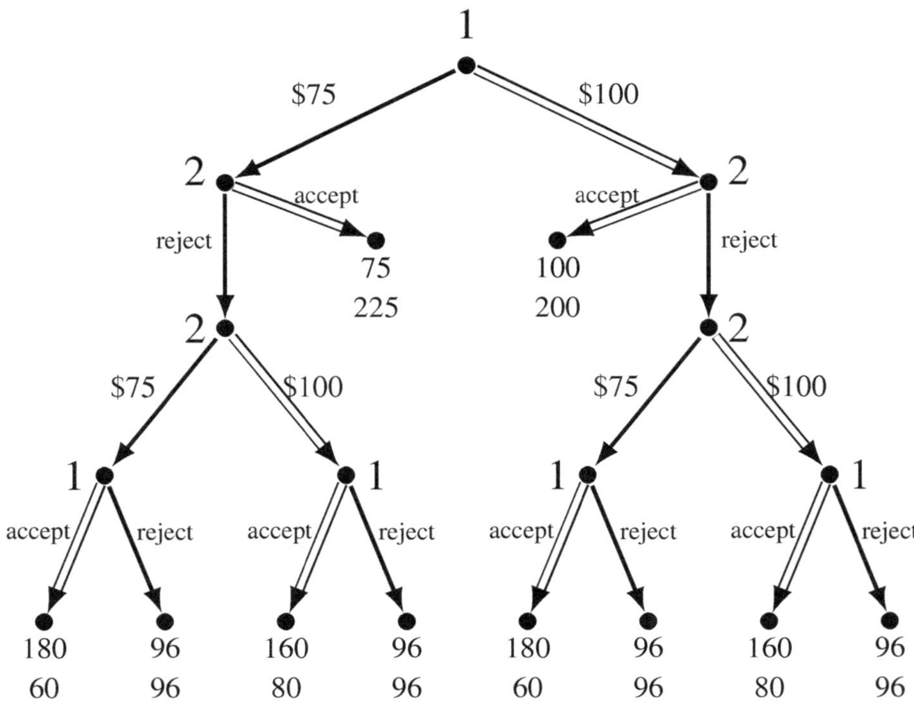


Figure 3.19: The extensive-form game of Exercise 3.14.

**Exercise 3.15** Dan wants to auction $30 to Ann and Barb. He tells them the following:

> Each of you can bid either $0 or $15 or $30. Ann goes first and states her bid openly, so that Barb can hear. Then Barb goes second and states her bid. I will then distribute the $30 as follows. If both of you bid $0, I will keep the $30. Otherwise, I will give the $30 to the one who bid the larger amount of money, if there is such person. In case of ties I will split the $30 equally between the two of you. However, **no matter what the outcome is, each of you has to pay her own bid to me (thus even if she is not a winner)**.

**(a)** Represent this situation as an extensive-form frame with perfect information, taking as outcomes the **net** amount of money that each player gets (what she receives minus what she pays). The players are only Ann and Barb (that is, do not treat Dan as a player).

**(b)** Consider the extensive-form game obtained from the frame of Part (a) by adding the assumption that both players are selfish and *spiteful*. Selfish means that in comparing two outcomes, say $x$ and $y$, a player prefers $x$ to $y$ if the net amount of money she gets in $x$ is larger then the net amount of money she gets in $y$. Spiteful means that if $x$ and $y$ are two outcomes in which she gets the same net amount of money, then she prefers $x$ to $y$ if in $x$ the other player gets less than in $y$.

(a) Draw the extensive-form game replacing the outcomes with payoffs. For each player use a utility function with values in the set $\{1,2,3,4,5,6\}$.

(b) Find the backward induction solution

**(c)** Find the backward induction solution of the extensive-form game obtained from the frame of Part (a) assuming that both players are selfish and *nice*. Nice means that if $x$ and $y$ are two outcomes in which she gets the same net amount of money, then she prefers $x$ to $y$ if in $x$ the other player gets more than in $y$.

**Solution to Exercise 3.15.**
**(a)** The extensive-form frame is shown in Figure 3.20.

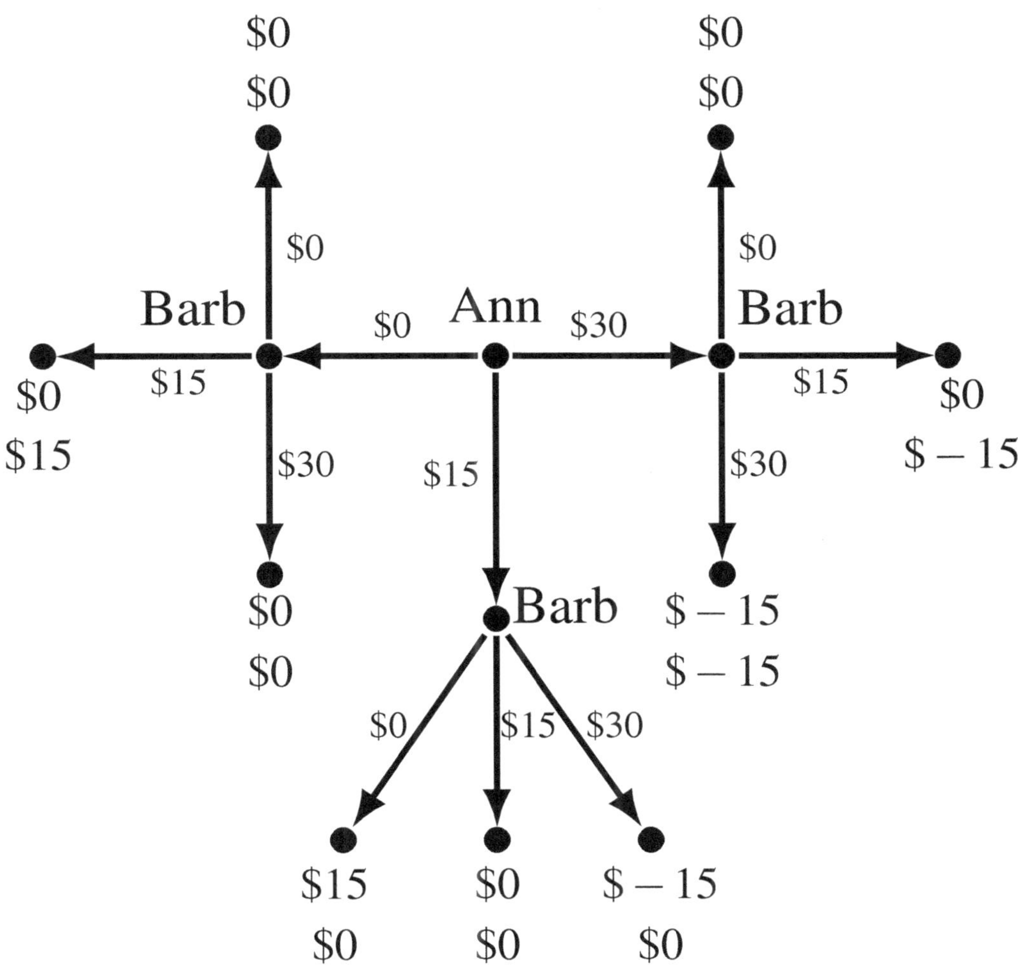


Figure 3.20: The extensive-form frame of Exercise 3.15.

**(b)** We can represent the players' preferences with the following utility functions.

- For Ann: $U(\$15,\$0) = 6, U(\$0,\$-15) = 5, U(\$0,\$0) = 4, U(\$0,\$15) = 3, U(\$-15,\$-15) = 2, U(\$-15,\$0) = 1$.

- For Barb: $V(\$0,\$15) = 6, V(\$-15,\$0) = 5, V(\$0,\$0) = 4, V(\$15,\$0) = 3, V(\$-15,\$-15) = 2, V(\$0,\$-15) = 1$.

The backward-induction solution of the corresponding game is shown as double edges in Figure 3.21.

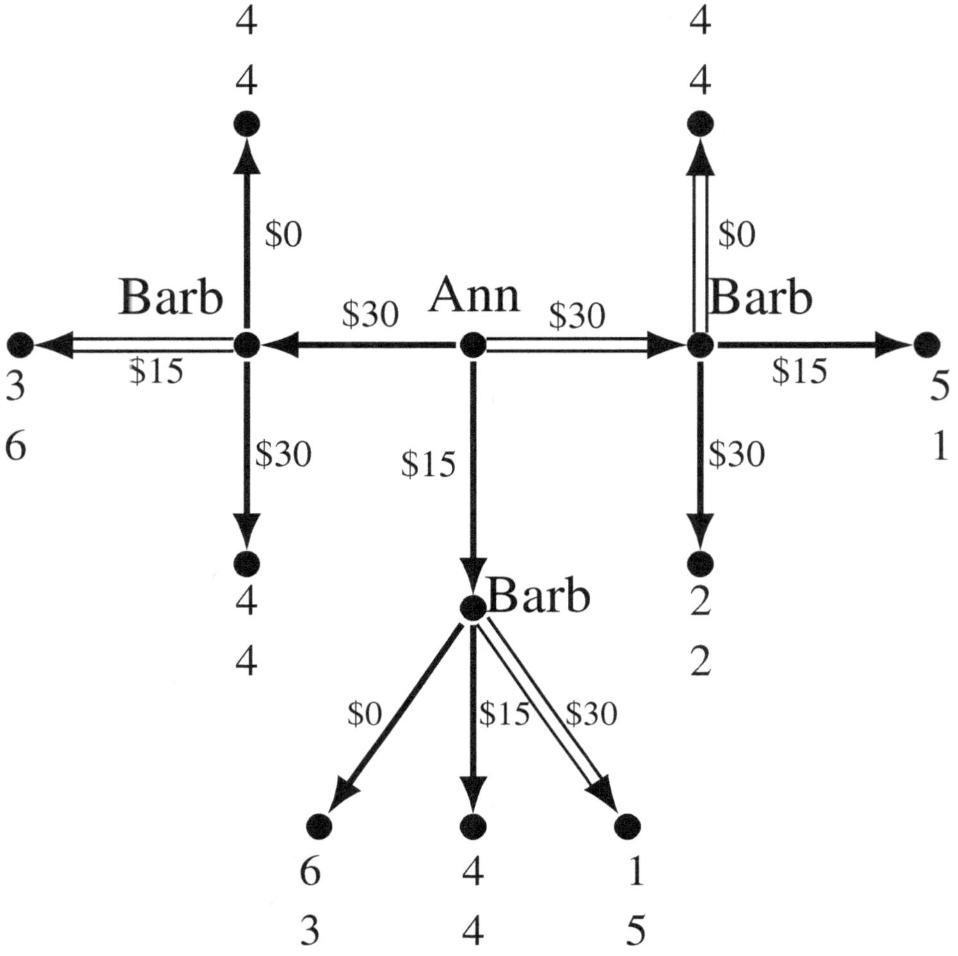


Figure 3.21: The extensive-form game of Part (b) of Exercise 3.15.

**(c)** The backward-induction solution of the corresponding game is as follows.

- At Barb's left node, since her preferences are $(\$0,\$15) \succ (\$0,\$0)$, she will choose to bid \$15.
- At Barb's bottom node, since her preferences are $(\$15,\$0) \succ (\$0,\$0) \succ (\$0,\$-15)$, she will choose to bid \$0.
- At Barb's right node, since her preferences are $(\$0,\$0) \succ (\$0,\$-15) \succ (\$-15,\$-15)$, she will choose to bid \$0.
- Thus Ann will bid \$15 since her preferences are $(\$15,0) \succ (\$0,\$15) \succ (\$0,\$0)$.

## 3.3 Strategies in perfect-information games

The exercises in this section deal with the notion of strategy in extensive-form games with perfect information (Definition 3.3.1, Section 3, Volume 1).

Exercise 3.16 Consider the perfect information game of Figure 3.22 (as usual, the payoffs are given in the following order, from top to bottom: Player 1, Player 2, Player 3).

**(a)** Write the corresponding strategic-form game.
**(b)** Are there values of $x$ for which Player 3 has a **strictly** dominant strategy?
**(c)** Does Player 1 have a weakly **dominated** strategy? If your answer is Yes, name a strategy that dominates it; if your answer is No prove your claim.
**(d)** What strategy profiles are Nash equilibria irrespective of the value of $x$?
**(e)** Find all the backward induction solutions for the case where $x = 0$.
**(f)** Find all the backward induction solutions for the case where $x = 10$.

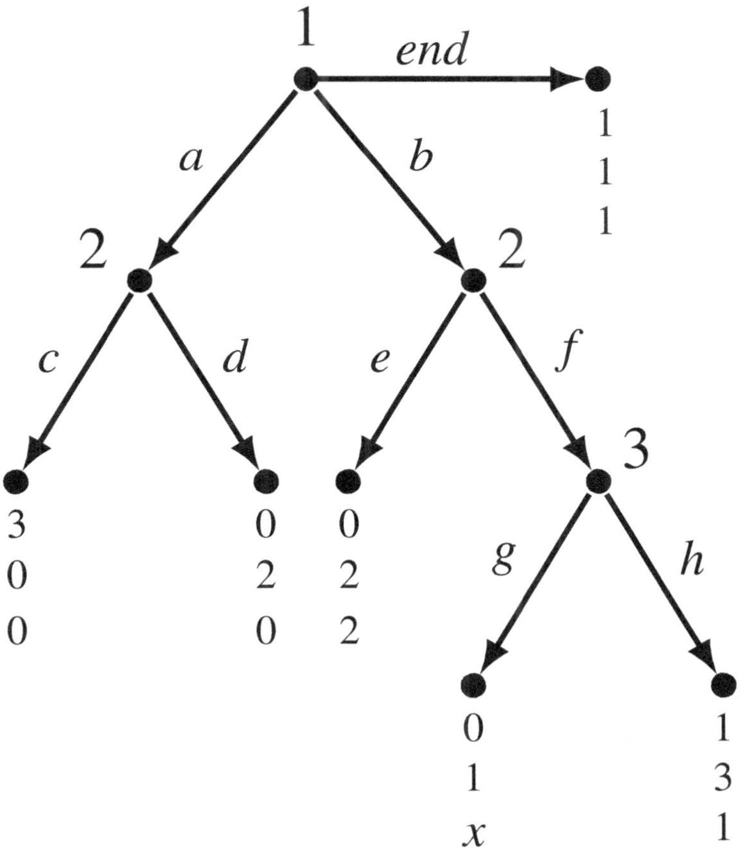


Figure 3.22: The game for Exercise 3.16.

**Solution to Exercise 3.16.**

**(a)** The strategic-form game is shown in Figure 3.23.

Player 3: $g$

| Player 1 \ Player 2 | $ce$ | $cf$ | $de$ | $df$ |
|---|---|---|---|---|
| $a$ | 3 0 0 | 3 0 0 | 0 2 0 | 0 2 0 |
| $b$ | 0 2 2 | 0 1 $x$ | 0 2 2 | 0 1 $x$ |
| $end$ | 1 1 1 | 1 1 1 | 1 1 1 | 1 1 1 |

Player 3: $h$

| Player 1 \ Player 2 | $ce$ | $cf$ | $de$ | $df$ |
|---|---|---|---|---|
| $a$ | 3 0 0 | 3 0 0 | 0 2 0 | 0 2 0 |
| $b$ | 0 2 2 | 1 3 1 | 0 2 2 | 1 3 1 |
| $end$ | 1 1 1 | 1 1 1 | 1 1 1 | 1 1 1 |

Figure 3.23: The strategic-form game corresponding to the extensive-form game of Figure 3.22.

**(b)** There are no values of $x$ for which Player 3 has a **strictly** dominant strategy (e.g. if Player 1 plays *end* then Player 3 gets the same payoff, namely 1, whatever her strategy).

**(c)** Yes: for Player 1, $b$ is weakly dominated by *end*.

**(d)** The following four strategy profiles: $(end, de, g)$, $(end, de, h)$, $(end, df, g)$ and $(end, df, h)$.

**(e)** When $x = 0$ there are two backward-induction solutions: $(end, df, h)$ and $(b, df, h)$.

**(f)** When $x = 10$ there is only one backward-induction solution: $(end, de, g)$.

Exercise 3.17 There are two players and two envelopes. One of the envelopes is marked "Player 1" and the other is marked "Player 2". At the beginning of the game, each envelope contains one dollar. Player 1 is given the choice between stopping the game and continuing. If she stops, then each player receives the money in her own envelope and the game ends. If Player 1 chooses to continue, then a dollar is removed from her envelope and two dollars are added to Player 2's envelope. Then Player 2 must choose between stopping the game and continuing. If she stops then the game ends and each player receives the money in her own envelope and the game ends. If Player 2 chooses to continue, then a dollar is removed from her envelope and two dollars are added to Player 1's envelope. Play continues in this fashion, alternating between the players, until either one of them decides to stop or $k$ rounds of play have elapsed (every time one of the players has made a choice, we say that one round has elapsed). If neither player has chosen to stop by the end of the $k^{th}$ round, then both players have to surrender their envelopes and thus get no money. Assume that each player is selfish and greedy, that is, she is only interested in the amount of money she obtains and prefers more money to less.

**(a)** Draw a perfect-information extensive form to represent this game for the case where $k = 5$.

**(b)** Find the backward-induction solution of the game of Part (a).

**(c)** How many strategies does Player 1 have in the game of Part (a)?

**(d)** How many strategies does Player 2 have in the game of Part (a)?

**(e)** For any positive integer $k$, describe the backward-induction solution of the game with $k$ rounds.

■

**Solution to Exercise 3.17.**

**(a)** Denote an outcomes as $(\$x, \$y)$, where $x$ is the amount Player 1 obtains and $y$ the amount Player 2 obtains. Then we can take as utility function of Player 1 the function $U_1(\$x, \$y) = x$ and as utility function of Player 2 the function $U_2(\$x, \$y) = y$. The extensive-form game is as shown in Figure 3.24, where '$s$' stands for 'stop' and '$c$' stands for 'continue'.

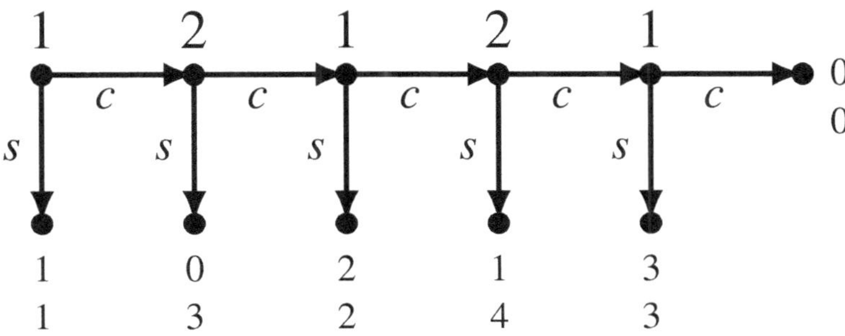


Figure 3.24: The game for Exercise 3.17.

**(b)** The backward-induction solution is shown as double edges in Figure 3.25.

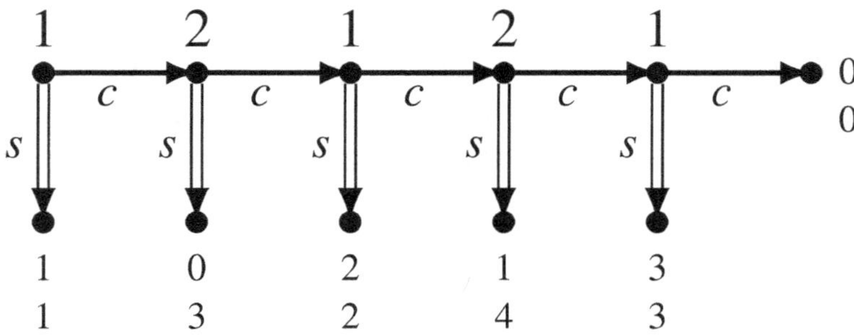


Figure 3.25: The backward-induction solution of the game of Figure 3.24.

**(c)** Player 1 has 8 strategies (one choice for each of her nodes). Her backward-induction strategy is $(s,s,s)$ (at each node, stop the game).

**(d)** Player 2 has 4 strategies: $(c,c)$, $(c,s)$, $(s,c)$ and $(s,s)$.

**(e)** For every positive integer $k$, the backward induction strategy profile has each player stopping the game at each of her decision nodes. Thus the outcome is that the game is stopped at the first move.

Exercise 3.18 Consider the extensive-form frame of Figure 3.26.

**(a)** How many strategies does Player 1 have? Write one possible strategy.

**(b)** How many strategies does Player 2 have? Write one possible strategy.

**(c)** How many cells are there in the table that represents the corresponding strategic-form frame?

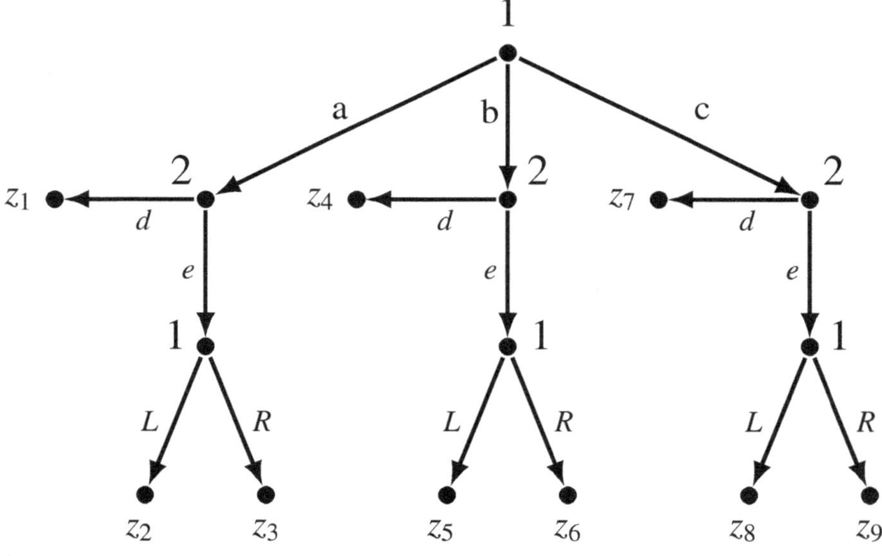


Figure 3.26: The game-frame for Exercise 3.18.

**Solution to Exercise 3.18.**

**(a)** Player 1 has 4 decision nodes and two choices at each decision node, thus she has $2^4 = 16$ possible strategies. One of them is: $(a, L, R, R)$, which says that she starts by choosing $a$ and then, if she has to play again (at her bottom left node), she plays $L$ and if, by any chance (perhaps by mistake), she starts with $b$ and then has to play again (at her bottom middle node) then she plays $R$ and if, by any chance (perhaps by mistake), she starts with $c$ and then has to play again (at her bottom right node) then she plays $R$.

**(b)** Player 2 has 3 decision nodes and two choices at each decision node, thus she has $2^3 = 8$ possible strategies. One of them is: $(d, e, d)$, which says that if Player 1 plays $a$ then he plays $d$ and if Player 1 plays $b$ then he plays $e$ and if Player 1 plays $c$ then he plays $d$.

**(c)** The table would have 32 rows and 8 columns, so 256 cells.

Exercise 3.19 Consider the extensive-form game of Figure 3.27.

**(a)** Are there values of $x$ for which Player 1 has a **strictly** dominant strategy? If your answer is Yes, say what values and what strategy, if your answer is No explain why not.

**(b)** Are there values of $y$ for which Player 3 has a **strictly** dominant strategy? If your answer is Yes, say what values and what strategy, if your answer is No explain why not.

**(c)** Does Player 2 have **weakly** dominated strategies? If your answer is Yes, name the strategies and the strategies that dominate them; if your answer is No explain why not.

**(d)** For what values of $y$ does Player 3 have a **weakly** dominated strategy?

**(e)** How many strategies does Player 2 have?

**(f)** Find all the backward-induction solutions when $x = 1$ and $y = 2$.

**(g)** Find the backward-induction solution when $x = 1$ and $y = 3$.

**(h)** Assume that $x = 1$ and $y = 1$. Explain why $(b, f, r)$ is not a Nash equilibrium of the associated strategic-form game.

**(i)** Assume that $x = 1$ and $y = 1$. Explain why $(b, f, r)$ is not a backward-induction solution.

**(j)** Assume that $x = 1$ and $y = 1$. Is there a Nash equilibrium where Player 1 plays $c$? ■

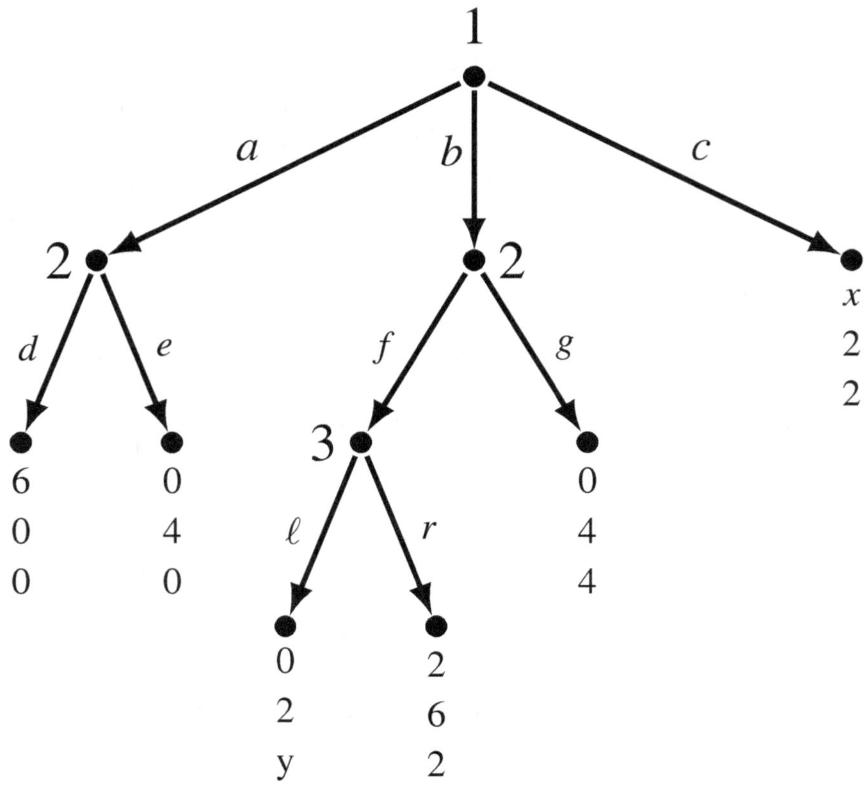


Figure 3.27: The game for Exercise 3.19.

**Solution to Exercise 3.19.**

**(a)** Yes, any $x > 6$ (and only those values of $x$). The strategy is $c$.
**(b)** No, because if Player 1 plays $c$ the payoff of Player 3 is the same (namely 2), no matter what strategy Player 3 chooses.
**(c)** Yes $(d, f)$ is weakly dominated by $(e, f)$ and $(d, g)$ is weakly dominated by $(e, g)$.
**(d)** For $y \neq 2$ (and only those values of $y$): if $y > 2$ then $\ell$ weakly dominates $r$ and if $y < 2$ then $r$ weakly dominates $\ell$.
**(e)** Four: $(d, f)$, $(d, g)$, $(e, f)$ and $(e, g)$.
**(f)** One solution is $(c, (e, g), \ell)$ and the other is $(b, (e, f), r)$.
**(g)** $(c, (e, g), \ell)$.
**(h)** Because it is not a strategy profile (the strategy of Player 2 is not complete: what would she do if Player 1 played $a$?).
**(i)** Again, because it is not a strategy profile: $(b, f, r)$ is the play (or outcome) associated with the backward-induction solution (namely $(b, (e, f), r)$) but it is not itself a backward-induction solution.
**(j)** Yes. Indeed there are three: $(c, (e, g), \ell)$, $(c, (e, g), r)$ and $(c, (e, f), \ell)$.

## 3.4 Relationship between backward induction and other solutions

The exercises in this section deal with the relationship between backward induction and other solution concepts, such as Nash equilibrium and the iterated deletion of weakly dominated strategies (IDWDS).

Exercise 3.20 Consider the extensive-form game shown in Figure 3.28 (as usual, the top number is the payoff of Player 1 and the bottom number is the payoff of Player 2).

**(a)** Write the corresponding strategic-form game.

**(b)** What strategy profiles survive the iterated deletion of weakly dominated strategies (IDWDS) in the strategic-form game of Part (a)?

**(c)** What are the Nash equilibria of the strategic-form game of Part (a)?

**(d)** Find the backward-induction solutions. Is each backward-induction solution also a Nash equilibrium? Is each backward-induction solution also a strategy profile that survives the IDWDS procedure?

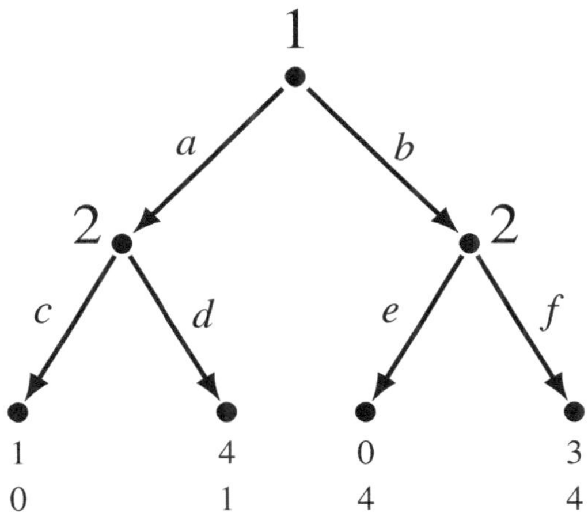


Figure 3.28: The game for Exercise 3.20.

**Solution to Exercise 3.20.**

**(a)** The strategic form is shown in Figure 3.29

Player 2

| | | $ce$ | | $cf$ | | $de$ | | $df$ | |
|---|---|---|---|---|---|---|---|---|---|
| Player 1 | $a$ | 1 | 0 | 1 | 0 | 4 | 1 | 4 | 1 |
| | $b$ | 0 | 4 | 3 | 4 | 0 | 4 | 3 | 4 |

Figure 3.29: The strategic form of the extensive-form game of Figure 3.28.

**(b)** For Player 2 both $ce$ and $cf$ are weakly dominated by $de$ (and also by $df$). After deleting $ce$ and $cf$ in the reduced game for Player 1 $b$ is strictly dominated by $a$. Thus the output of the IDWDS procedure is $\{(a,de),(a,df)\}$.

**(c)** There are three Nash equilibria: $(b,cf)$, $(a,de)$ and $(a,df)$.

**(d)** The backward-induction solutions are $(a,de)$ and $(a,df)$. Both of them are Nash equilibria and both of them survive the IDWDS procedure.

## 3.5 Perfect-information games with two players

The exercises in this section deal with two-player extensive-form games with perfect information and only two outcomes: win and lose. In such games one of the players has a winning strategy, that is, a strategy that guarantees that she will win no matter how the opponent plays (Theorem 3.5.1, Volume 1, Chapter 3, Section 5).

Exercise 3.21 Consider the following two-player game with perfect information. There are 28 red tiles and 16 blue tiles on the table. Players take turns removing tiles, with Player 1 making the first move. The rules are as follows: each player must remove at least one and at most two red tiles and, at the same time, can remove either 0 or 1 or 2 blue tiles. The player who removes the **last red tile** loses (that is, if there is at least one red tile before the player moves and no more red tiles after she moves, then she loses) and the other player wins.

**(a)** Determine which player has a winning strategy and give a detailed description of the winning strategy.

**(b)** Suppose we modify the game as follows: each player is given one (and only one) chance to pass, that is, if it is her turn to move then she can exercise her option to pass and skip her turn (so that it is the other player's turn to move again). For the case considered above where there are 28 red tiles and 16 blue tiles, determine which player has a winning strategy and give a detailed description of the winning strategy.

■

**Solution to Exercise 3.21.**

**(a)** A player whose turn it is to move when there is only one red tile left on the table loses. On the other hand, the number of remaining blue tiles is clearly irrelevant. Let $R$ denote the number of red tiles that are on the table. Thus $R=1$ is a losing position. It follows that $R=4$ is a losing position (the player who has to move when $R=4$ must change the position to either $R=3$, in which case the other player can remove 2 tiles and take her to the losing position of $R=1$, or change the position to $R=2$, in which case the other player can remove one tile and take her to the losing position $R=1$). Continuing this way, $R=7$ is a losing position and so are $R=10,13,16,19,22,25,28$. Thus at the beginning of the game Player 1 is in a losing position. Hence Player 2 has a winning strategy. The winning strategy is as follows: if Player 1 has just removed $n$ red tiles, Player 2 should remove $(3-n)$ red tiles (and any allowed number of blue tiles, including zero).

**(b)** Since both players have a chance to pass, Player 2 can neutralize Player 1's choice of "pass" by exercising her option as soon as Player 1 does. Thus it remains true that Player 2 has a winning strategy and the strategy is the same as before with the additional clause: "if (and only if) Player 1 at any time passes then pass immediately after him (thus restoring the turn to him)".

Exercise 3.22 Consider the map shown in Figure 3.30. Two players play the following game. Player 2 is at the location marked "Start" and her objective is to move to the location marked "End". She can travel along any of the edges marked 1 to 7 to a contiguous location. Player 2 wins if she reaches the node marked "End". Player 1 wins if he manages to prevent Player 2 from reaching that node. The game is played as follows. Player 1 starts and removes one of the seven edges, then Player 2 moves to a contiguous node using **one** of the remaining edges, then Player 1 removes another edge and Player 2 moves to a contiguous node using one of the remaining edges, and so on. Player 2 can move in any direction she likes; for example, from node $B$ she can go back to Start, if she so wishes (and if edge 2 is still available). The game ends either when Player 2 has reached the node "End" or when that node has become unreachable.

**(a)** Explain why in this game, as well as in any similar game with a finite number of nodes and a finite number of edges, one of the players has a strategy that allows her/him to win, no matter what the other player does.

**(b)** For the game described above, determine which player has a winning strategy and describe that strategy.

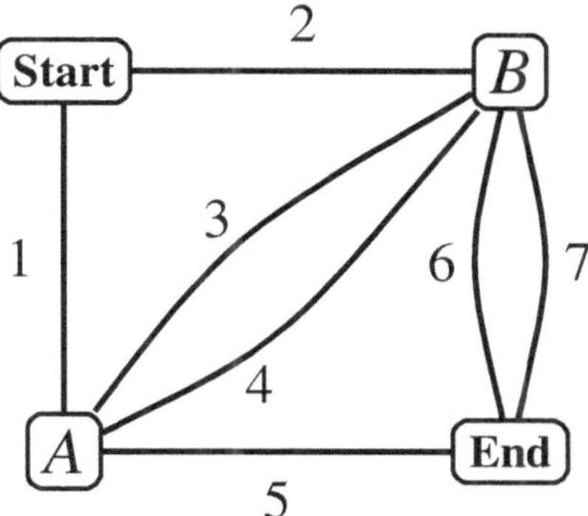


Figure 3.30: The map for the game of Exercise 3.22.

**Solution to Exercise 3.22.**

**(a)** This is a finite game with perfect information with only two outcomes (Player 1 wins or Player 2 wins). Thus, by Theorem 3.5.1 (Volume 1, Chapter 3, Section 5), one of the two players has a winning strategy.

**(b)** For the game described above, Player 1 is the one who has a winning strategy, which is as follows. First remove edge 7. Then

- if Player 2 goes to node $B$ delete edge 6 and then, after the next move of Player 2, delete edge 5 (now "End" has become unreachable),
- if, instead, Player 2 goes to node $A$ delete edge 5 and then, after the next move of Player 2, delete edge 6 (now "End" has become unreachable).

Exercise 3.23 Consider the following game, played on an $n \times n$ chess-like board (Figure 3.31 shows a $6 \times 6$ board, that is, a board with 6 rows and 6 columns). A coin is put in the bottom-right cell, marked as 'start'. The players take turns moving, with Player 1 making the first move. At each turn the player whose turn it is to move must shift the coin either one square to the left (if possible) or one square up (if possible), as indicated by the arrows in Figure 3.31; shifting the coin diagonally is not allowed. The game ends when the coin reaches the top-left cell (marked as 'end'). The player who moves the coin to that cell **loses** (in other words, the last player to move loses). Thus there are only two outcomes (Player 1 wins and Player 2 wins) and each pleyer prefers winning herself to having the other player win.

**(a)** For the cases $n = 2$ and $n = 3$, determine which player has a winning strategy.

**(b)** For the cases $n = 2$ and $n = 3$, explain why this is a trivial game.

**(c)** Now modify the game by allowing also a diagonal move, so that the coin can also be shifted one square diagonally in the up-left direction. For the cases $n = 3$ and $n = 4$, determine which player has a winning strategy and construct that strategy. Is the game still trivial?

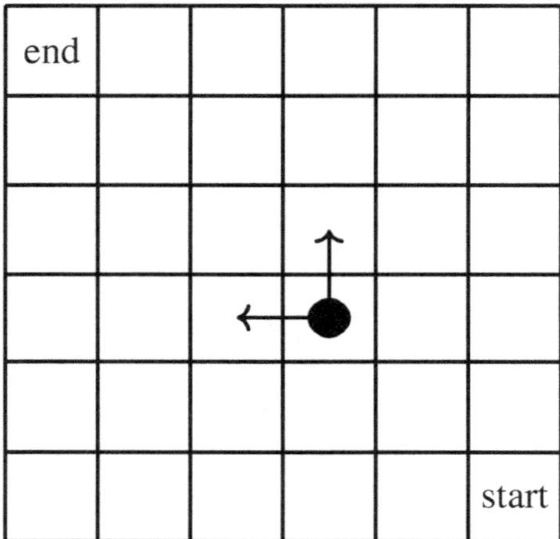


Figure 3.31: The $6 \times 6$ board for the game of Exercise 3.23.

**Solution to Exercise 3.23.**

**(a)** The case $n = 2$ is very simple: wherever Player 1 moves the coin, Player 2 is forced to move the coin to 'end' and loses. Thus Player 1 has a winning strategy and, indeed, both strategies of Player 1 are winning strategies.

The case $n = 3$ is shown in Figure 3.32, where each square is marked either as 'L' or 'W'. If the square is marked as 'L' then the player whose turn it is to move when the coin is in that square will lose, while if the square is marked as 'W' then the player whose turn to move when the coin is in that square will win. It is clear from Figure 3.32 that Player 1 is again the one with a winning strategy, and – again – any strategy of Player 1 is a winning strategy.

**(b)** The game is trivial because, for the player with a winning strategy, any strategy will ensure a win and, for the other player, any strategy will guarantee losing.

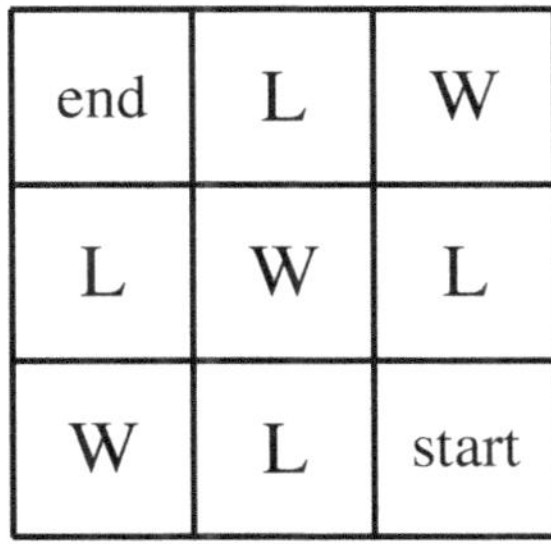

| end | L | W |
|---|---|---|
| L | W | L |
| W | L | start |

Figure 3.32: The case where $n = 3$.

**(c)** The modified with $n = 3$ can be analyzed the same way as done in Figure 3.32, but with an important difference: now all three squares adjacent to 'start' are marked 'W', so that it is now Player 2 who has a winning strategy, which is as follows: (1) if Player 1 moves the coin up then move it diagonally to the square marked 'L' on the right of 'end', (2) if Player 1 moves the coin left then move it diagonally to the square marked 'L' below 'end' and (3) if Player 1 moves the coin diagonally to the center square then either move it up or move it left. The game is not trivial any more, because if Player 2 plays badly then Player 1 can win (for example, if Player 1 moves the coin up and Player 2 also moves it up, then Player 1 will win).

The modified with $n = 4$ can be analyzed as shown in Figure 3.33. In this case it is Player 1 who has a winning strategy, which is as follows: (1) move the coin diagonally to the only square adjacent to 'start' marked 'L' and then (2) wherever Player 2 moves the coin, move it to an adjacent square marked 'L'. Again, the game is not trivial, because if Player 1 plays badly then Player 2 can win.

| end | L | W | L |
|---|---|---|---|
| L | W | W | W |
| W | W | L | W |
| L | W | W | start |

Figure 3.33: The modified game with $n = 4$.

## 3.6 More difficult exercises

The exercises in this section are more difficult and challenging than the previous ones.

Exercise 3.24 Ten pirates (ranked according to age, with 1 being the youngest and 10 the oldest) jointly own 100 gold coins. They have to decide on an allocation $(x_1, x_2, \ldots, x_{10})$ where $x_i \in \{0, 1, 2, \ldots, 99, 100\}$ for all $i = 1, ..., 10$ and $x_1 + x_2 + \cdots + x_{10} = 100$. They have agreed on the following decision rule. The oldest pirate (pirate 10) is the first to propose an allocation. If at least half of the pirates (including the pirate who makes the proposal) approve of this allocation, it is enforced. Otherwise pirate 10 is forced to "take early retirement". The nine-player version of the game is then played, where pirate 9 makes a proposal. For approval of this allocation the vote of at least four pirates among 1, 2, ..., 8 is needed. If the proposal is accepted, it is enforced, otherwise pirate 9 is forced to "take early retirement". And so on.

Please note: if the number of pirates in the game is odd, a strict majority is necessary; if the number is even, the one who made the proposal breaks ties. Also look carefully at the admissible values of $x_i$ : any integer between 0 and 100 (*including* 0 and 100).

Each pirate is selfish and greedy, that is, only cares about how many coins he gets and prefers getting more coins to getting fewer coins. We also need to make some hypothesis as to how a pirate reacts to a proposal $P$ which would give him the same number of coins as the proposal $P'$ which he expects to be offered in response, should $P$ be rejected. We make two alternative hypotheses:

- **Spiteful** hypothesis: a pirate will vote **against** a proposal if he is offered the same as what he expects to get if the proposal is rejected.
- **Altruistic** hypothesis: a pirate will vote **in favor** of a proposal if he is offered the same as what he expects to get if the proposal is rejected.

**(a)** Find the backward induction solution under the spiteful hypothesis.
**(b)** Find the backward induction solution under the altruistic hypothesis.
**(c)** Now consider the case where each pirate – while still being selfish and greedy – is neither spiteful nor altruistic, in the sense that he is indifferent between any two outcomes where he gets the same number of coins.
  **(1)** What are all the backward induction outcomes described in terms of what pirate 10 gets?
  **(2)** Restricting attention to the backward induction solutions, what is the minimum number of coins that pirate 10 obtains?
  **(3)** Restricting attention to the backward induction solutions, what is the maximum number of coins that pirate 10 obtains?

■

**Solution to Exercise 3.24.**

**(a)** If only pirates 2 and 1 are left, pirate 2 keeps 100 coins for he does not need pirate 1's approval. Think next of pirate 3 submitting a proposal $x = (x_1, x_2, x_3)$. Under the spiteful assumption, he expects that pirate 1 will reject $x$ unless $x_1 > 0$. Hence pirate 3's optimal proposal is (1,0,99). Keeping the pessimistic assumption, pirate 4's optimal proposal is (0,1,0,99), since to obtain pirate 2's approval he must offer him at least one coin. Repeating the argument we end up with the backward-induction allocation (0, 1, 0, 1, 0, 1, 0, 1, 0, 96), which gives most of the coins to the oldest pirate (pirate 10).

**(b)** Under the altruistic assumption (that any pirate approves of a proposal if he does not expect more by rejecting it), the backward-induction solution gives all the coins to pirate 10.

**(c)** Let us now turn to the indifference hypothesis.

**(1)** There are more backward-induction solutions besides the two found above. At these backward-induction solutions pirate 10 gets either 96 or 97 or 98 or 99 or 100 coins.

**(2)** The minimum number of coins that pirate 10 gets is 96.

**(3)** The maximum number of coins that pirate 10 gets is 100.

Note: why are pirates 1 to 9, exploited by pirate 10 at any backward induction solution? Because they cannot credibly make cooperative promises. For instance, when pirate 3 proposes (99, 0, 1), pirate 2 would like to propose to pirate 1 "Let us reject this proposal; I then will offer you 50 coins in the next round". Yet pirate 1 does not believe this promise, since nothing will prevent pirate 2 from reneging on his promise (and he will have every incentive to do so).

Exercise 3.25 The board of directors of a company consists of five women: A, B, C, D and E; they need to elect one of its members as chairwoman by a procedure involving successive majority votes. To start with, A will be paired with B; the one getting the majority in this vote will then be paired with C, the winner paired with D, and the winner of that with E. The winner of that fourth ballot will then be chairwoman.
The members know each other well and there is no uncertainty about anybody's preferences. It is known to everybody that their rankings are as follows (in particular, everybody wants to be chairperson):

A's ranking: $A \succ B \succ C \succ D \succ E$
B's ranking: $B \succ A \succ E \succ D \succ C$
C's ranking: $C \succ D \succ A \succ E \succ B$
D's ranking: $D \succ B \succ A \succ E \succ C$
E's ranking: $E \succ D \succ C \succ B \succ A$

According to the rules, a member who leaves before the vote may, but does not have to, give her voting rights to another member, who then casts two votes on each ballot. In other words, if X leaves and gives her voting rights to Y, then Y's vote will count as two; note that X cannot instruct Y on how to vote: all she can do is give her right to vote to Y. If on account of absence the number of voters is even, a tie on any ballot is resolved by picking the candidate that comes second in alphabetical order.
A is considering how her absence would affect the outcome. She reasons that, if she stays, she can't win, that B can beat C but not D, and that D will beat E on the last ballot. But if A leaves without giving her vote to somebody, D and E will be tied, and, by the tie-breaking rule, E will win. A has decided to leave and give her voting rights to somebody. She expects that her colleagues will vote strategically, that is, not necessarily in accordance with their true preferences.
In answering the following questions assume that both A and E are naïve in the following

sense: each thinks that if she leaves nobody else will leave (thus A believes that, after she leaves, the remaining four voters will stay and vote, and E believes that, after she leaves too, the remaining three voters will stay and vote).

**(a)** A announces that she must leave and she names the person to whom she transfers her voting rights. Who is this person?

**(b)** There is a long pause while each of the remaining members study the implications of A's departure and plan their own strategies. Eventually E announces that she is content to let her three colleagues settle the election in her absence, and she leaves without bothering to give anybody her vote. Why did E leave?

■

**Solution to Exercise 3.25.**

**(a)** When A left, she named E as her representative. Under this new situation (where E has two votes) A beats B, C, D and E, and becomes elected.
The reasoning is as follows. In the last vote nobody can gain by voting insincerely (that is, not in accordance with their true preferences) The possibilities for the last vote are: (A,E), (B,E), (C,E) and (D,E). With E having two votes and with sincere voting, (A,E) would result in A being elected and the others in E being elected. Since B, C and D all prefer A to E, it is in their interest to carry A to the last ballot, i.e. to vote for A throughout. Thus the winner will be A, which is E's worst outcome.

**(b)** If E leaves without giving her voting rights to another member and the decision is taken by B, C and D, then D is elected, whom E prefers to A.
The reasoning is similar. In the last ballot there will be sincere voting, so that: $(A,E) \to A$, $(B,E) \to B$, $(C,E) \to E$, $(D,E) \to D$. Since C cannot win, there is no point voting for her ever. Having removed C, D becomes the top choice of both C and D. Hence they will make sure that D is the outcome. This can be achieved by voting for D as soon as D appears on the ballot. Thus the winner is D.

Now the complete analysis.

CASE 1: all voters are present. Then in the last stage $(A,E) \to A$, $(B,E) \to B$, $(C,E) \to E$, $(D,E) \to D$. Thus in the second-to-last stage $(A,D) \to D$, $(B,D) \to D$ and $(C,D) \to D$ (in this vote the voters realize that voting for C is really voting for E, because in the last stage E would beat C; thus voters A, C and D will vote for D). That is, no matter what the previous votes, D is the winner. Hence when all voters are present, D wins .

CASE 2: A leaves and gives her vote to B. Then in the last stage $(A,E) \to A$, $(B,E) \to B$, $(C,E) \to E$. $(D,E) \to E$. Thus in the second-to-last stage $(A,D) \to A$ (in this vote the voters realize that voting for D is really voting for E, because in the last stage E would beat D; thus voters B, C and D will vote for A), $(B,D) \to B$ (once again, voting for D is really voting for E so that voters B and D will vote for B) and $(C,D) \to D$ (again, on the understanding that a vote for D is really a vote for E). In the second vote $(A,C) \to A$ (a vote for A is truly a vote for A, while a vote for C is a vote that will ultimately lead to E through the chain $(C,D) \to D$ and $(D,E) \to E$), $(B,C) \to B$ (same reasoning). Thus in the first vote $(A,B) \to B$: B wins .

CASE 3: A leaves and gives her vote to C. Then in the last stage $(A,E) \to A$, $(B,E) \to B$, $(C,E) \to E$, $(D,E) \to D$. Thus in the second-to-last stage $(A,D) \to D$, $(B,D) \to D$ and $(C,D) \to D$. That is, no matter what the previous votes, D is the winner.

CASE 4: A leaves and gives her vote to D. Then in the last stage $(A,E) \to A$, $(B,E) \to B$,

$(C,E) \to E$, $(D,E) \to D$. Thus in the second-to-last stage $(A,D) \to D$, $(B,D) \to D$ and $(C,D) \to D$. . That is, no matter what the previous votes, D is the winner.

CASE 5: A leaves and gives her vote to E. This was analyzed above: A wins.

Hence A is better off leaving and giving her vote to E. What happens when E leaves was explained above.

Exercise 3.26 It is common knowledge between a Seller (he) and a Buyer (she) that the Seller owns an object that is worth \$0 to him and is worth \$$v$ to the Buyer, with $v > 0$. We will consider two versions of the game.

**Version 1** (two-stage game). In stage 1 the Seller announces an integer price $p_1 \geq 0$, which the Buyer accepts or rejects. If the Buyer accepts, then she gets the object by paying $p_1$ to the Seller and her payoff is $v - p_1$, while the Buyer's payoff is $p_1$. If the Buyer rejects, then in stage 2 the Seller makes a second offer (again, an integer) $p_2 \geq 0$, which the Buyer accepts or rejects. If the Buyer accepts then she gets the object by paying $p_2$ to the Seller and her payoff is $v - 1 - \varepsilon - p_2$, with $0 < \varepsilon < 1$, while the Buyer's payoff is $p_2$. If the Buyer rejects, the game ends with no sale and both players get a payoff of 0. Assume that if, at any stage, the Buyer is indifferent between accepting and rejecting, then she will accept and this fact is common knowledge between Buyer and Seller.

**(a)** Find the backward-induction solution.

**(b)** Are there any Nash equilibria that differ from the backward-induction solution? If Yes, find one, if No explain why not.

**Version 2** ($n$-stage game with $n \geq 3$). The first two stages are the same as in Version 1, however, if the Buyer rejects the second-stage offer of $p_2$ then in stage 3 the Seller makes a third offer (again, an integer) $p_3 \geq 0$, which the Buyer accepts or rejects. And so on, up to the last stage $n$. If in the last stage the Buyer rejects then there is no sale and both players get a payoff of 0. If the Buyer accepts in stage $k$ with $2 \leq k \leq n$ then her payoff is $v - (k-1) - \varepsilon$, with $0 < \varepsilon < 1$, while the Buyer's payoff is $p_k$.

**(c)** Find the backward-induction solution.

■

**Solution to Exercise 3.26.**

**(a)** In stage 2 the Buyer will accept any $p_2 \leq v - 2$ (only such values of $p_2$ give the Buyer a positive payoff; recall that $p_2$ must be an integer) and reject any other offer. Thus at the beginning of period 2 the Seller will offer $p_2 = v - 2$ and the Buyer will accept. Hence in period 1 the Seller knows that he can guarantee himself a payoff of $v - 2$ in period 2, leaving the Buyer with a payoff of $v - 1 - \varepsilon - (v - 2) = 1 - \varepsilon$ (recall that $0 < \varepsilon < 1$ so that $0 < 1 - \varepsilon < 1$). In period 1 the Buyer will reject any $p_1 \geq v$ and accept any $p_1 \leq v - 1$. Hence the Seller will offer $p_1 = v - 1$ and the Buyer will accept; the payoffs are $v - 1$ for the Seller and 1 for the Buyer.

**(b)** Yes there are many other Nash equilibria, based on incredible threats by the Buyer. For example if the Buyer's strategy (which is not credible) is to reject in every period any price greater than 1, then the best reply for the Seller is to offer 1 in period 1 (and to plan to offer 1 again in period 2).

**(c)** The reasoning here is an extension of the reasoning in Part (a). In period $k \geq 2$ the Buyer will accept any $p_k \leq v - k$ and reject any other offer; thus at the beginning of period $k$ the Seller would offer $p_k = v - k$. Hence in period 1 the Seller will offer $p_1 = v - 1$ and the Buyer will accept; the payoffs are $v - 1$ for the Seller and 1 for the Buyer.

Exercise 3.27 Two players, 1 and 2, can contribute to the provision of a public good. The good will be provided if and only if the sum of their contributions exceeds some critical level $X > 0$. In this case, each player obtains a benefit $B > 0$. However, contributing is costly and contributions are non-refundable. The cost of contributing $x \geq 0$ is $c(x) = x$. The payoff of a player who contributes $x$ is $B - x$ if the public good is provided and $-x$ if the public good is not provided. Assume that $2B > X$ so that providing the good is efficient. There is no discounting, so that these payoffs do not depend on the period in which the public good is provided (if it is provided). The players contribute sequentially, with Player 1 contributing in period 1 and Player 2 in period 2. Assume that contributions are observable and that the good is provided if and only if, and as soon as, the total contribution level is at least $X$.

**(a)** Suppose first that $B > X$. Find the backward-induction solution(s). Is the public good provided in the first period, the second period, or never?

**(b)** Suppose now that $B = X$.

**(1)** Find the backward-induction solution(s). Is the public good provided in the first period, the second period, or never?

**(2)** Find a Nash equilibrium that is not a backward-induction solution.

**(c)** Suppose now that $B < X$. Find the backward-induction solution(s). Is the public good provided in the first period, the second period, or never?

**Solution to Exercise 3.27.**

**(a)** Assume that $B > X$. Let $x_i \geq 0$ be the contribution of player $i$. The backward-induction strategy of Player 2 is

$$x_2(x_1) = \begin{cases} X - x_1 & \text{if } x_1 \leq X \\ 0 & \text{if } x_1 > X. \end{cases}$$

Thus the backward-induction strategy of Player 1 is to choose $x_1 = 0$. Hence the public good is provided in period 2.

**(b)** Assume that $B = X$.

**(1)** The best-reply function of Player 2 is

$$x_2(x_1) = \begin{cases} \text{either } 0 \text{ or } X & \text{if } x_1 = 0 \\ X - x_1 & \text{if } 0 < x_1 \leq X \\ 0 & \text{if } x_1 > X. \end{cases}$$

There is only one backward-induction solution: $x_1 = 0$ together with Player 2's strategy given above with $x_2(0) = X$ . Thus the public good is provided in period 2.

If Player 2 responds to $x_1 = 0$ with $x_2 = 0$ then Player 1 would prefer to contribute a positive amount $\hat{x}_1$ to which Player 2 would respond with $x_2(\hat{x}_1) = X - \hat{x}_1$; the problem is that Player 1 would want to choose the smallest such positive amount and it does not exist if the strategy sets are continuous. If the strategy sets were discrete then we would have a second backward-induction solution at which Player 1 chooses the smallest positive contribution.

**(2)** There are several Nash equilibria that are not backward-induction solutions. One of them is: $x_1 = \frac{X}{2}$ and Player 2's strategy is to choose $x_2 = \frac{X}{2}$ if $x_1 = \frac{X}{2}$ and $x_2 = 0$ otherwise.

**(c)** Assume that $B < X$. Note that, since (by hypothesis) $2B > X$, $B > X - B$. The best-reply function of Player 2 is

$$x_2(x_1) = \begin{cases} 0 & \text{if } 0 \le x_1 < X - B \text{ (i.e. if } X - x_1 > B) \\ \text{either } 0 \text{ or } B & \text{if } x_1 = X - B \\ X - x_1 & \text{if } x_1 > X - B \text{ (i.e. if } X - x_1 < B). \end{cases}$$

The backward-induction solution is $x_1 = X - B$ together with Player 2's strategy given above with $x_2(X - B) = B$ (the same logic applies here as in (1) of Part (b) concerning Player 1's best reply to Player 2's strategy with $x_2(X - B) = 0$). Thus the public good is provided in period 2.

Exercise 3.28 [Note: this exercise requires the use of calculus.]

There are two customer service workers; worker $i$ ($i \in \{1,2\}$) chooses effort level $E_i \in [0,\infty)$ to exert in helping a customer. The cost of effort is $c_1(E_1) = \frac{(E_1)^2}{15}$ for worker 1 and $c_2(E_2) = \frac{(E_2)^2}{10}$ for worker 2. The customer's satisfaction level equals the total effort that the workers exert, up to a maximum satisfaction level of 10, that is, it is equal to $\min\{E_1 + E_2, 10\}$. Each worker's payoff is the difference between the customer's satisfaction level and his own effort cost. We will consider two different games.

**Game 1**. First Player 1 (Worker 1) selects his effort level, then Player 2 (Worker 2) – after observing Player 1's choice – chooses her own effort level.

**(a)** Find the backward-induction solution.

**Game 2**. The initial sequence of moves is the same as in Game 1, but after Player 2 moves, Player 1 observes her choice and decides whether to exert some additional effort $\delta_1 \in [0, 10 - E_1]$, so that his own total effort is $E_1 + \delta_1$. Also, in this game, $E_1, E_2 \in [0, 10]$ (instead of $[0,\infty)$)

**(b)** Explain what a strategy for Player 1 is and and what a strategy for Player 2 is.

**(c)** Find all the backward-induction solutions.

**Solution to Exercise 3.28.**

**(a)** Player 2's best reply function is

$$E_2(E_1) = \begin{cases} 5 & \text{if } E_1 \leq 5 \\ 10 - E_1 & \text{if } 5 < E_1 \leq 10 \\ 0 & \text{if } E_1 > 10 \end{cases}$$

Thus Player 1 chooses $E_1$ to maximize

$$U_1(E_1) = \begin{cases} E_1 + 5 - \frac{(E_1)^2}{15} & \text{if } E_1 \leq 5 \\ 10 - \frac{(E_1)^2}{15} & \text{if } E_1 > 5 \end{cases}$$

The function $E_1 + 5 - \frac{(E_1)^2}{15}$ is increasing up to the point $E_1 = 7.5$ and thus achieves a maximum at $E_1 = 5$ in the range $[0,5]$. The function $0 - \frac{(E_1)^2}{15}$ is decreasing in the range $(5, \infty)$. Thus the payoff-maximizing value of $E_1$ is 5 and the actual choices at the backward-induction solution are $E_1 = E_2 = 5$.

**(b)** A strategy of Player 1 specifies his initial effort level, as well as his additional effort level as a function of both his initial effort level and Player 2's effort level. A strategy of Player 2 specifies her effort level as a function of Player 1's initial effort level.

**(c)** The payoff function of Player 1 is

$$U_1(E_1, \delta_1, E_2) = \begin{cases} E_1 + \delta_1 + E_2 - \frac{(E_1+\delta_1)^2}{15} & \text{if } E_1 + E_2 + \delta_1 \leq 10 \\ 10 - \frac{(E_1+\delta_1)^2}{15} & \text{if } E_1 + E_2 + \delta_1 > 10 \end{cases}$$

At the end of the game, as shown above, it is optimal for Player 1 to target the sum of his efforts to $E_1 + \delta_1 = 7.5$ if $E_2 \leq 2.5$ and to target the total effort to $E_1 + E_2 + \delta_1 = 10$ if $E_2 > 2.5$. Hence, Player 1's optimal strategy at the end of the game is

$$\delta_1(E_1, E_2) = \begin{cases} 7.5 - E_1 & \text{if } E_2 \leq 2.5 \text{ and } E_1 \leq 7.5 \\ 10 - E_1 - E_2 & \text{if } E_2 > 2.5 \text{ and } E_1 + E_2 \leq 10 \\ 0 & \text{if } E_1 > 7.5 \text{ or } E_1 + E_2 > 10 \end{cases}$$

In the middle of the game, Player 2 realizes that if she chooses $E_2 \geq 2.5$, Player 1's additional effort will bring the total effort up to at least 10. Thus Player 2 should never choose an effort level higher than 2.5. If Player 1 chose an effort $E_1 \leq 7.5$ in period 1, then this is exactly what Player 2 should do. But if Player 1 chose a higher effort level, then Player 2 need only exert enough effort to bring the total effort up to 10. Thus, Player 2's strategy in any subgame perfect equilibrium is:

$$E_2(E_1) = \begin{cases} 2.5 & \text{if } E_1 \leq 7.5 \\ 10 - E_1 & \text{if } 7.5 < E_1 \leq 10 \\ 0 & \text{if } E_1 > 10 \end{cases}$$

In the initial period, Player 1 should never choose an effort higher than 7.5, since choosing 7.5 is enough to guarantee that the total effort will be 10. Any $E_1 \in [0, 7.5]$ will cause Player 2 to choose $E_2 = 2.5$, which Player 1 will follow by choosing $\delta_1 = 7.5 - E_1$. Thus any $E_1 \in [0, 7.5]$ can be chosen initially in a backward-induction solution so that there are many backward-induction solutions.

Exercise 3.29 You won \$200M (\$200,000,000) playing the lottery. Two charities have approached you for a donation. You prefer to keep all the money for yourself, but you don't want to appear to be a selfish person, so you summon the directors of the two charities and inform them that they will play the following game. First, the director of Charity 1 (call her $D_1$) will make a public request, which can be any integer dollar amount, starting from \$1 and up to \$200M. Then the director of Charity 2 (call him $D_2$) – having heard $D_1$'s request – will announce his own request, which, again, can be any integer dollar amount starting from \$1 and up to \$200M. You inform them that one of your objectives is to punish greed and thus you will proceed as follows. Let $d_1$ denote the amount requested by $D_1$ and $d_2$ the amount requested by $D_2$. If $d_1 > d_2$ then $D_1$ – being the greedy one – will get nothing, while $D_2$ will be rewarded with two times what he asked for, up to \$200M; that is, $D_2$ will get $\min\{2d_2, 200M\}$. Symmetrically, if $d_1 < d_2$ then $D_2$ will get nothing, while $D_1$ will get $\min\{2d_1, 200M\}$. Finally, if $d_1 = d_2$ then each charity will get what it requested, up to \$100M, that is, each charity will get $\min\{d_1, 100M\}$.

**(a)** Suppose that it is common knowledge that each director is selfish and greedy, in the sense that he/she only cares about how much money his/her own charity gets and prefers more money to less. Find all the backward induction solutions.

**(b)** Suppose now that it is common knowledge that each director ranks the outcomes exclusively on the basis of the total amount of money that goes to the two charities (the more the better). Find all the backward induction solutions.

Now let us change the game. It is no longer a sequential game; instead, each director submits a written request, not knowing how much money the other director is requesting (that is, the game is simultaneous).

**(c)** Assume that the preferences are as described in Part (a) (both directors are selfish and greedy). Write the payoff functions of the two players and find all the Nash equilibria. Prove that what you propose are Nash equilibria and that there are no other Nash equilibria.

**(d)** Assume now that the preferences are as described in Part (b). Find all the Nash equilibria. Prove that what you propose are Nash equilibria and that there are no other Nash equilibria.

**(e)** Assume now that each director has the following lexicographic preferences: outcomes are ranked first on the basis of the total amount of money that goes to the two charities (the more the better) and then on the basis of how much money his/her own charity gets (the more the better). Find all the Nash equilibria. Prove that what you propose are Nash equilibria and that there are no other Nash equilibria.

■

**Solution to Exercise 3.29.**

**(a)** First we determine the optimal response of Player 2 to any possible $d_1$. Choose an arbitrary $d_1$. If Player 2 chooses $d_2 > d_1$ he gets 0; if he chooses $d_2 = d_1$ he gets $\min\{d_2, 100M\} > 0$ and if he chooses $d_2 = d_1 - 1$ he gets $\min\{2(d_1 - 1), 200M\}$ (choosing $d_2 < d_1 - 1$ is worse than choosing $d_2 = d_1 - 1$). Note that $2(d_1 - 1) > d_1$ if and only if $d_1 > 2$, that is, if and only if $d_1 \geq 3$. Thus the best reply of Player 2, denoted by $d_2(d_1)$, is as follows:

$$d_2(d_1) = \begin{cases} d_2 = 1 & \text{if } d_1 = 1 \\ d_2 = 1 \text{ or } d_2 = 2 & \text{if } d_1 = 2 \\ d_2 = d_1 - 1 & \text{if } 3 \leq d_1 \leq 100M \\ \text{any } d_2 \text{ with } 100M \leq d_2 \leq d_1 - 1 & \text{if } d_1 > 100M \end{cases}$$

Hence there are two sets of backward induction solutions:

**(1)** Player 1's strategy is $d_1 = 1$ and Player 2's strategy is

$$\begin{cases} d_2 = 1 & \text{if } d_1 = 1 \text{ or } d_1 = 2 \\ d_2 = d_1 - 1 & \text{if } 3 \leq d_1 < 100M \\ d_2 = \hat{d}_2 & \text{if } d_1 \geq 100M \end{cases}$$

for an arbitrary $\hat{d}_2$ with $100M \leq \hat{d}_2 \leq d_1 - 1$; all these solutions give rise to the outcome where each charity gets \$1!

**(2)** Player 1's strategy is $d_1 = 2$ and Player 2's strategy is

$$\begin{cases} d_2 = 1 & \text{if } d_1 = 1 \\ d_2 = 2 & \text{if } d_1 = 2 \\ d_2 = d_1 - 1 & \text{if } 3 \leq d_1 < 100M \\ d_2 = \hat{d}_2 & \text{if } d_1 \geq 100M \end{cases}$$

for an arbitrary $\hat{d}_2$ with $100M \leq \hat{d}_2 \leq d_1 - 1$; all these solutions give rise to the outcome where each charity gets \$2!

**(b)** Let us compute the best reply of Player 2. Consider an arbitrary $d_1$. If Player 2 responds with any $d_2 \geq d_1$ then the total contribution to both charities is $\min\{2d_1, 200M\}$; if Player 2 responds with $d_2 = d_1 - 1$ then the total contribution to both charities is $\min\{2(d_1 - 1), 200M\}$. Since $\min\{2(d_1 - 1), 200M\} < \min\{2d_1, 200M\}$ if and only if $d_1 \leq 100M$, the best reply of Player 2 is as follows:

$$d_2(d_1) = \begin{cases} \text{any } d_2 \geq d_1 & \text{if } d_1 \leq 100M \\ \text{any } d_2 \text{ with } d_2 \geq d_1 \text{ or } 100M \leq d_2 \leq d_1 - 1 & \text{if } d_1 > 100M \end{cases}$$

Thus the backward induction solutions are as follows. Player 1's strategy is any $d_1 \geq 100M$ and Player 2's strategy is any function obtained from $d_2(d_1)$ given above.

**(c)** The payoff functions are

$$\pi_1(d_1, d_2) = \begin{cases} \min\{2d_1, 200M\} & \text{if } d_1 < d_2 \\ \min\{d_1, 100M\} & \text{if } d_1 = d_2 \\ 0 & \text{if } d_1 > d_2 \end{cases}$$

$$\pi_2(d_1, d_2) = \begin{cases} 0 & \text{if } d_1 < d_2 \\ \min\{d_1, 100M\} & \text{if } d_1 = d_2 \\ \min\{2d_2, 200M\} & \text{if } d_1 > d_2 \end{cases}$$

Any $(d_1, d_2)$ with $d_1 < d_2$ is not a Nash equilibrium because Player 2 can get more by choosing $d_2 = d_1$. Similarly for any $(d_1, d_2)$ with $d_1 > d_2$. So we are left with pairs of the form $d_1 = d_2 = d$. For Player 1 deviating to $d_1 < d-1$ is worse than deviating to $d_1 = d-1$ (and similarly for Player 2). Since $2(d-1) > d$ if and only if $d > 2$, the Nash equilibria are (\$2,\$2) (where each charity gets \$2) and (\$1,\$1) (where each charity gets \$1).

**(d)** In this case

- every pair $(d_1, d_2)$ with $d_1 = d_2 = d$ is a Nash equilibrium: the total amount given to charity is $\min\{2d, 200M\}$ and it remains the same if Player 1 increases $d_1$ and it remains the same or decreases if Player 1 reduces $d_1$ (similarly for Player 2); furthermore
- every pair $(d_1, d_2)$ with $d_1 < d_2$ and $2d_1 \geq 200M$, that is, $d_1 \geq 100M$ is a Nash equilibrium, and
- every pair $(d_1, d_2)$ with $d_1 > d_2$ and $d_2 \geq 100M$ is a Nash equilibrium.

There are no other Nash equilibria.

**(e)** Consider a pair $(d_1, d_2)$ with $d_1 = d_2 = d$ and $2d \geq 200M$, that is, $d \geq 100M$ (so that the total is $\min\{2d, 200M\} = 200M$ and each player gets $\min\{d, 100M\} = 100M$). If Player 1 increases $d_1$ then the total does not change (it remains $\min\{2d, 200M\} = 200M$), but Player 1 gets 0 and thus is worse off. If Player 1 switches to $d_1 = d-1$ then the total is $\min\{2(d-1), 200M\}$; thus (1) if $2(d-1) < 200M$, that is, $d \leq 100M$ then the total goes down and Player 1 is worse off and (2) if $2(d-1) \geq 200M$, that is, $d > 100M$ then the total does not change and Player 1 gets $\min\{2(d-1), 200M\} = 200M$ and thus he is better off. The same reasoning applies to Player 2. Hence of all the such pairs only $(100M, 100M)$ is a Nash equilibrium.

Now consider a pair $(d_1, d_2)$ with $d_1 = d_2 = d$ and $2d < 200M$, that is, $d < 100M$ (so that the total is $\min\{2d, 200M\} = 2d$ and each player gets $\min\{d, 100M\} = d$). If Player 1 increases $d_1$ then the total does not change (it remains $\min\{2d, 200M\} = 2d$), but Player 1 gets 0 and thus is worse off. If Player 1 switches to $d_1 = d-1$ (assuming that this is possible, that is, that $d \geq 2$) then the total goes down to $\min\{2(d-1), 200M\} = 2(d-1) < 2d$ and thus Player 1 is worse off. Hence every such pair is a Nash equilibrium.

Now consider any pair $(d_1, d_2)$ with $d_1 < d_2$. Then the total is $\min\{2d_1, 200M\}$ and Player 2 gets 0. If Player 2 switched to $d_2 = d_1$ then the total would remain $\min\{2d_1, 200M\}$, but Player 2 would get a positive amount, namely $\min\{d_1, 100M\}$. Hence no such pair is a Nash equilibrium. Similar reasoning for pairs of the form $(d_1, d_2)$ with $d_1 > d_2$.

In conclusion, the Nash equilibria are all the pairs $(d_1, d_2)$ with $d_1 = d_2 = d \leq 100M$.

**Exercise 3.30** [This question requires the use of calculus.]

Consider the following two-player game. Player 1 chooses $x_1 \in [0,\infty)$ and Player 2 chooses $x_2 \in [0,\infty)$. The payoff functions are as follows: $\pi_1(x_1,x_2) = \min\{A, x_1+x_2\} - \frac{(x_1)^2}{20}$ and $\pi_2(x_1,x_2) = \min\{A, x_1+x_2\} - \frac{(x_2)^2}{12}$, where $A > 0$.

The game is sequential: Player 1 moves first and Player 2 moves second after having observed Player 1's choice.

**(a)** For the case where A = 12, find all the backward-induction solutions.

**(b)** For the case where A = 8, find all the backward-induction solutions.

**(c)** For the case where A = 4, find all the backward-induction solutions.

**Solution to Exercise 3.30.**

**(a)** Let $A = 12$ The best reply of Player 2 to every choice of $x_1$ is:

$$x_2(x_1) = \begin{cases} 6 & \text{if } x_1 \le 6 \\ 12 - x_1 & \text{if } 6 < x_1 \le 12 \\ 0 & \text{if } x_1 > 12 \end{cases}$$

This is the backward induction strategy of Player 2. Thus Player 1 chooses $x_1$ to maximize $\pi_1(x_1) = \min\{x_1 + x_2(x_1), 12\} - \frac{(x_1)^2}{20}$. The maximum is achieved at $x_1 = 6$. Thus the backward induction solution is given by $(6, x_2(x_1))$ with actual choices $x_1 = 6$ and $x_2 = 6$.

**(b)** Let A = 8. The best reply of Player 2 to every choice of $x_1$ is:

$$x_2(x_1) = \begin{cases} 6 & \text{if } x_1 \le 2 \\ 8 - x_1 & \text{if } 2 < x_1 \le 8 \\ 0 & \text{if } x_1 > 8 \end{cases}$$

This is the backward induction strategy of Player 2. Thus Player 1 chooses $x_1$ to maximize $\pi_1(x_1) = \min\{x_1 + x_2(x_1), 8\} - \frac{(x_1)^2}{20}$. The maximum is achieved at $x_1 = 2$. Thus the backward induction solution is given by $(2, x_2(x_1))$ with actual choices $x_1 = 2$ and $x_2 = 6$.

**(c)** Let A = 4. The best reply of Player 2 to every choice of $x_1$ is:

$$x_2(x_1) = \begin{cases} 4 - x_1 & \text{if } x_1 \le 4 \\ 0 & \text{if } x_1 > 4 \end{cases}$$

This is the backward induction strategy of Player 2. Thus Player 1 chooses $x_1$ to maximize $\pi_1(x_1) = \min\{x_1 + x_2(x_1), 12\} - \frac{(x_1)^2}{20}$. The maximum is achieved at $x_1 = 0$. Thus the backward induction solution is given by $(0, x_2(x_1))$ with actual choices $x_1 = 0$ and $x_2 = 4$.

Exercise 3.31 Consider the following two-stage game. There are two rooms and in each room there are 5 balloons: 4 white and 1 red. In stage 1, three players, Ann, Bob and Carla are taken to the first room and take turns moving in alphabetical order (first Ann, then Bob, then Carla and then again, if needed, in the same order). Each player has to pop at least 1 and at most 2 balloons. The game ends as soon as somebody pops the red balloon; the player who pops the red balloon is eliminated and gets nothing, while the other two players get $10 each and then they proceed to the second stage. In the second stage the two remaining players are taken to the second room where they proceed in the same way: they take turns moving, with the first in alphabetical order moving first (Ann if she was not eliminated, otherwise Bob); again, at each turn, the player who is moving has to pop at least 1 and at most 2 balloons. The player who pops the red balloon gets no additional money, while the other player gets and additional $50. Each player is selfish (only cares about how much money he/she gets) and greedy (prefers getting more money to getting less money).

How much money will the three players end up with at any backward-induction solution of this two-stage game? ■

**Solution to Exercise 3.31.** Let us begin with the second stage. The player who moves first has a strategy that guarantees that he/she will get the additional $50. The strategy is as follows: start by popping 1 white balloon and then

- if the second player pops 1 white balloon then pop 2 white balloons (so that the other player is left only with the red balloon),
- if the second player pops 2 white balloons then pop 1 white balloon (so that the other player is left only with the red balloon).

Now consider the first stage. Bob's objective is to have Ann eliminated, so that in the second stage he can apply the strategy described above and get the additional $50. So Bob should employ the following strategy:

- if Ann pops 1 white balloon then Bob should pop 2 white balloons so that Carla is left with the choice of (1) either popping 2 balloons (the red one being one of the two) and be eliminated or (3) popping the red balloon and be eliminated, or (3) popping the only remaining white balloon, in which case the move goes back to Ann who is forced to pop the only remaining balloon, which is the red one, and be eliminated; by choosing options (1) or (2) Carla ends up with nothing, while with option (3) she ends up with at least $10.
- if Ann pops 2 white balloons then Bob should pop 1 white balloon putting Carla in the same situation described above.

All the backward-induction solutions are obtained by Bob and Carla employing the strategies described above. They only differ in the initial move of Ann in stage 1 and Carla's move in stage 2.

So the final outcome is: Ann gets nothing, Bob gets $60 and Carla gets $10.

# 4. General Dynamic Games

## 4.1 Imperfect information

The exercises in this section deal with the notion of *finite extensive form (or frame) with perfect recall* (Definition 4.1.1, Volume 1, Chapter 4, Section 1).

Exercise 4.1 Consider the following situation involving a cat (C), a mouse (M), and a dog (D).

[C's objective is to catch M while avoiding D; M wants to tease C while avoiding getting caught; D just wants some peace and is unhappy when he is disturbed; however, for the purpose of this exercise, their objectives do not matter, because we are only interested in the structure of moves, irrespective of payoffs.]

In the morning, C and M simultaneously decide what activity to engage in. C can either nap ($n$) or roam ($r$), where roaming involves moving D's bone. M can either hide ($h$) or play ($p$). If nap and hide are chosen, then the game ends (outcome $z_1$). The game also ends immediately if roam and play are chosen, in which case C captures M (outcome $z_7$). On the other hand, if nap and play are chosen, then M observes that C is napping and must decide whether to move ($m$) D's bone or leave it where it is ($\ell$). If M chooses to leave the bone where it is, then the game ends (outcome $z_2$). Finally, in the event that D's bone was moved (either by C choosing to roam, or by M while C naps), then D learns that his bone was moved but does not observe who moved it; in this case, D must choose whether to punish C ($pc$) or punish M ($pm$). After D's move, the game ends (outcome $z_3$ if M moved the bone and D punishes C, outcome $z_4$ if M moved the bone and D punishes M, outcome $z_5$ if C moved the bone and D punishes C, outcome $z_6$ if C moved the bone and D punishes M).

Draw an extensive-form game-frame that represents this situation. ■

**Solution to Exercise 4.1.** The extensive-form game-frame is as follows (an alternative representation is one where the order of moves at the beginning is reversed: first M chooses and then C chooses; whenever moves are simultaneous, the order of play constitutes an arbitrary choice).

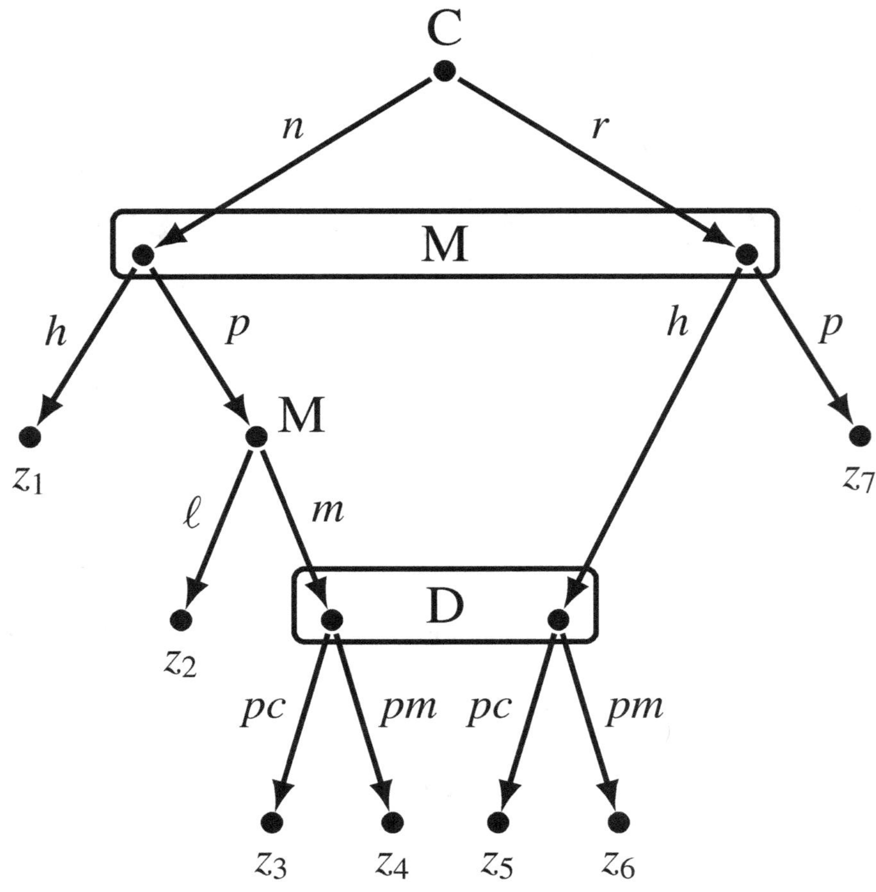


> Exercise 4.2 Al and Bob are playing the following game. Bob puts a tracking device on Al 's ankle, that Al cannot remove but he can turn off. Today Al is at his home. He makes two consecutive decisions: first whether or not to turn off the tracking device; then whether to stay in his home or go to his parents' house. Tomorrow Bob first checks if the tracking device is sending a signal and, if it is, from which location; then – whether or not the tracking device is sending a signal – decides whether to go to Al's house or to the house of Al's parents. If Bob goes to the location where Al is, Bob wins, otherwise Al wins
>
> Draw the extensive-form game-frame that represents the situation described above. ■

**Solution to Exercise 4.2.** The extensive-form game-frame is shown in Figure 4.3, where 'H' means 'stay at home', 'P' means 'got to parents' house' and each terminal node is labeled with the name of the winner.

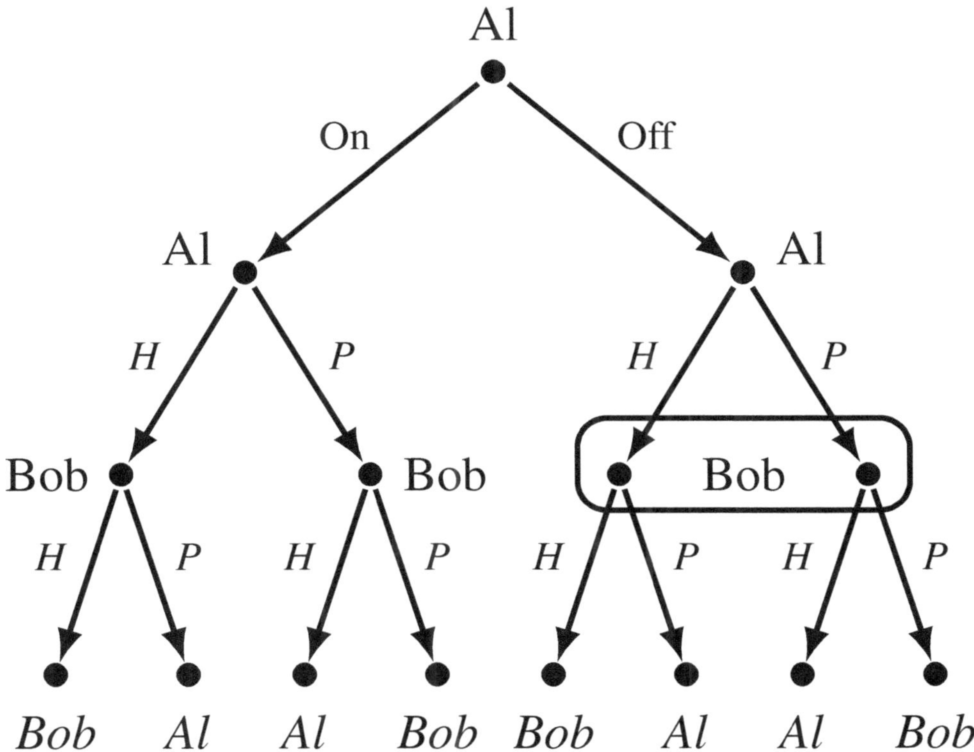


Figure 4.1: The extensive-form frame for Exercise 4.2.

Exercise 4.3 Consider the following three-player game. Player 1 moves first and chooses between A and S.

- If Player 1 chooses S then Player 2 is informed of this and asked to choose either A or S; if Player 2 also chooses S then the game ends and the outcome is $z_5$; if Player 2 chooses A, then Player 3 is asked to choose either A – in which case the game ends and the outcome is $z_3$ – or S, in which case the game ends and the outcome is $z_4$.
- If Player 1 chooses A, then Player 2 is not consulted and Player 3 is asked to choose either A – in which case the game ends and the outcome is $z_1$ – or S, in which case the game ends and the outcome is $z_2$.
- If asked to make a decision (between A and S), Player 3, is not informed whether or not Player 2 was consulted: all she knows is that either Player 1 chose A or Player 1 chose S and then Player 2 chose S.

Draw the extensive-form game-frame that represents the situation described above. ■

**Solution to Exercise 4.3.** The extensive-form game-frame is as follows.

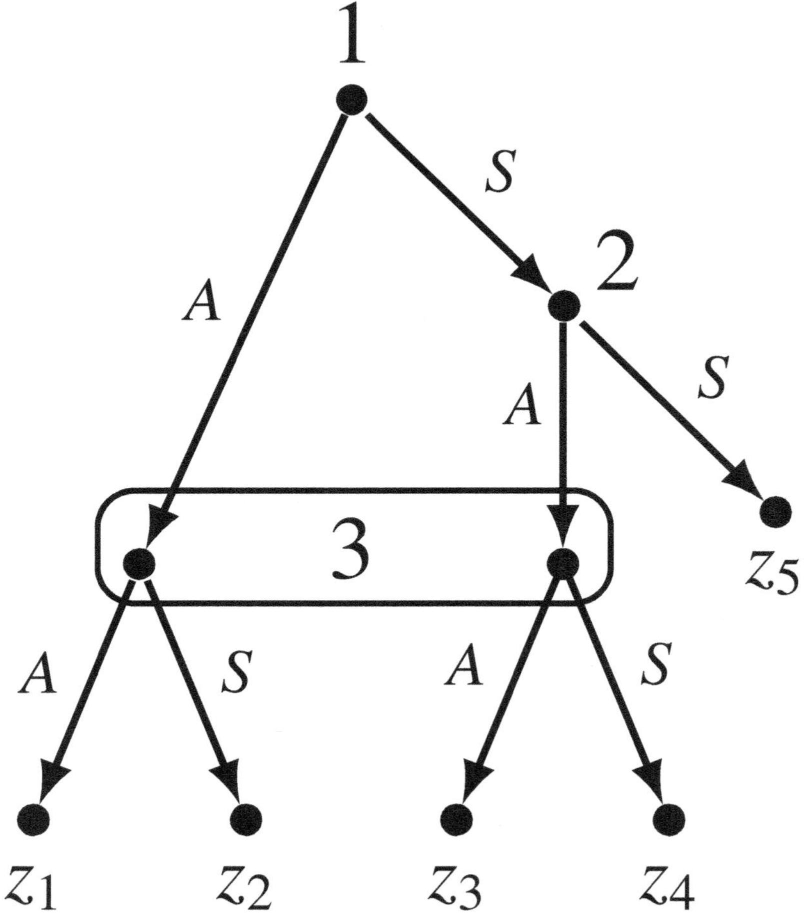


Exercise 4.4 Players 1 and 2 are placed in separate rooms.

In Player 1's room there are two boxes, labeled A and B. Player 1 is given \$100 and a piece of paper. She makes two consecutive decisions: first she decides in which box to put the \$100 and then she writes on the piece of paper a note that says either "The money is in box A" or "The money is in box B"; then she leaves the note on a table and exits the room.

Player 2 is then brought into this room and is shown what the note says. It is common knowledge between the two players that the statement in the note may be true or may be false. Player 2 then decides which box to open: if she finds the money in the box that she opens, she keeps it, otherwise the \$100 is given to Player 1.

Draw an extensive-form game-frame to represent this situation. ■

**Solution to Exercise 4.4.** The extensive-form game-frame is shown in Figure 4.2. At each terminal node, the top number is the amount of money given to Player 1 and the bottom number is the amount of money given to Player 2.

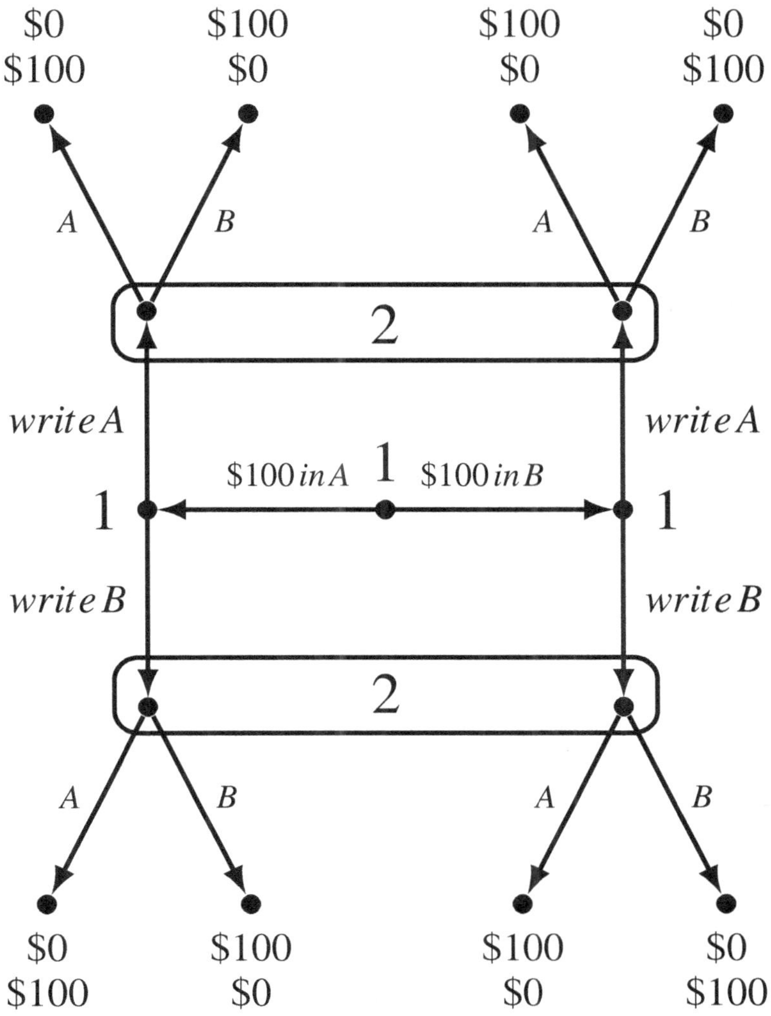


Figure 4.2: The extensive-form game-frame for Exercise 4.4.

Exercise 4.5 Sam has five dollars to distribute to his children, Ann and Ben. He asks them to play the following game. Ann must go to the kitchen and write either 'Ann' or 'Ben' on a piece of paper and give it to her father. Ben must go to the garage and write either 'Ann' or 'Ben' on a piece of paper and give it to his father. If they both wrote 'Ann', then Ann gets \$3 and Ben gets \$2. If they both wrote 'Ben', then Ben gets \$3 and Ann gets \$2. If they wrote different names, then they both get nothing.

Draw two different extensive-form games that represent this situation. ■

**Solution to Exercise 4.5.** The two extensive-form representations differ in terms of which player moves first, as shown below.

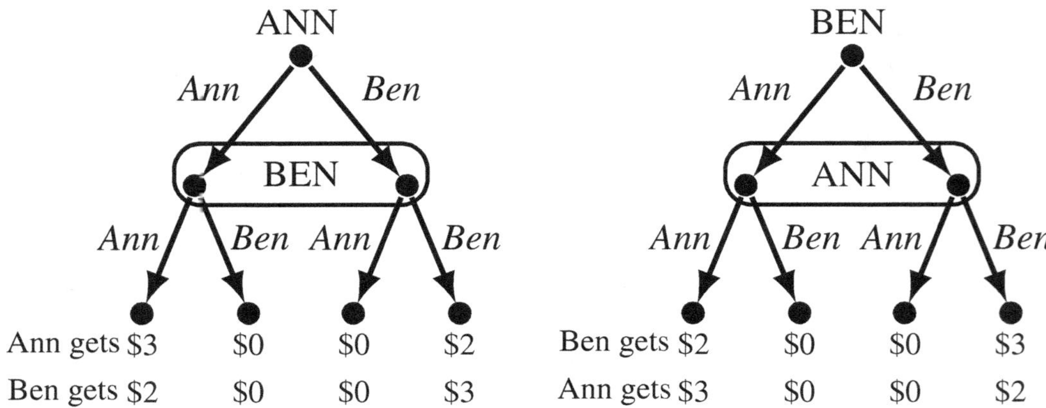


## 4.2 Strategies

The exercises in this section deal with the notion of *strategy* (Definition 4.2.1, Volume 1, Chapter 4, Section 2) in general extensive-form games and the strategic-form game associated with an extensive-form game.

**Exercise 4.6** Consider the following game. Janet is a contestant in a popular game show and her task is to guess behind which door Liz, another contestant, is hiding. With Janet out of the room, Liz chooses a door behind which to hide: either door A or door B. The host, Monty, observes this choice. Janet, not having observed Liz's choice, enters the room. Monty says to Janet either "Red milk" or "Green coffee" (which sounds silly, of course, but it is as good as anything else you see on TV these days!). After hearing Monty's statement, Janet picks a door: either A or B. If she picks the correct door, that is, if Liz is hiding behind that door, then she wins $100. If she picks the wrong door then she wins nothing. Liz wins $100 if Janet picks the wrong door and nothing if Janet picks the correct door. Thus, Liz would like to hide from Janet, while Janet would like to find Liz. What about Monty? Well, he likes the letter A. If Janet selects door A, then Monty gets $10, while if Janet selects door B then he gets nothing.
All players are selfish and greedy, that is, care only about how much money they themselves get and prefer more money to less.

**(a)** Draw an extensive-form game that represents this exciting situation.

**(b)** Write the corresponding strategic form. Let Liz be the row player, Monty the column player and Janet the remaining player.

**(c)** Find the Nash equilibria. ■

**Solution to Exercise 4.6.** For each player we can take as utility of an outcome (which specifies how much money each of the three gets) a number equal to the amount of money given to that player, that is, if the outcome is $(\$x, \$y, \$z)$, where $\$x$ is what Liz gets, $\$y$ is what Monty gets and $\$z$ is what Janet gets, then $U_{Liz}(\$x, \$y, \$z) = x$, $U_{Monty}(\$x, \$y, \$z) = y$ and $U_{Janet}(\$x, \$y, \$z) = z$.

**(a)** The extensive-form game is as follows, where at each terminal node the top number is Liz's utility, the middle number Monty's utility and the bottom number Janet's utility.

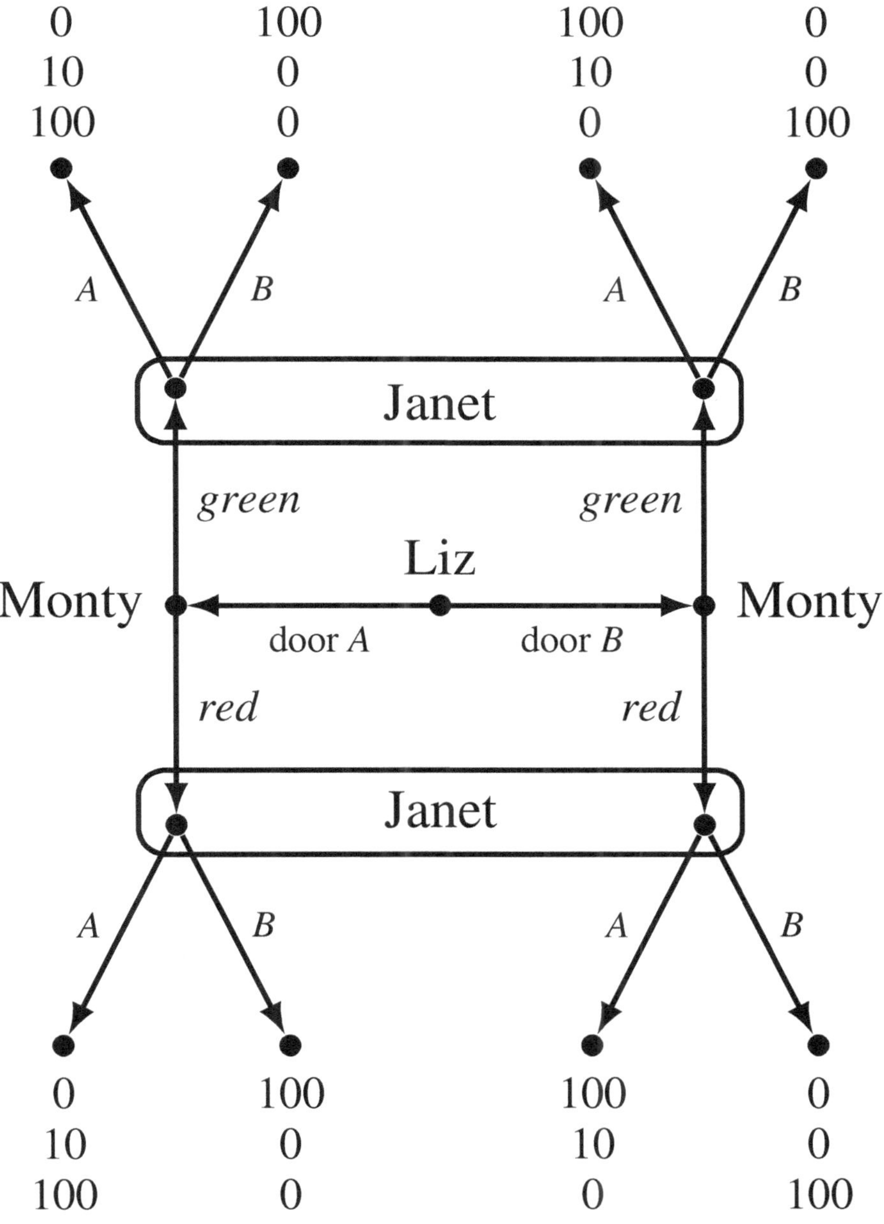


**(b)** The strategic form is as follows (for Monty we go left to right, thus RG means 'Red if A and Green if B'; for Janet we go top to bottom, thus BA means 'B if Green and A if Red').

Monty

Janet: AA

| Liz | RR | | | RG | | | GR | | | GG | | |
|---|---|---|---|---|---|---|---|---|---|---|---|---|
| *A* | 0 | 10 | 100 | 0 | 10 | 100 | 0 | 10 | 100 | 0 | 10 | 100 |
| *B* | 100 | 10 | 0 | 100 | 10 | 0 | 100 | 10 | 0 | 100 | 10 | 0 |

Monty

Janet: BA

| Liz | RR | | | RG | | | GR | | | GG | | |
|---|---|---|---|---|---|---|---|---|---|---|---|---|
| *A* | 0 | 10 | 100 | 0 | 10 | 100 | 100 | 0 | 0 | 100 | 0 | 0 |
| *B* | 100 | 10 | 0 | 0 | 0 | 100 | 100 | 10 | 0 | 0 | 0 | 100 |

Monty

Janet: AB

| Liz | RR | | | RG | | | GR | | | GG | | |
|---|---|---|---|---|---|---|---|---|---|---|---|---|
| *A* | 100 | 0 | 0 | 100 | 0 | 0 | 0 | 10 | 100 | 0 | 10 | 100 |
| *B* | 0 | 0 | 100 | 100 | 10 | 0 | 0 | 0 | 100 | 100 | 10 | 0 |

Monty

Janet: BB

| Liz | RR | | | RG | | | GR | | | GG | | |
|---|---|---|---|---|---|---|---|---|---|---|---|---|
| *A* | 100 | 0 | 0 | 100 | 0 | 0 | 100 | 0 | 0 | 100 | 0 | 0 |
| *B* | 0 | 0 | 100 | 0 | 0 | 100 | 0 | 0 | 100 | 0 | 0 | 100 |

**(c)** There are two Nash equilibria: $(A, RG, BA)$ and $(A, GR, AB)$.

Exercise 4.7 Consider the extensive-form game (based on the game-frame of Figure 4.3) shown in Figure **??**.

**(a)** How many strategies does Al have?

**(b)** How many strategies does Bob have?

**(c)** Without writing the strategic-form game associated with this extensive-form game, identify strategies that are weakly dominated for Bob.

■

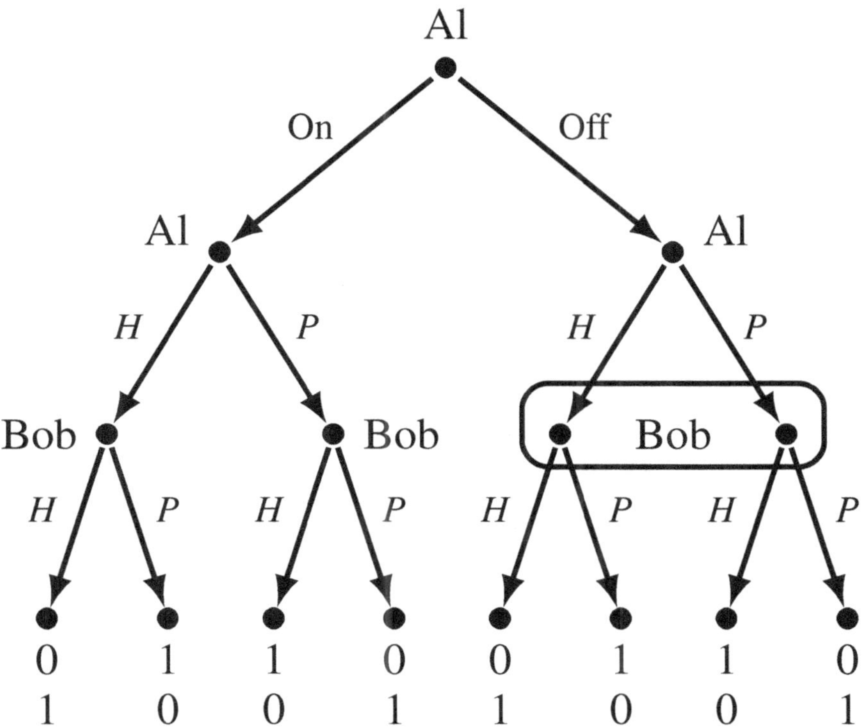


Figure 4.3: The game for Exercise 4.2.

**Solution to Exercise 4.7.**

**(a)** Al has three information set (all singleton sets) and two choices at each; thus Al has $2^3 = 8$ strategies.

**(b)** Bob has three information sets (the two singleton nodes and the information set on the right) and two choices at each; thus also Bob has $2^3 = 8$ strategies.

**(c)** Denote a strategy of Bob as a triple $(x,y,z)$ where $x \in \{H,P\}$ is his choice at the left node (after history "On-H"), $y \in \{H,P\}$ is his choice at the second node from the left (after history "On-P"), and $z \in \{H,P\}$ is his choice at the information set on the right. Then, for Bob,

- all of the following strategies are weakly dominated by $(H,P,H)$: $(H,H,H)$, $(P,H,H)$ and $(P,P,H)$
- all of the following strategies are weakly dominated by $(P,P,P)$: $(H,H,P)$, $(P,H,P)$ and $(P,P,P)$.

Exercise 4.8 Write the strategic-form game associated with the extensive-form game shown in Figure 4.4. ■

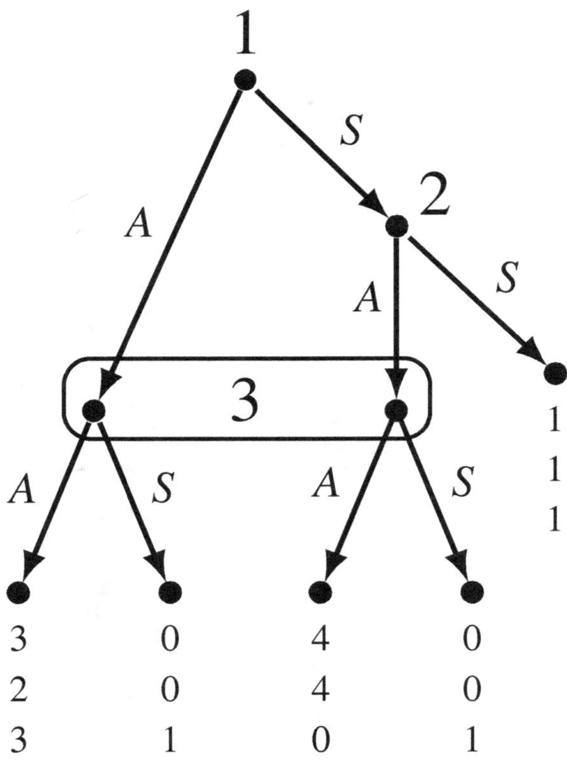


Figure 4.4: The game for Exercise 4.8.

**Solution to Exercise 4.8.** The strategic-form game is as follows.

Player 3: $A$

| Player 1 \ Player 2 | $A$ | $S$ |
|---|---|---|
| $A$ | 3 2 3 | 3 2 3 |
| $S$ | 4 4 0 | 1 1 1 |

Player 3: $S$

| Player 1 \ Player 2 | $A$ | $S$ |
|---|---|---|
| $A$ | 0 0 1 | 0 0 1 |
| $S$ | 0 0 1 | 1 1 1 |

## 4.3 Subgames

The exercises in this section deal with the notion of *proper subgame* and of *minimal* subgame (Definitions 4.3.1 and 4.3.2, Volume 1, Chapter 4, Section 3).

Exercise 4.9 Consider the extensive-form game-frame shown in Figure 4.5.
**(a)** How many *proper* subgames does the game have?
**(b)** What are the minimal subgames? ■

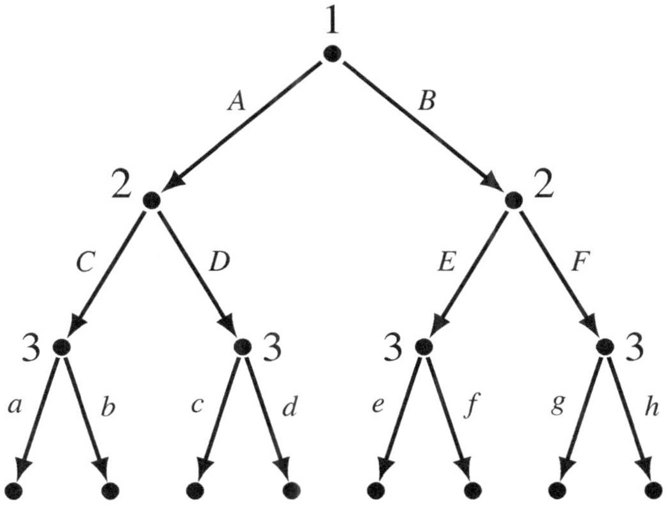

Figure 4.5: The game for Exercise 4.9.

**Solution to Exercise 4.9.**

**(a)** Six: the two that start at the decision nodes of Player 2 and the four that start at the decision nodes of Player 3.
**(b)** Four: the ones that start at the decision nodes of Player 3

Exercise 4.10 Consider the extensive-form game-frame shown in Figure 4.6.
**(a)** How many *proper* subgames does the game have?
**(b)** What are the minimal subgames? ■

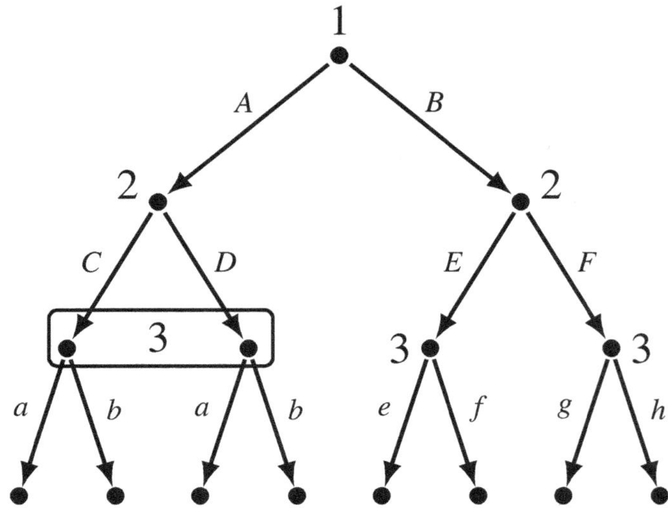


Figure 4.6: The game for Exercise 4.10.

**Solution to Exercise 4.10.**

**(a)** Four: the two that start at the decision nodes of Player 2 and the two that start at the two decision nodes of Player 3 on the right (which constitute singleton information sets).

**(b)** Three: the one that starts at the left decision node of Player 2 and the ones that start at the decision nodes of Player 3 on the right.

Exercise 4.11 Consider the extensive-form game-frame shown in Figure 4.7.

**(a)** How many *proper* subgames does the game have?

**(b)** What are the minimal proper subgames?

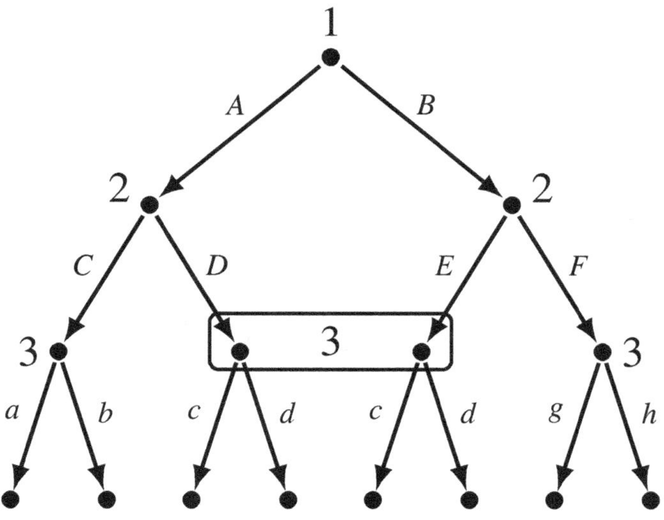


Figure 4.7: The game for Exercise 4.11.

**Solution to Exercise 4.11.**

**(a)** Two: the one that starts at the left-most decision node of Player 3 and the one that starts at the right-most decision node of Player 3 (which constitute singleton information sets).

**(b)** Two: both proper subgames are minimal.

Exercise 4.12 Consider the extensive-form game-frame shown in Figure 4.8.

How many *proper* subgames does the game have?

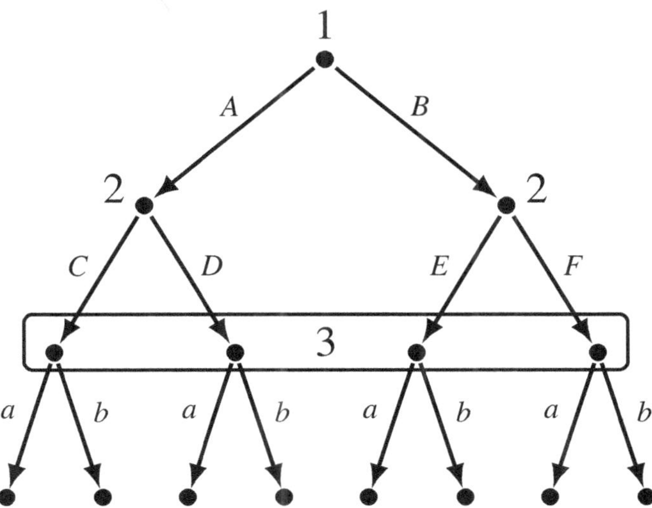


Figure 4.8: The game for Exercise 4.12.

**Solution to Exercise 4.12.** This game does not have any proper subgames.

## 4.4 Subgame-perfect equilibria

The exercises in this section deal with the notion of *subgame-perfect equilibrium* (Definition 4.4.1, Volume 1, Chapter 4, Section 4).

Exercise 4.13 Consider the game shown in Figure 4.9. As usual, at each terminal node the top number is Player 1's payoff, the middle number is Player 2's payoff and the bottom number is Player 3's payoff.

**(a)** Write all the strategies of Player 2.

**(b)** Write all the strategies of Player 3.

**(c)** Find all the subgame-perfect equilibria.

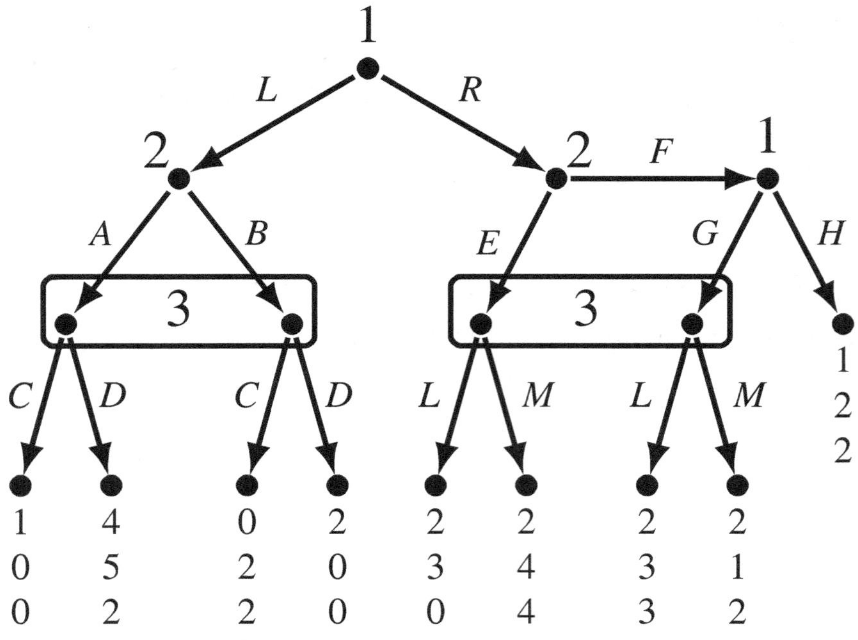


Figure 4.9: The game for Exercise 4.13.

**Solution to Exercise 4.13.**

**(a)** Player 2 has four strategies: *AE*, *AF*, *BE* and *BF*.

**(b)** Player 3 has four strategies: *CL*, *CM*, *DL* and *DM*.

**(c)** The game has two proper subgames that start at the nodes of Player 2. First we solve the subgame on the left. Its strategic form is as follows:

| | | Player 3 | |
|---|---|---|---|
| | | *C* | *D* |
| Player 2 | *A* | 0 0 | 5 2 |
| | *B* | 2 2 | 0 0 |

This game has two Nash equilibria: $(B,C)$ and $(A,D)$.

Next we solve the subgame on the right. Its strategic form is as follows:

Player 2

| Player 1 | E | F |
|---|---|---|
| G | 2 3 0 | 2 3 3 |
| H | 2 3 0 | 1 2 2 |

Player 3: $L$

Player 2

| Player 1 | E | F |
|---|---|---|
| G | 2 4 4 | 2 1 2 |
| H | 2 4 4 | 1 2 2 |

Player 3: $M$

This game has three Nash equilibria: $(G,F,L)$, $(G,E,M)$ and $(H,E,M)$.

Thus there are six subgame-perfect equilibria, which are as follows.

- There are three subgame-perfect equilibria that yield outcome $LAD$ with payoffs $(4,5,2)$, namely $((L,G),(A,E),(D,M))$, $((L,H),(A,E),(D,M))$, and $((L,G),(A,F),(D,L))$.

- There are two subgame-perfect equilibria that yield outcome $REM$ with payoffs $(2,4,4)$, namely $((R,G),(B,E),(C,M))$ and $((R,H),(B,E),(C,M))$.

- There is one subgame-perfect equilibrium that yields outcome $RFGL$ with payoffs $(2,3,3)$, namely $((R,G),(B,F),(C,L))$.

Exercise 4.14 Consider the following situation. A firm is managed by two individuals: Ann and Bill. Claudia has applied for a job with the firm. The application goes first to Ann, who can either veto it, in which case Claudia will not be hired, or approve it, in which case Claudia will be hired. Ann can also decide not to do any of the above and forward the file to Bill, asking him to make the hiring decision. In this case Bill can either hire Claudia or not hire her. If Claudia is hired, she is not told who made the hiring decision. If Claudia is hired, she can either work hard or shirk. The rankings of the outcomes are as follows.

- Claudia's favorite outcome is to be hired by Bill and shirk (she used to be in love with him and he rejected her!). The next best outcome is to be hired by Ann and shirk. The least favorite outcome is to be hired by Bill and work hard. Of the remaining outcomes, her most preferred is the one where she is hired by Ann and works hard and her least preferred is the one where Bill made the decision not to hire her; the remaining outcome is ranked between these two.
- Each of the two managers has the following preferences. The best outcome is one where he/she made the decision to hire Claudia and Claudia works hard (the hiring manager gets the credit for making a good hiring decision). The least favorite outcome is one where the other manager hired Claudia and she works hard (the other manager will get the credit). The outcome immediately above this (i.e. second to last) is the one where he/she made the decision to hire Claudia and Claudia shirks (he/she will take the blame for making a bad hiring decision). Of the remaining outcomes, his/her most favorite is the one where Claudia was hired by the other manager and shirks (the other manager will be blamed for making a bad decision and he/she will be able to take advantage of it). If Claudia is not hired, each manager prefers the other manager to have made the decision (no feelings of guilt).

For each player use a utility function whose values are consecutive integers with the lowest being 0.

**(a)** Draw an extensive-form game that represents the situation described above.

**(b)** Write the corresponding strategic form and find the Nash equilibria.

**(c)** Are any of the Nash equilibria not subgame perfect?

■

**Solution to Exercise 4.14.**
Name the outcomes as follows:

$z_1$ Ann hires Claudia and Claudia works hard
$z_2$ Ann hires Claudia and Claudia shirks
$z_3$ Bill hires Claudia and Claudia works hard
$z_4$ Bill hires Claudia and Claudia shirks
$z_5$ Bill makes the decision not to hire Claudia
$z_6$ Ann makes the decision not to hire Claudia

Ann's ranking is: $z_1 \succ z_4 \succ z_5 \succ z_6 \succ z_2 \succ z_3$.

Bill's ranking is: $z_3 \succ z_2 \succ z_6 \succ z_5 \succ z_4 \succ z_1$.

Claudia's ranking is: $z_4 \succ z_2 \succ z_1 \succ z_6 \succ z_5 \succ z_3$.

**(a)** The extensive-form game is as follows:

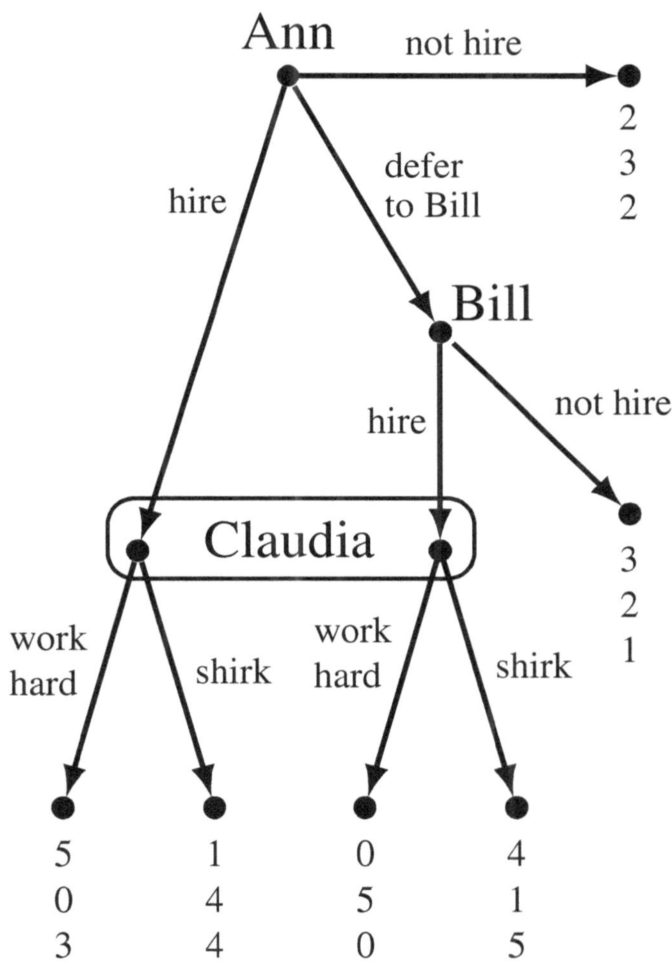


**(b)** The corresponding strategic-form is as follows:

CLAUDIA: work hard (BILL: columns; ANN: rows)

| ANN \ BILL | not hire | hire |
|---|---|---|
| not hire | 2 3 2 | 2 3 2 |
| hire | 5 0 3 | 5 0 3 |
| defer | 3 2 1 | 0 5 0 |

CLAUDIA: shirk (BILL: columns; ANN: rows)

| ANN \ BILL | not hire | hire |
|---|---|---|
| not hire | 2 3 2 | 2 3 2 |
| hire | 1 4 4 | 1 4 4 |
| defer | 3 2 1 | 4 1 5 |

There is only one Nash equilibrium given by: (defer, not hire, shirk).

**(c)** No, because there are no proper subgames.

Exercise 4.15 Two investors have each deposited \$100 in a bank. The bank has invested these deposits in a long-term project. If the bank is forced to liquidate the investment before the project matures, a total of \$160 can be recovered. However, if the bank allows the investment to reach maturity, the project will pay out a total of \$300.

There are two dates at which the investors can make withdrawals from the bank: date 1, before the bank's investment matures, and date 2, after its maturity. If both investors make withdrawals at date 1 then each receives \$80 and the game ends. If only one investor makes a withdrawal at date 1 then that investor receives \$100, the other receives \$60, and the game ends. If neither investor makes a withdrawal at date 1 then the project matures and the investors make withdrawal decisions at date 2. If both investors make withdrawals at date 2 then each receives \$150. If only one investor makes a withdrawal at date 2 then that investor receives \$200 and the other receives \$100. Finally, if neither investor makes a withdrawal at date 2 then the bank returns \$150 to each investor. At each date the decision whether or not to make a withdrawal is made simultaneously by both investors. Each investor is selfish and greedy, that is, cares only about his own wealth and prefers more money to less.

**(a)** Represent this situation as an extensive-form game.

**(b)** How many proper subgames are there?

**(c)** Find the pure-strategy subgame-perfect equilibria.

**(d)** Convert the extensive-form game of Part (a) into a strategic-form game.

**(e)** Find all the pure-strategy Nash equilibria of the game of part (d).

**(f)** Are all the pure-strategy Nash equilibria subgame perfect?

■

**Solution to Exercise 4.15.**
Denote an outcome as $(\$x, \$y)$ where $\$x$ is the amount of money given to Player 1 and $\$y$ the amount of money given to Player 2. Then we can take the following utility functions: $U_1(\$x, \$y) = x$ for Player 1 and $U_2(\$x, \$y) = y$ for Player 2.

**(a)** The extensive-form game is as follows ('W' means 'withdraw' and 'N' means 'Not withdraw')

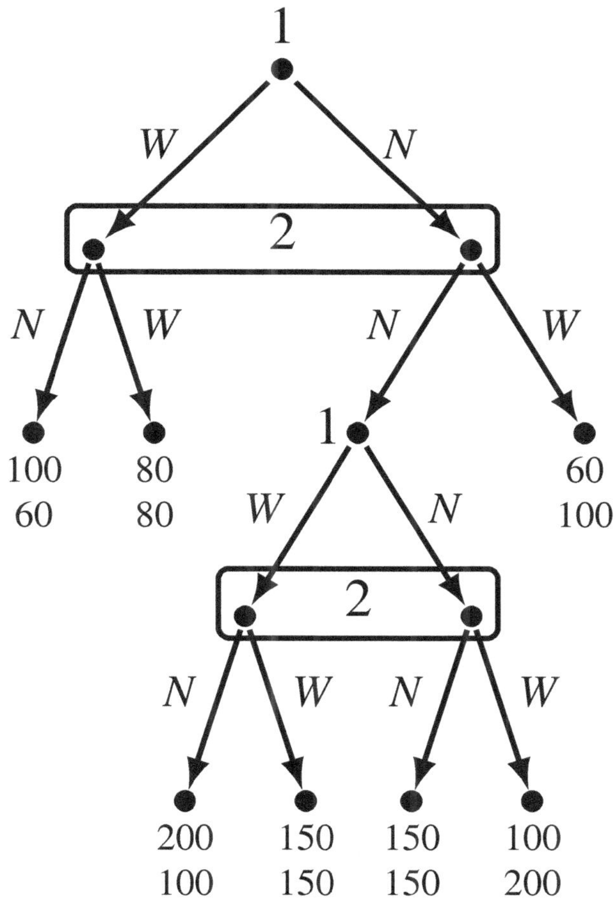


**(b)** There is only one proper subgame, namely the second-period game.

**(c)** Let us first solve the second-period game. Its strategic form is as follows:

| | | Player 2 | |
|---|---|---|---|
| | | W | N |
| Player 1 | W | 150 150 | 200 100 |
| | N | 100 200 | 150 150 |

W is a strictly dominant strategy for each player. Thus there is a unique Nash equilibrium: (W,W). Hence the extensive-form game simplifies to the following:

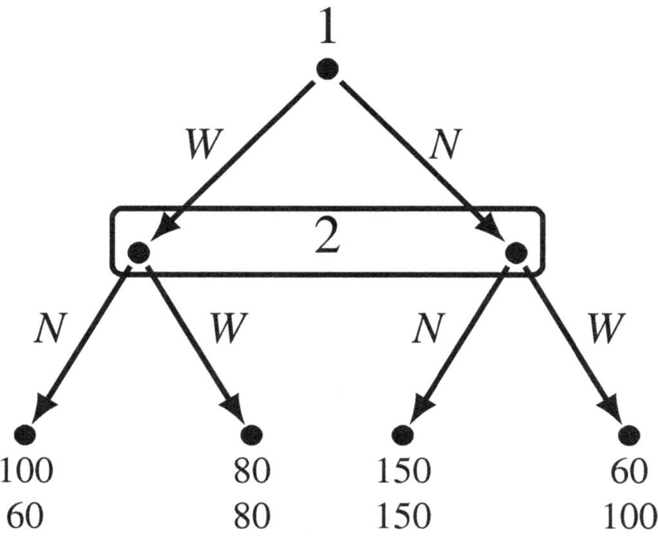


The corresponding strategic form is as follows:

| | | Player 2 | |
|---|---|---|---|
| | | W | N |
| Player 1 | W | 80 80 | 100 60 |
| | N | 60 100 | 150 150 |

This game has two Nash equilibria: (W,W) and (N,N). Thus there are two subgame-perfect equilibria of the original game: (WW, WW) and (NW, NW).

**(d)** Each player has four strategies. The strategic form is as follows:

| | | Player 2 WW | | WN | | NW | | NN | |
|---|---|---|---|---|---|---|---|---|---|
| | WW | 80 | 80 | 80 | 80 | 100 | 60 | 100 | 60 |
| Player | WN | 80 | 80 | 80 | 80 | 100 | 60 | 100 | 60 |
| 1 | NW | 60 | 100 | 60 | 100 | 150 | 150 | 200 | 100 |
| | NN | 60 | 100 | 60 | 100 | 100 | 200 | 150 | 150 |

**(e)** There are 5 Nash equilibria: (WW,WW), (WW,WN), (WN,WW), (WN,WN) and (NW,NW).

**(f)** As we saw above, only (WW, WW) and (NW, NW) are subgame-perfect.

Exercise 4.16 Two players are involved in a two-stage repeated Prisoners' Dilemma game. In each stage they play the game shown in Figure 4.10, where '$c$' stands for 'cooperation' and '$d$' for defection. More precisely, in the first stage, they play this game simultaneously (each player chooses between $c$ and $d$). In the second stage, after having been informed of what the other player did in the first stage, they simultaneously play the game a second time (again, each player chooses between $c$ and $d$). For each player his total payoff is the sum of the payoffs he receives in the two stages.

**(a)** Draw an extensive-form game that represents this two-stage game.

**(b)** How many proper subgames does the game have?

**(c)** Write down one of the strategies of Player 1.

**(d)** How many strategies does Player 1 have?

**(e)** How many strategies does Player 2 have?

**(f)** Find the subgame-perfect equilibrium of the extensive-form game of Part (a).

| | | Player 2 $c$ | | $d$ | |
|---|---|---|---|---|---|
| Player | $c$ | 5 | 5 | 0 | 6 |
| 1 | $d$ | 6 | 0 | 1 | 1 |

Figure 4.10: The stage game for Exercise 4.16.

**Solution to Exercise 4.16.**

**(a)** The extensive-form game is as follows:

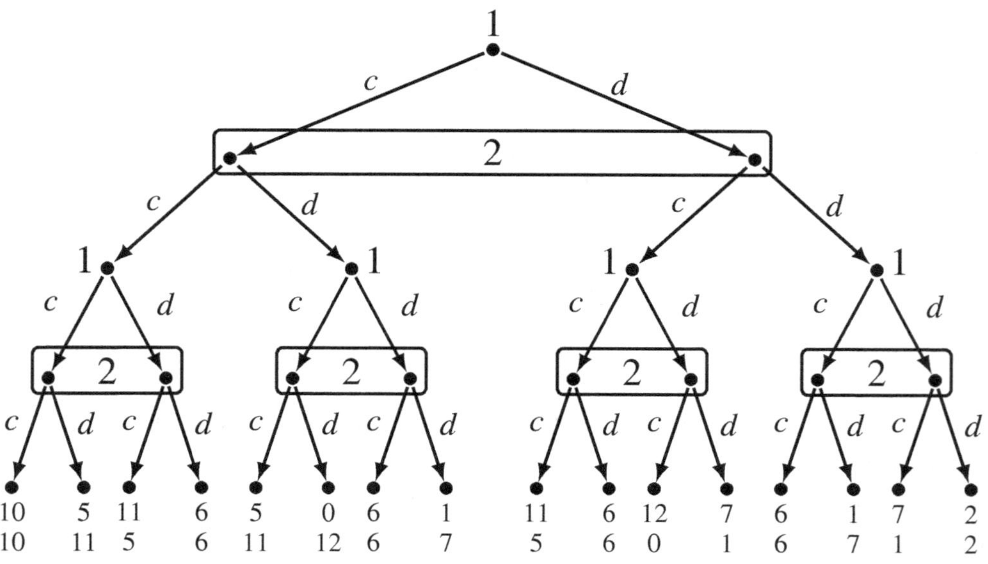


**(b)** There are four proper subgames (the second-stage ones).

**(c)** Player 1 has 5 information sets; hence a strategy for Player 1 has to specify five instructions. For example a possible strategy is:

**(1)** start with $c$ and then

**(2)** if I played $c$ and 2 played $c$ then play $d$,

**(3)** if I played $c$ and 2 played $d$ then play $c$,

**(4)** if I played $d$ and 2 played $c$ then play $c$,

**(5)** if I played $d$ and 2 played $d$ then play $d$.

**(d)** Player 1 has $2^5 = 32$ strategies.

**(e)** Also Player 2 has $2^5 = 32$ strategies.

**(f)** In each proper subgame (that is, in each second-stage game), playing $d$ is a strictly dominant choice for every player. Thus the unique Nash equilibrium of each second-stage game is $(d,d)$. Hence the game simplifies to the following:

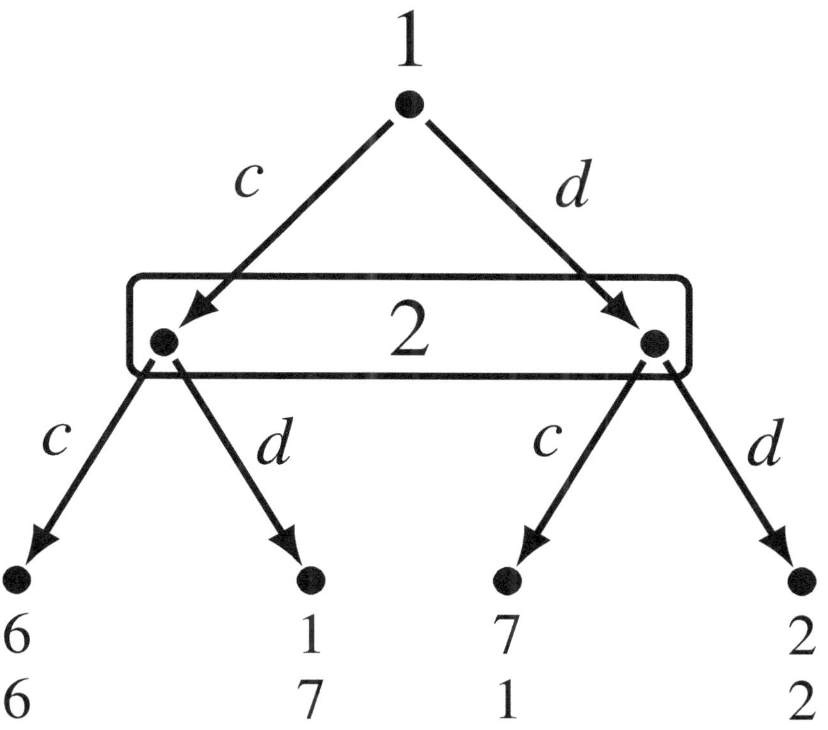


In this simplified game, $d$ is again a strictly dominant strategy for each player. Thus the subgame-perfect equilibrium is: $((d,d,d,d,d),(d,d,d,d,d))$.

Exercise 4.17 Consider the extensive-form game shown in Figure 4.11.

**(a)** List all the strategies of Player 1.

**(b)** List all the strategies of Player 2.

**(c)** Find the subgame-perfect equilibrium.

**(d)** Find two Nash equilibria that are not subgame-perfect.

■

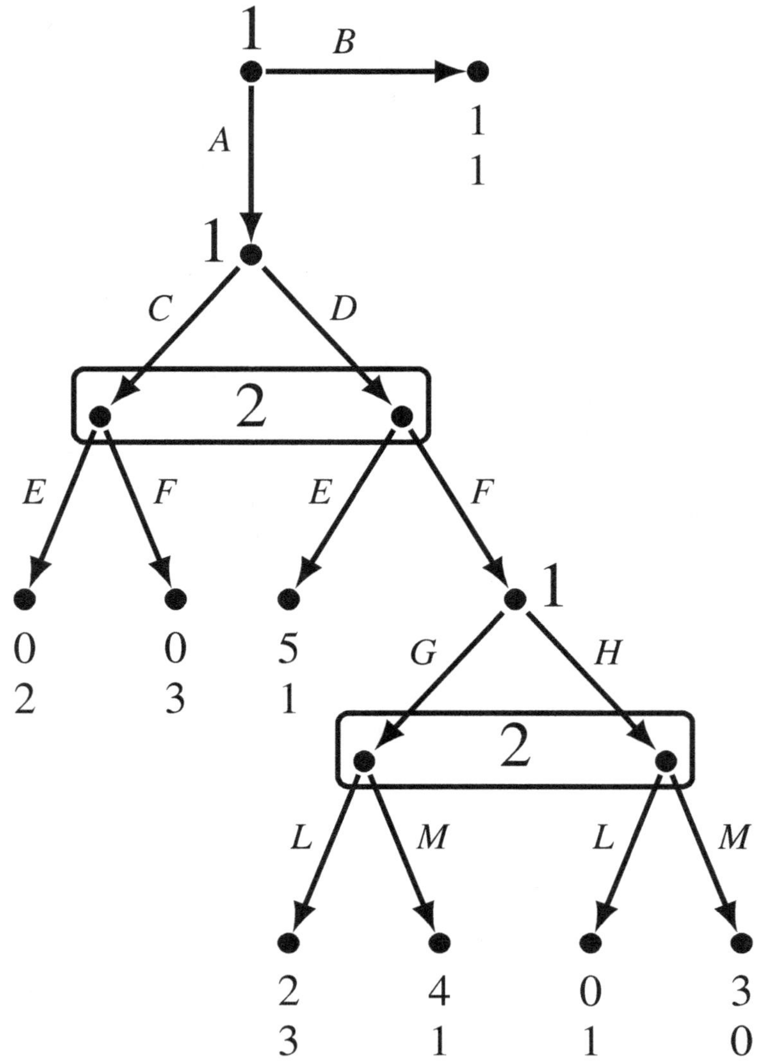


Figure 4.11: The game for Exercise 4.17.

**Solution to Exercise 4.17.**

**(a)** Player 1 has 8 strategies:

$$(A,C,G),(A,C,H),(A,D,G),(A,D,H),(B,C,G),(B,C,H),(B,D,G),(B,D,H).$$

**(b)** Player 2 has 4 strategies: $(E,L),(E,M),(F,L),(F,M)$.

**(c)** First we solve the minimal subgame that starts at Player 1's third node (after history $ADF$). In this game $G$ is a strictly dominant choice for Player 1 and $L$ is a strictly dominant choice for Player 2. Thus this game has a unique Nash equilibrium, namely $(G,L)$. Replacing the subgame with the corresponding payoffs, namely $(2,3)$, we get the following reduced game:

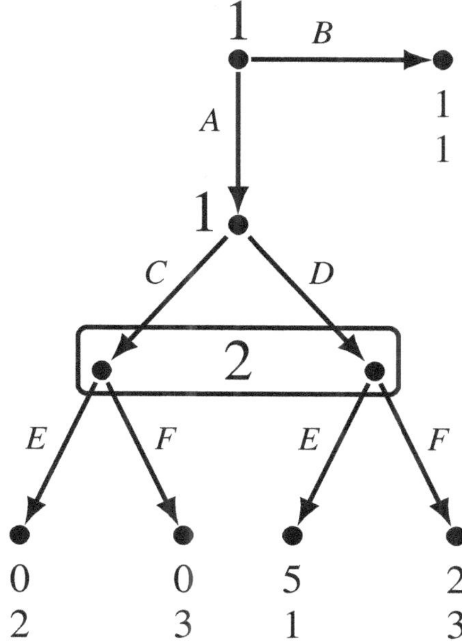


In the reduced game, let us solve the proper subgame that starts at Player 1's second node (after his initial choice of $A$). In this game, for Player 2, $F$ is a strictly dominant choice. Thus there is a unique Nash equilibrium given by $(D,F)$. Hence the original game has a unique subgame-perfect equilibrium given by $((A,D,G),\ (F,L))$.

**(d)** Two Nash equilibria that are not subgame perfect are $((A,D,H),\ (E,L))$ and $((A,D,H),\ (E,M))$. The reason why they are not subgame-perfect equilibria is that the restriction of these strategy profiles to the minimal subgame, namely $(H,L$ and $(H,M)$, respectively, are not Nash equilibria of the subgame. Intuitively, the threat by Player 1 to play $H$ in the that subgame is not credible (since $H$ is strictly dominated by $H$).

Exercise 4.18 Consider the following game. Firm $A$ decides whether to enter firm $B$'s industry, where currently firm $B$ is a monopoly. If firm $A$ enters, then the two firms simultaneously decide whether to advertise. If firm $A$ does not enter, then firm $B$ is informed of that decision and decides alone whether to advertise. With two firms in the market, the firms earn profits of \$3 each if they both advertise and \$5 each if they both do not advertise. If only one firm advertises, then it earns \$6 while the other firm earns \$1. When firm $B$ is alone in the industry it earns \$4 if it advertises and \$3.5 if it does not advertise. Firm $A$ earns zero profits if it does not enter.

**(a)** Represent this situation as a game in extensive form.

**(b)** Write the corresponding strategic form.

**(c)** Find all the pure-strategy Nash equilibria.

**(d)** Find the subgame-perfect equilibrium.

**Solution to Exercise 4.18.**

**(a)** The extensive form is as follows:

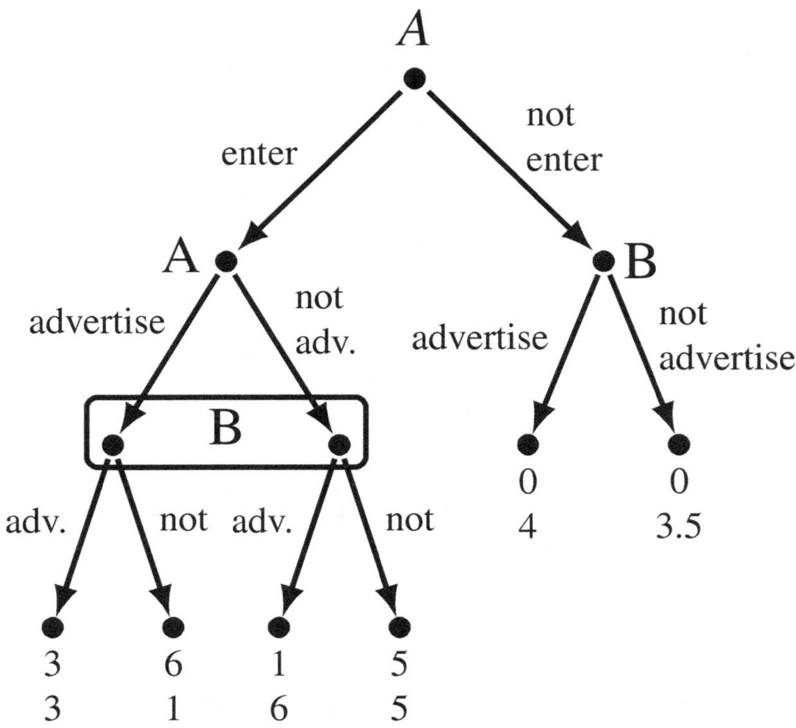


**(b)** The corresponding strategic form is as follows:

| | | Firm B | | | |
|---|---|---|---|---|---|
| | | *aa* | *an* | *na* | *nn* |
| Firm A | *ea* | 3 3 | 3 3 | 6 1 | 6 1 |
| | *en* | 1 6 | 1 6 | 5 5 | 5 5 |
| | *na* | 0 4 | 0 3.5 | 0 4 | 0 3.5 |
| | *nn* | 0 4 | 0 3.5 | 0 4 | 0 3.5 |

**(c)** The pure-strategy Nash equilibria are: (1) (enter/advertise, advertise/advertise) and (2) (enter/advertise, advertise/not advertise).

**(d)** There are two proper subgames. The simultaneous game on the left (following entry) and the one-person decision problem on the right. The latter has a unique Nash equilibrium where firm *B* advertises. The former has the following strategic form:

| | | Firm B | |
|---|---|---|---|
| | | $a$ | $n$ |
| Firm A | $a$ | 3 3 | 6 1 |
| | $n$ | 1 6 | 5 5 |

which has a unique Nash equilibrium: (advertise, advertise). Thus the unique subgame-perfect equilibrium of the entire game is: (enter/advertise, advertise/advertise).

Exercise 4.19 Consider the following game. Player 1 moves first and chooses between $A$, $B$ and $C$. If he plays $A$ or $B$ then Player 2 has to choose between $E$ and $F$ only knowing that Player 1 played either $A$ or $B$ (thus not knowing which of $A$ or $B$ Player 1 chose). If Player 1 plays $C$ then this becomes common knowledge between Players 1 and 2 and the two players play a simultaneous game where Player 1 chooses between $G$ and $H$ and Player 2 chooses between $K$ and $L$. The payoffs are as follows:

| Play: | 1's payoff | 2's payoff |
|---|---|---|
| $AE$ | 5 | 9 |
| $AF$ | 3 | 3 |
| $BE$ | 3 | 3 |
| $BF$ | 9 | 5 |
| $CGK$ | 2 | 9 |
| $CGL$ | 0 | 3 |
| $CHK$ | 0 | 3 |
| $CKL$ | 6 | 5 |

**(a)** Draw an extensive-form game that represents the situation described above.

**(b)** Find the subgame-perfect equilibria.

**(c)** Find a Nash equilibrium which is not subgame-perfect.

■

**Solution to Exercise 4.19.**

**(a)** There are two different ways to represent the game in an extensive form, depending on who moves first in the simultaneous game after choice $C$ of Player 1. In the following figure we made Player 1move first.

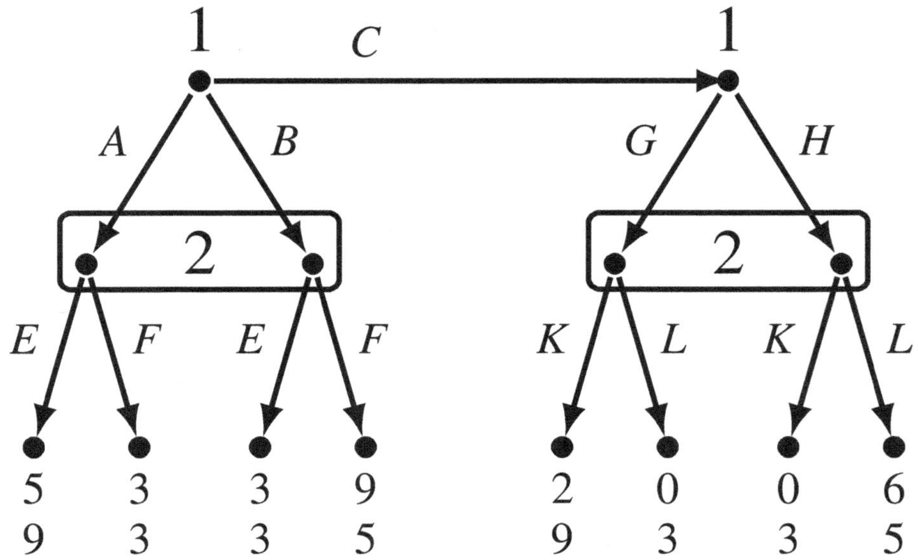


**(b)** There is only one proper subgame, namely the one that starts at Player 1's second node (after choice $C$). Its strategic form is as follows:

| | | Player 2 | |
|---|---|---|---|
| | | $K$ | $L$ |
| Player 1 | $G$ | 2   9 | 0   3 |
| | $H$ | 0   3 | 6   5 |

The subgame has two Nash equilibria: $(G,K)$ and $(H,L)$.

Let us first select the Nash equilibrium $(G,K)$ and replace the subgame with the corresponding payoffs, namely $(2,9)$. Then the extensive game reduces to the following:

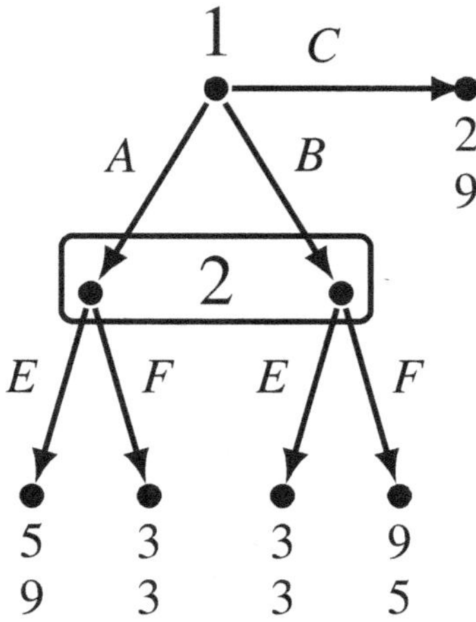


whose strategic form is as follows:

Player 2

| Player 1 | | $E$ | | $F$ | |
|---|---|---|---|---|---|
| | $A$ | 5 | 9 | 3 | 3 |
| | $B$ | 3 | 3 | 9 | 5 |
| | $C$ | 2 | 9 | 2 | 9 |

This game has two Nash equilibria: $(A,E)$ and $(B,F)$.

Thus we have found two subgame-perfect equilibria of the initial game: $((A,G),(E,K))$ and $((B,G),(F,K))$

Next we select the other Nash equilibrium of the subgame, namely $(H,L)$, and replace the subgame with the corresponding payoffs: $(6,5)$. Then the extensive game reduces to the following:

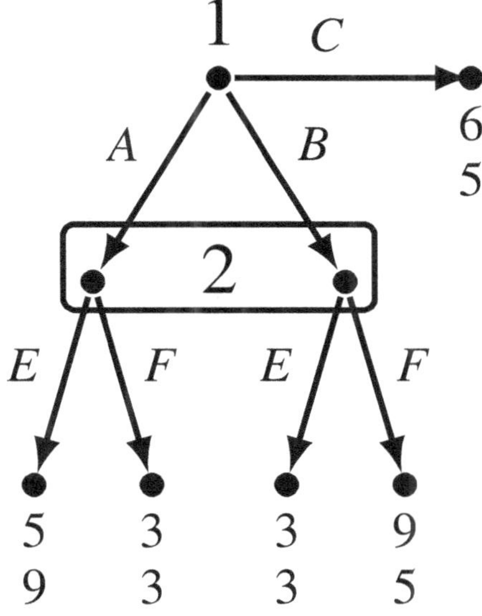


whose strategic form is as follows:

Player 2

| Player 1 | | $E$ | | $F$ | |
|---|---|---|---|---|---|
| | $A$ | 5 | 9 | 3 | 3 |
| | $B$ | 3 | 3 | 9 | 5 |
| | $C$ | 6 | 5 | 6 | 5 |

This game has two Nash equilibria: $(C,E)$ and $(C,F)$.

Thus we have found two more subgame-perfect equilibria of the initial game: $((C,H),\ (E,L))$ and $((C,H),\ (F,L))$

In conclusion, the initial game has four subgame-perfect equilibria: $((A,G),\ (E,K))$, $((B,G),\ (F,K))$, $((C,H),\ (E,L))$ and $((C,H),\ (F,L))$.

**(c)** There are several. For example, $((A,H),\ (E,K))$. It is a Nash equilibrium because, given Player 2's strategy, Player 1 cannot get more than 5 with any other strategy and Player 2 is getting her largest payoff. It is not subgame-perfect because its restriction to the subgame, namely $(H,K)$ is not a Nash equilibrium of the subgame.

## 4.5 Games with Chance moves

The exercises in this section deal with games that incorporate Chance moves (or moves by Nature) and the notions of money lottery and of risk neutrality (Definitions 4.5.1 and 4.5.2, Volume 1, Chapter 4, Section 5).

Exercise 4.20 Consider the following cases. Assume throughout that the firms are selfish and greedy (that is, each firm cares only about its own profits and prefers more money to less) and **risk-neutral**. In all of the cases place the choices of Nature (or Chance) at the beginning of the game.

**CASE 1.** Tomorrow, with equal probability, it will either rain or shine. Firms 1 and 2 have to make their production plans today not knowing what the weather will be like tomorrow. They simultaneously choose whether to produce umbrellas or sun lotion. Their profits are as follows: if they make the same choice (both choose umbrellas or both choose lotion) then they make a profit of \$1 each, irrespective of the weather. If they make different choices, then the firm that makes the choice appropriate to the weather makes a profit of \$4, while the other one makes a profit of \$0 (e.g. if Firm 1 chooses lotion, Firm 2 chooses umbrellas and the weather turns out to be rain, then Firm 1 makes a profit of \$0 and Firm 2 makes a profit of \$4).

**(a)** Draw the extensive-form game.

**(b)** Write the corresponding strategic-form game and find all the Nash equilibria.

**CASE 2.** Consider now the alternative situation where it is common knowledge that Firm 1 knows with certainty what the weather will be like (e.g. because it has access to an accurate weather forecast), while Firm 2 does not know. The production decisions are still made "simultaneously", that is, in ignorance of the choice of the other firm. [You can think of this situation as equivalent to the situation where production decisions are made after the realization of the weather and Firm 1 gets to see what the weather is like, while Firm 2 does not.]

**(d)** Draw the extensive-form of this game.

**(e)** Write the corresponding strategic-form game and find all the Nash equilibria.

**CASE 3.** Now let us go back to Case 1, where both firms are uncertain about the weather, but modify it as follows: Firm 1 chooses first (not knowing what the weather will be like) and then Firm 2 learns what Firm 1 chose and makes its choice second – knowing what Firm 1 did, but not knowing what the weather will be like.

**(f)** Draw the extensive-form of this game.
**(g)** Write the corresponding strategic-form game and find all the Nash equilibria.

**CASE 4.** Now modify Case 2 (where Firm 1 knows the weather before it makes its decision) as follows: Firm 2, when it makes its own decision, knows Firm 1's decision but not what the weather is like.
**(h)** Draw the extensive-form of this game.
**(i)** Write the corresponding strategic-form game and find all the Nash equilibria.

**COMPARISON.**
**(j)** Now compare Case 1 and Case 2: does Firm 1 gain from the additional information it has in Case 2 (the state of the weather) as compared to Case 1? Does Firm 2 gain or suffer from the additional information given to Firm 1?
**(k)** Now compare Case 3 and Case 4: does Firm 1 gain from the additional information it has in Case 4 (the state of the weather) as compared to Case 3? Does Firm 2 gain or suffer from the additional information given to Firm 1? Try to explain why this is the case.

**Solution to Exercise 4.20.**

**(a)** **CASE 1.** The extensive form is as follows: ($U$ stands for umbrellas and $L$ for lotion; the top payoff is Firm 1's profit and the bottom payoff is Firm 2's profit).

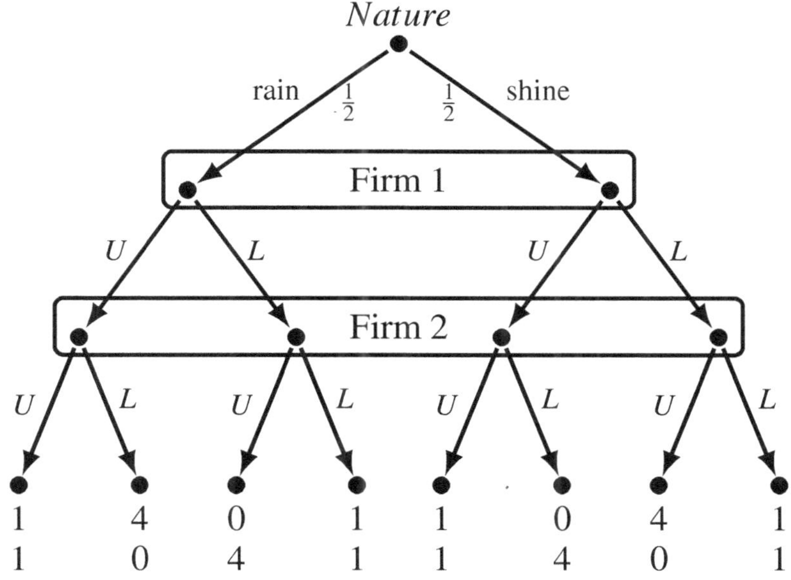


**(b)** The strategic form is as follows:

| Firm 1 \ Firm 2 | $U$ | $L$ |
|---|---|---|
| $U$ | 1 1 | 2 2 |
| $L$ | 2 2 | 1 1 |

There are two pure-strategy Nash equilibria: $(L,U)$ and $(U,L)$. The two equilibria are identical in terms of expected payoffs: each firm has an expected profit of 2 in both equilibria.

**(c)** **CASE 2.** The extensive form is as follows:

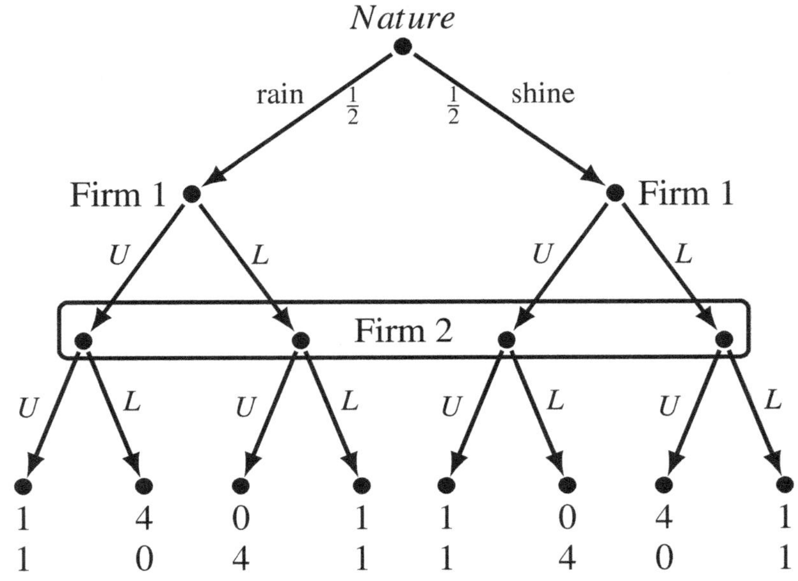


**(d)** Firm 1 now has two information sets and thus has four strategies. The strategic form is as follows:

| Firm 1 \ Firm 2 | $U$ | $L$ |
|---|---|---|
| $U$ always | 1 1 | 2 2 |
| $U$ if rain, $L$ if shine | 2.5 0.5 | 2.5 0.5 |
| $U$ if shine, $L$ if rain | 0.5 2.5 | 0.5 2.5 |
| $L$ always | 2 2 | 1 1 |

There are two pure-strategy Nash equilibria: ($U$ if rain and $L$ if shine, $U$) and ($U$ if rain and $L$ if shine, $L$). The two equilibria are identical in terms of expected payoffs: Firm 1 has an expected profit of 2.5 in both equilibria and Firm 2 has an expected profit of 0.5 in both equilibria.

**(e)** **CASE 3.** The extensive game is as follows:

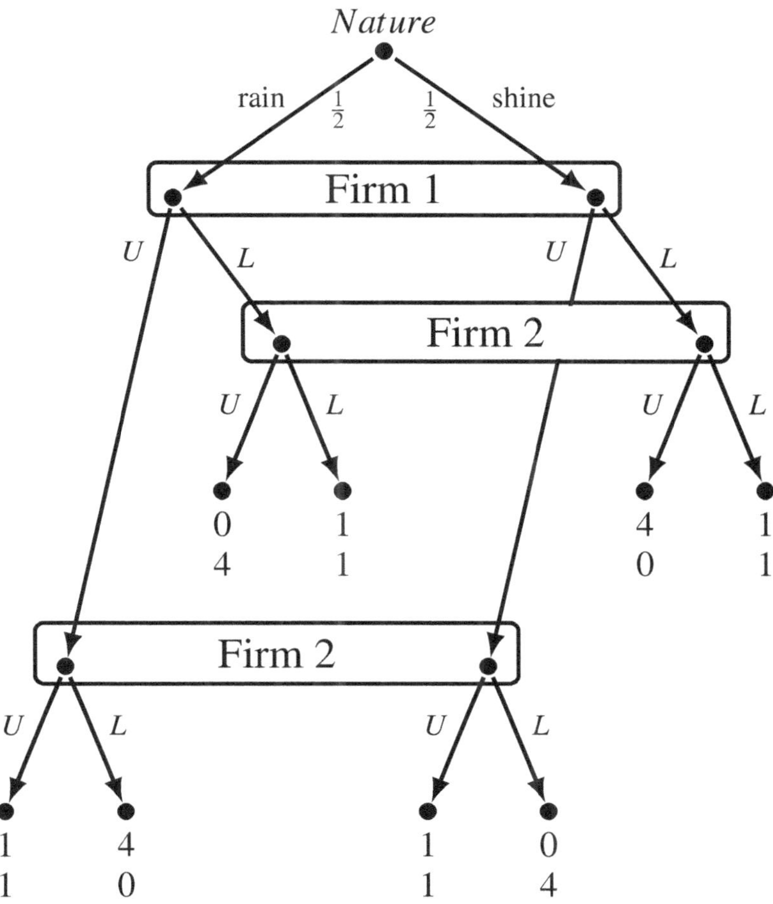


**(f)** Now Firm 2 has two information sets and therefore four strategies. The strategic form is as follows:

Firm 2

| Firm 1 | $U$ always | $L$ always | $U$ if $U$, $L$ if $L$ | $U$ if $L$, $L$ if $U$ |
|---|---|---|---|---|
| $U$ | 1 1 | 2 2 | 1 1 | 2 2 |
| $L$ | 2 2 | 1 1 | 1 1 | 2 2 |

There are four pure-strategy Nash equilibria: ($U$, $L$ always), ($U$, $U$ if $L$ and $L$ if $U$), ($L$, $U$ always) and ($L$, $U$ if $L$ and $L$ if $U$). The four equilibria are identical in terms of expected payoffs: both firms have an expected profit of 2.

**(g) CASE 4.** The extensive game is as follows:

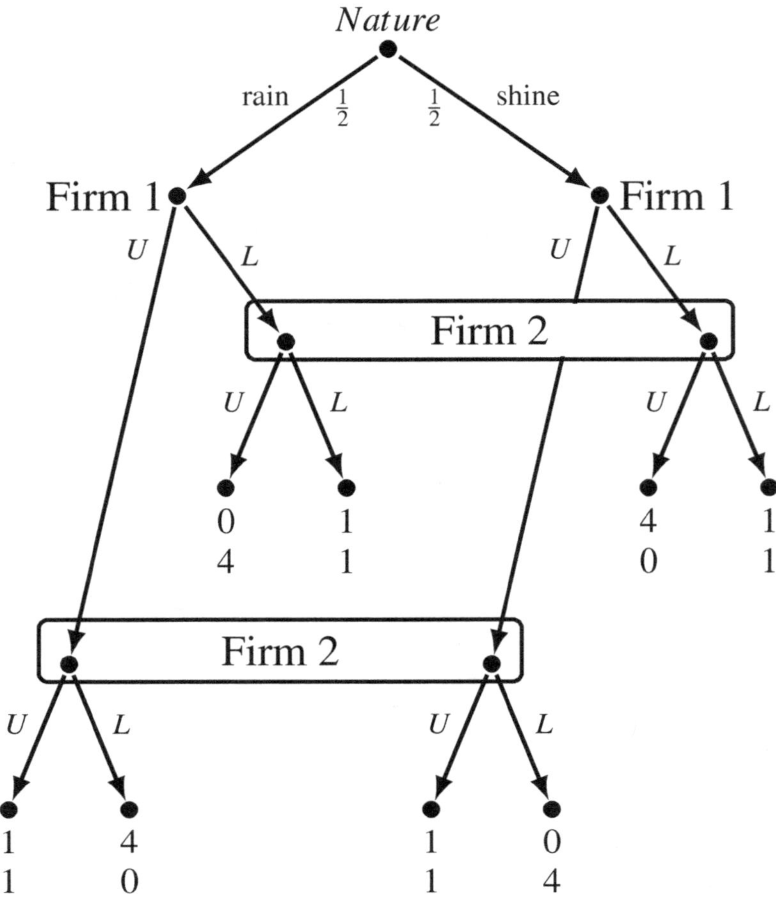


**(h)** Now each firm has two information sets and therefore four strategies. The strategic form is as follows:

| | | Firm 2: *U* always | Firm 2: *L* always | Firm 2: *U* if *U*, *L* if *L* | Firm 2: *U* if *L*, *L* if *U* |
|---|---|---|---|---|---|
| Firm 1 | *U* always | 1 1 | 2 2 | 1 1 | 2 2 |
| Firm 1 | *U* if rain, *L* if shine | 2.5 0.5 | 2.5 0.5 | 1 1 | 4 0 |
| Firm 1 | *U* if shine, *L* if rain | 0.5 2.5 | 0.5 2.5 | 1 1 | 0 4 |
| Firm 1 | *L* always | 2 2 | 1 1 | 1 1 | 2 2 |

Now there is only one pure-strategy equilibrium, namely
($U$ if rain and $L$ if shine , $U$ if $U$ and $L$ if $L$) with expected profit of 1 for each firm.

(i)
- Comparing Case 2 to Case 1, the additional information given to Firm 1 in Case 2 (the state of the weather) is advantageous to Firm 1: its expected equilibrium payoff is 2.5 while it was only 2 in Case 1. On the other hand, Firm 2 is hurt by the extra information given to Firm 1, because its expected equilibrium payoff is reduced from 2 to 0.5.

- Comparing Case 4 to Case 3 we get the surprising result that the additional information given to Firm 1 in Case 4 (the state of the weather) hurts both firms, in particular also Firm 1: the expected payoff of each firm is reduced from 2 to 1.

  This may seem paradoxical, since in one-person contexts more information is always advantageous. In interactive situations this is no longer true, as the above example shows: more information may be detrimental. The intuitive reason why this is happening here is as follows. Once Firm 1 is informed about the weather, it becomes a dominant strategy for it to produce umbrellas if it rains and lotion if it shines. Since Firm 2 knows that Firm 1 knows what the weather will be like and observes Firm 1's choice, Firm 2 can infer from Firm 1's choice what the weather will be like and the best reply to Firm 1's choice is also to produce umbrellas if it rains (that is, if Firm 1 chose to produce umbrellas) and to produce lotion if it shines (that is, if Firm 1 chose to produce lotion). Thus both firms end up doing the same and making a profit of 1.

  [The difference between Case 4 and Case 2 is that in Case 4 Firm 2 can adjust its choice to Firm 1's choice, which it observes, while in Case 2 Firm 2 cannot do so because it does not observe Firm 1's choice.]

## 4.6 More difficult exercises

The exercises in this section are more difficult and challenging than the previous ones.

Exercise 4.21 Construct an extensive-form game with three proper subgames but only one Nash equilibrium. ■

**Solution to Exercise 4.21.** The simplest such game is one with perfect information, as shown below. The only Nash equilibrium is $(C,(F,H,K))$.

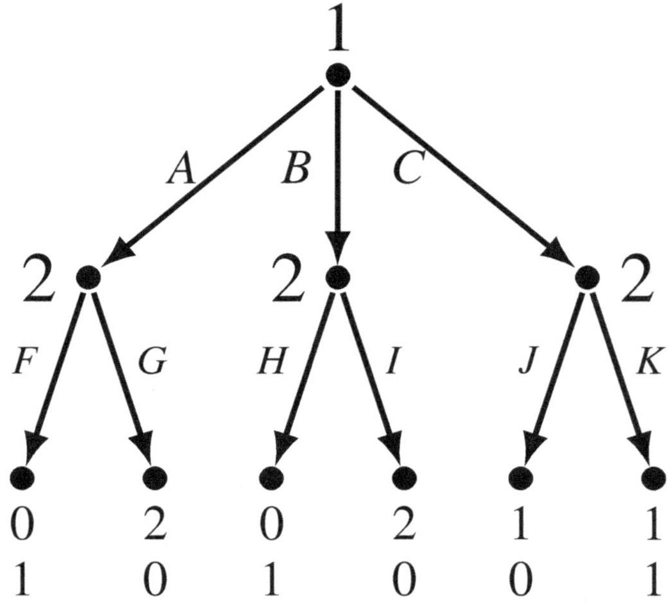


Exercise 4.22 You have been arrested for allegedly committing a serious crime. Three judges, Alice, Beth and Clara (from now on referred to as A, B and C) have heard all the evidence and are now going to vote on the verdict. There are three possible verdicts: Acquittal, Life sentence and Death penalty. During the trial it has become common knowledge among everybody (in particular, you and the three judges) that the judges' preferences over the possible verdicts are as follows:

| | A's ranking | B's ranking | C's ranking |
|---|---|---|---|
| best | *Death* | *Life* | *Acquittal* |
| | *Life* | *Acquittal* | *Death* |
| worst | *Acquittal* | *Death* | *Life* |

The legal system allows you to choose the procedure by which the judges reach a verdict. There are three different procedures. All of them involve two stages: in Stage 1 the judges simultaneously and secretly vote on a first issue; at the end of the first stage, it is made public what the first-stage votes were (i.e. which issue each judge voted for) and in Stage 2 there is a second simultaneous and secret vote on the second issue. In each stage the corresponding issue is decided by majority voting. It is common knowledge that the judges are strategic, that is, they do not necessarily vote according to their true preferences and the objective of each judge is to try to bring about an outcome which is best according to her true ranking.

**(1) Innocent/Guilty procedure (IG).** In this procedure the first vote is on the issue of whether you are innocent or guilty. If a majority of the judges votes for innocence in the first stage then the outcome is that you are acquitted and the matter ends there; otherwise the second vote is on which of the two punishments (death or life) should be applied to you.

(2) **Sequential Punishment procedure (SP).** In this procedure the first vote is on the issue of whether you should be sentenced to death; if a majority of the judges votes Yes, then the outcome is the death penalty. If a majority votes No in the first stage, then in the second stage the vote is on the issue of whether you should be given a life sentence; if a majority votes Yes, then the outcome is life in prison, otherwise the outcome is acquittal.

(3) **The Punishment Assessment procedure (PA).** In this procedure the first vote is on which punishment is appropriate for the crime (independently of whether you committed it or not): life sentence or death penalty. The second vote is whether you are innocent (so that the outcome is acquittal) or guilty (in which case the outcome is the punishment picked in the first vote by a majority of the judges).

You should attempt to answer the following questions without going through the laborious task of drawing the extensive-form games that represent the above procedures, although it might help to sketch the structure of such games.

**(a)** For the IG procedure find a subgame-perfect equilibrium of the extensive-form game that represents it, by using – whenever possible – either the notion of dominant-strategy profile (DSP) or the notion of Iterated Deletion of Weakly Dominated Strategies (IDWDS).

**(b)** For the SP procedure find a subgame-perfect equilibrium of the extensive-form game that represents it, by using – whenever possible – either the notion of DSP or the notion of IDWDS.

**(c)** For the PA procedure find two subgame-perfect equilibria of the extensive-form game that represents it, by using – whenever possible – either the notion of DSP or the notion of IDWDS.

**(d)** Which procedure should you ask the panel of judges to use?

**Solution to Exercise 4.22.** The key observation is that – under majority voting – when there are only two choices, voting sincerely (that is, according to one's true preferences) is a weakly dominant choice.

**(a)** In the IG procedure the first vote is between 'guilty' and 'innocent'. If you are found innocent, then you are acquitted. If, instead, you are found guilty, then there is a second vote where each judge has to vote either for 'life' (L) or 'death' (D). In the second-stage vote – if the outcome of the first stage was 'guilty' (there are four proper subgames corresponding to this stage) – for judge A voting 'D' is a dominant choice, for judge B voting 'L' is a dominant choice and for judge C voting 'D' is a dominant choice, that is, voting sincerely (i.e. according to one's true ranking) is a dominant strategy for every judge. Hence everybody can predict that the outcome of the second-stage vote will be 'death'. Incorporating this prediction into the first-stage vote we get the following ('g' means 'guilty', 'i' means 'innocent', 'A' means 'acquittal' and 'D' mead death penalty'):

Judge C: $i$

| Judge A \ Judge B | $i$ | $g$ |
|---|---|---|
| $i$ | $A$ | $A$ |
| $g$ | $A$ | $D$ |

Judge C: $g$

| Judge A \ Judge B | $i$ | $g$ |
|---|---|---|
| $i$ | $A$ | $D$ |
| $g$ | $D$ | $D$ |

In this game for judge A voting ‘g’ is a dominant strategy, for judge B voting ‘i’ is a dominant strategy and for judge C voting ‘i’ is a dominant strategy. Thus the outcome of the IG procedure is that you are acquitted.

**(b)** In the SP procedure the first vote is between ‘death’ and ‘no death’. If a majority of judges vote for ‘death’ then you are executed. If, instead, the outcome of the first vote is ‘no death’, then there is a second vote where each judge has to vote either for ‘life’ (L) or ‘acquittal’ (A). Thus the second-stage vote – if the outcome of the first stage was ‘no death’(there are four proper subgames corresponding to this second-stage vote) – is as follows:

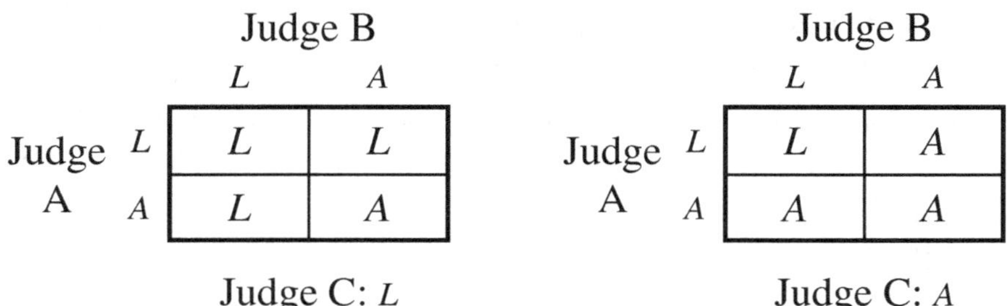

Judge C: $L$

| Judge A \ Judge B | $L$ | $A$ |
|---|---|---|
| $L$ | $L$ | $L$ |
| $A$ | $L$ | $A$ |

Judge C: $A$

| Judge A \ Judge B | $L$ | $A$ |
|---|---|---|
| $L$ | $L$ | $A$ |
| $A$ | $A$ | $A$ |

At this stage, for judge A voting ‘L’ is a dominant choice, for judge B voting ‘L’ is a dominant choice and for judge C voting ‘A’ is a dominant choice (once again, truthful voting is a dominant strategy). Thus everybody can predict that the outcome of the second-stage vote will be ‘life’. On the basis of this prediction, the first-stage vote reduces to one where "death" means "death" and “no death” means "life", that is, the game reduces to the following (‘D’ stands for ‘death’ and ‘N’ for ‘no death’):

Judge C: $D$

| Judge A \ Judge B | $D$ | $N$ |
|---|---|---|
| $D$ | $D$ | $D$ |
| $N$ | $D$ | $L$ |

Judge C: $N$

| Judge A \ Judge B | $D$ | $N$ |
|---|---|---|
| $D$ | $D$ | $L$ |
| $N$ | $L$ | $L$ |

In this game for judge A voting ‘D’ is a dominant strategy, for judge B voting ‘L’ is a dominant strategy and for judge C voting 'D' is a dominant strategy. Thus the outcome of the SP procedure is that you are sentenced to death.

**(c)** In the PA procedure the first vote is between 'death' and 'life'. If the majority voted for 'death' then the next vote will be between 'acquittal' (a) and 'guilty' (g), where 'guilty' will mean 'death sentence'. If, instead, the outcome of the first vote was 'life', then the next vote will be between 'acquittal' (a) and 'guilty' (g), where 'guilty' will mean 'life sentence'. Thus there are two different second-stage games, depending on the outcome of the first-stage vote.

**CASE 1**: the outcome of the first-stage vote was death. In this case the second-stage game (there are four proper subgames corresponding to this) becomes:

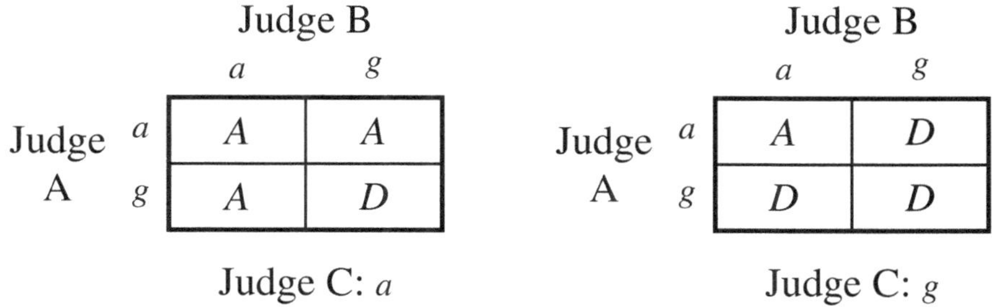

Judge B
a
g
Judge A
a
A
A
g
A
D
Judge C: a
Judge B
a
g
Judge A
a
A
D
g
D
D
Judge C: g


In this game, for judge A voting 'g' is a dominant choice while for judges B and C voting 'a' is a dominant choice. Thus everybody can predict that the outcome of this second-stage vote will be 'acquittal'.

**CASE 2**: the outcome of the first-stage vote was life. In this case the second-stage game (there are four proper subgames corresponding to this) becomes:

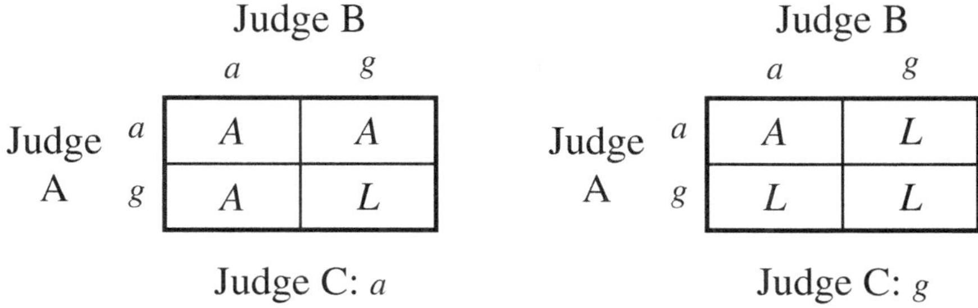

Judge B
a
g
Judge A
a
A
A
g
A
L
Judge C: a
Judge B
a
g
Judge A
a
A
L
g
L
L
Judge C: g


In this game, for judges A and B voting 'g' is a dominant choice while for judge C voting 'a' is a dominant choice. Thus everybody can predict that the outcome of this second-stage vote will be 'guilty'.

Thus there are two second-stage games. If life is chosen as the applicable penalty, then in the second stage there will be a majority for the life sentence. If death is the predetermined penalty, then in the second stage there will be a majority in favor of acquittal. Thus the first-stage vote reduces to one where "life" means "life" and "death" means "acquittal". The game thus reduces to one where the judges have to choose between 'death', D, and 'life', L:

Judge C: $D$

| | | Judge B: $D$ | Judge B: $L$ |
|---|---|---|---|
| Judge A | $D$ | $A$ | $A$ |
| | $L$ | $A$ | $L$ |

Judge C: $L$

| | | Judge B: $D$ | Judge B: $L$ |
|---|---|---|---|
| Judge A | $D$ | $A$ | $L$ |
| | $L$ | $L$ | $L$ |

In this game for judges A and B voting 'L' is a dominant strategy, while for judge C voting 'D' is a dominant strategy. Thus the outcome of the PA procedure is that you are sentenced to life in prison.

**(d)** Obviously, the IG procedure.

Exercise 4.23 Draw an extensive-form game-frame with imperfect information that satisfies the following properties:

**(1)** There are two players.
**(2)** Player 1 moves first.
**(3)** Player 1 has two information sets: one consists of only one node and the other contains three nodes.
**(4)** Player 2 has two information sets, each containing two nodes.
**(5)** For any game based on this game-frame, the set of Nash equilibria coincides with the set of subgame-perfect equilibria.

**Solution to Exercise 4.23.** For example, the following game. Requirement (5) is satisfied because the game does not have any proper subgames.

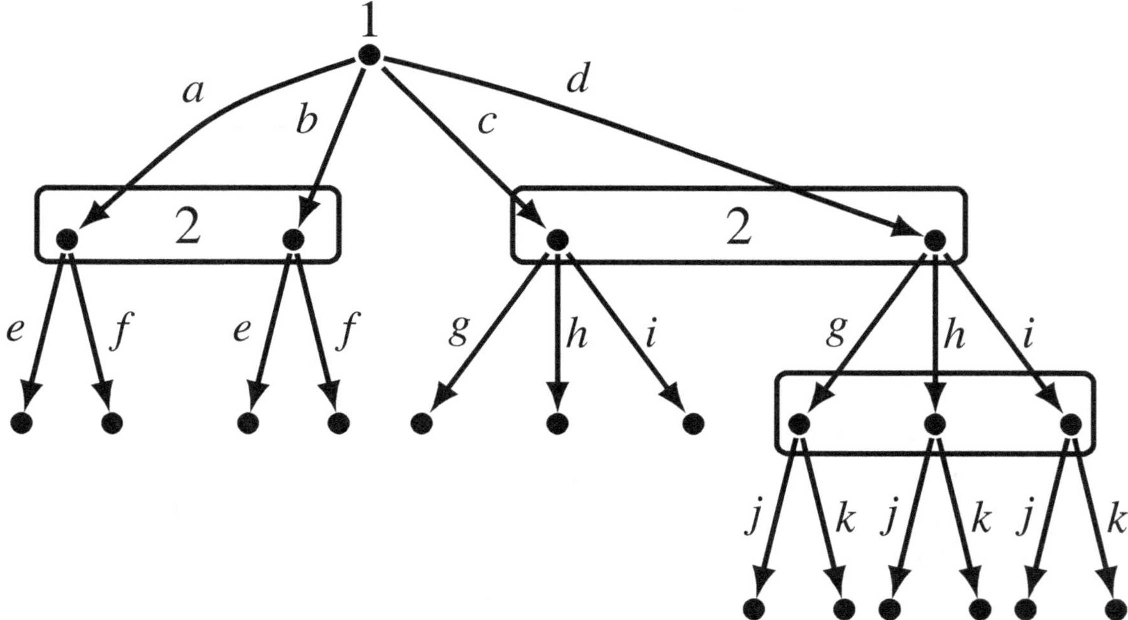

# Games with Cardinal Payoffs

# 5. Expected Utility Theory

## 5.1 Money lotteries and attitudes to risk

The exercises in this section deal with the notions of *money lottery*, *risk aversion*, *risk neutrality* and *risk love* (Definition 5.1.1, Volume 1, p. 168).

> Exercise 5.1 Sophie prefers more money to less and her preferences are transitive. She is asked to choose between the following options.
>
> - Option 1: get \$40 for sure.
> - Option 2: roll a symmetric 6-face die and get \$150 if it shows the face 1 or the face 2 and get nothing if it shows any of the other four faces.
>
> She says that she is indifferent between the two options, that is, she considers Option 1 to be just as good as Option 2.
>
> Is Sophie risk neutral, risk averse or risk loving? ■

**Solution to Exercise 5.1.**
Since the expected value of the lottery

$$L = \begin{pmatrix} \$0 & \$150 \\ \frac{2}{3} & \frac{1}{3} \end{pmatrix}$$

is \$50 and she considers the lottery to be just as good as \$40, she is risk averse (if she were risk neutral she would consider the lottery to be equivalent to \$50 and hence she would prefer the lottery to getting \$40 for sure; if she were risk loving she would prefer $L$ to \$50 and, since she prefers \$50 to \$40, she would prefer $L$ to \$40).

Exercise 5.2 Carla prefers more money to less and is risk neutral. How does she rank the following three lotteries?

- Lottery A: she gets \$1 with probability ½ and \$121 with probability ½.
- Lottery B: she gets \$25 with probability ½ and \$49 with probability ½.
- Lottery C: she gets \$9 with probability ⅓, \$36 with probability ⅓ and \$81 with probability ⅓.

**Solution to Exercise 5.2.**
Since Carla is risk neutral, she ranks lotteries according to their expected value. $\mathbb{E}[A] = 61$, $\mathbb{E}[B] = 37$ and $\mathbb{E}[C] = 42$. Thus she prefers $A$ to $C$ and $C$ to $B$.

Exercise 5.3 Kevin prefers more money to less and his preferences are transitive. He is asked to choose between the following options.

- Option 1: get \$78 for sure.
- Option 2: play the following lottery, call it $L$. Get \$16 with probability ¼, \$36 with probability ⅓ and \$144 with probability 5⁄12.

He says that he is indifferent between the two options. Is Kevin risk neutral, risk averse or risk loving?

**Solution to Exercise 5.3.**
Since Kevin prefers more money to less, he prefers \$78 to \$76. Since $\mathbb{E}[L] = 76$, if Kevin were risk averse he would prefer \$76 for sure to lottery $L$ and thus, by transitivity, he would prefer \$78 to $L$. If Kevin were risk neutral he would be indifferent between \$76 for sure and lottery $L$ and thus, by transitivity, he would prefer \$78 to $L$. Thus, Kevin is risk loving.

Exercise 5.4 A casino in Las Vegas offers the following deal. You pay \$9 and a fair coin is tossed repeatedly, up to ten times, until it shows heads for the first time. If it shows heads for the first time on the $k^{th}$ toss, you win \$$2k$. If it never shows heads, you get nothing. If you are risk neutral, will you play this game?

**Solution to Exercise 5.4.**
The game corresponds to the following lottery:

| Coin sequence | $H$ | $TH$ | $TTH$ | $TTTH$ | ... | $TT\ldots TH$ |
|---|---|---|---|---|---|---|
| Prize | \$2 | \$4 | \$8 | \$16 | ... | \$$2^{10}$ |
| Probability | $\frac{1}{2}$ | $\frac{1}{4}$ | $\frac{1}{8}$ | $\frac{1}{16}$ | ... | $\frac{1}{2^{10}}$ |

The expected value of this lottery is $\underbrace{1+1+\cdots+1}_{\text{ten times}} = 10$. Thus any risk neutral person should play the game (expecting to gain \$1).

Exercise 5.5 Colin is looking for a job. There are three firms that might be interested in hiring Colin. One pays a salary of \$16,000, the other a salary of \$25,000 and the remaining firm pays a salary of \$34,000. Colin has no wealth and his von Neumann-Morgenstern utility-of-money function is as follows: $U(\$16,000) = 1.2$, $U(\$25,000) = 2.3$ and $U(\$34,000) = 3.8$. Is Colin risk averse, risk neutral or risk loving? [Hint: construct a lottery using (some of) the sums of money above and then apply the definition of risk aversion/neutrality/love.] ■

**Solution to Exercise 5.5.**
Consider the lottery, call it $A$, that pays \$16,000 with probability 0.5 and \$34,000 with probability 0.5. The expected value of this lottery is $\mathbb{E}[A] = 25,000$. The utility of the expected value of $A$ is $U(\$25,000) = 2.3$. The expected utility of $A$ is $\mathbb{E}[U(A)] = 0.5U(\$16,000) + 0.5U(\$34,000) = 0.5(1.2) + 0.5(3.8) = 2.5$. Since the utility of the expected value of $A$ is less than the expected utility of $A$, Colin is risk-loving.

Exercise 5.6 Henrietta is celebrating her $16^{th}$ birthday today. Her boyfriend wants to give her a present. He has \$50 to spend. He can either give this sum of money to her, or he can buy a lottery ticket which costs \$50; the lottery is as follows:

$$L = \begin{pmatrix} \text{prize} & \$0 & \$30 & \$80 & \$100 \\ \text{probability} & \frac{1}{5} & \frac{2}{5} & \frac{1}{10} & \frac{3}{10} \end{pmatrix}$$

Of the two alternatives, he wants to choose the one that Henrietta prefers. Unfortunately, he does not know which of the two she would prefer. However, he does know something about her, namely the values that her von Neumann-Morgenstern utility-of-money function takes at \$10, \$15 and \$50. These are: $U(\$10) = 6$, $U(\$15) = 9$ and $U(\$50) = 28$. Extrapolating from this information, should he give her \$50 or the lottery ticket? [Hint: you need to be creative and construct a hypothetical lottery that will allow you to decide what her attitude to risk is.] ■

**Solution to Exercise 5.6.**
The expected value of $L$ is $\mathbb{E}[L] = 50$. Thus Henrietta should be given the lottery ticket if she is risk-loving and \$50 in cash if she is risk-averse. We have to try to infer from the information that we have whether she is risk-averse or risk-loving (or risk-neutral). What we know is that $U(\$10) = 6$, $U(\$15) = 9$ and $U(\$50) = 28$. Construct a lottery of the form $M = \begin{pmatrix} \$10 & \$50 \\ p & 1-p \end{pmatrix}$ whose expected value is 15. We need $10p + 50(1-p) = 15$, that is, $p = \frac{7}{8}$. The expected utility of this lottery is $\mathbb{E}[U(M)] = \frac{7}{8}6 + \frac{1}{8}28 = 8.75$. Since the expected utility of the lottery is less than the utility of the expected value of the lottery (that is, $8.75 < 9$), Henrietta is risk-averse relative to lottery $M$. Extrapolating from this, that is, assuming that she is risk averse relative to any money lottery, we conclude that she would prefer to be given \$50 in cash.

## 5.2 Expected utility: theorems

The exercises in this section deal with Expected Utility Theory (Theorem 5.2.1 and Theorem 5.2.2, Volume 1, pp. 170 and 172).

Exercise 5.7 Robert has developed a knee injury and is consulting his doctor on what to do. The doctor tells him that he has two options: (1) rest for a few months, or (2) have knee surgery. In the doctor's experience with similar injuries, resting leads to complete recovery (outcome $a$) in 30 out of a 100 cases, while in the remaining 70 cases the knee remains in bad shape (outcome $b$). As for Option 2, the doctor can report success in 65 out of 100 surgeries; success means that the knee will return to normal after a lengthy period of physical therapy (outcome $c$), while failure of surgery leads to the knee being in a worse state than it is now (outcome $d$). Robert ranks the possible outcomes as follows ($x \succ y$ means that outcome x is preferred to outcome y): $a \succ c \succ b \succ d$. Suppose that Robert's preferences satisfy the axioms of expected utility theory. He decides that he prefers rest to surgery. For each of the following utility functions determine whether it could be a von Neumann-Morgenstern utility function that represents Robert's preferences.

**(a)** $U_1(a) = 100$ $U_1(b) = 60$ $U_1(c) = 80$ $U_1(d) = 0$

**(b)** $U_2(a) = -1$ $U_2(b) = -4$ $U_2(c) = -2$ $U_2(d) = -5.$

**(c)** $U_3(a) = 10$ $U_3(b) = 2$ $U_3(c) = 8$ $U_3(d) = 0.$

**(d)** $U_4(a) = -1$ $U_4(b) = -3$ $U_4(c) = -2$ $U_4(d) = -4.$

**(e)** $U_5(a) = 10$ $U_5(b) = 6$ $U_5(c) = 8$ $U_5(d) = 0.$ ■

**Solution to Exercise 5.7.**

**(a)** The expected utility of rest is $0.3 \times 100 + 0.7 \times 60 = 72$ while the expected utility of surgery is $0.65 \times 80 + 0.35 \times 0 = 52$. Thus rest is better and $U_1$ could be Robert's von Neumann-Morgenstern utility function.

**(b)** The expected utility of rest is $-3.1$ while the expected utility of surgery is $-3.05$. Thus surgery is better and $U_2$ could **not** be Robert's von Neumann-Morgenstern utility function.

**(c)** The expected utility of rest is 4.4 while the expected utility of surgery is 5.2. Thus surgery is better and $U_3$ could **not** be Robert's von Neumann-Morgenstern utility function.

**(d)** The expected utility of rest is $-2.4$ while the expected utility of surgery is $-2.7$. Thus rest is better and $U_4$ could be Robert's von Neumann-Morgenstern utility function.

**(e)** Since $U_5 = \frac{1}{10}U_1$, $U_5$ and $U_1$ represent the same preferences, thus $U_5$ could be Robert's von Neumann-Morgenstern utility function.

Exercise 5.8 Berta's von Neumann-Morgenstern utility-of-money function is $U(\$m) = \sqrt{m}$. How does she rank the following lotteries?

A: she gets \$9 with probability ⅓, \$36 with probability ⅓ and \$81 with probability ⅓.

B: she gets \$25 with probability ½ and \$49 with probability ½.

C: she gets \$1 with probability ½ and \$121 with probability ½.

**Solution to Exercise 5.8.**
Berta is indifferent among all three lotteries, since the expected utility of each is the same, namely 6:

$\mathbb{E}[U(A)] = \frac{1}{3}\sqrt{9} + \frac{1}{3}\sqrt{36} + \frac{1}{3}\sqrt{81} = \frac{1}{3}3 + \frac{1}{3}6 + \frac{1}{3}9 = \frac{18}{3} = 6,$

$\mathbb{E}[U(B)] = \frac{1}{2}\sqrt{25} + \frac{1}{2}\sqrt{49} = \frac{5+7}{2} = 6$

$\mathbb{E}[U(B)] = \frac{1}{2}\sqrt{1} + \frac{1}{2}\sqrt{121} = \frac{1+11}{2} = 6.$

Exercise 5.9 When asked whether he would prefer to go on a 3-week tour of England and Ireland or a 1-week tour of France and Italy, Anton says that he strictly prefers the 3-week tour. He also strictly prefers the 1-week tour to staying at home.

When asked to choose between A and B below, Anton chooses B (he strictly prefers B to A).

**A.** A 50% chance of winning a 3-week tour of England and Ireland (a coin is tossed and if it comes up Heads, Anton will get to go on the 3-week tour, if it comes up Tails he won't go anywhere).

**B.** A 1-week tour of France and Italy for sure.

When asked to choose between C and D below, Anton chooses C (he strictly prefers C to D).

**C.** A 5% chance of winning a 3-week tour of England and Ireland and a 95% chance of staying at home.

**D.** A 10% chance of winning a 1-week tour of France and Italy and a 90% chance of staying at home.

Are Anton's choices compatible with Expected Utility Theory?

**Solution to Exercise 5.9.** Let

$z_1$ = 3-week tour of England and Ireland
$z_2$ = 1-week tour of France and Italy
$z_3$ = staying at home.

The information we have is that for Anton $z_1 \succ z_2 \succ z_3$. Suppose that there exists a von Neumann-Morgenstern utility function over $\{z_1, z_2, z_3\}$ that represents Anton's preferences. Let us use the normalized utility function so that $U(z_1) = 1$, $U(z_2) = a$ and $U(z_3) = 0$, with $0 < a < 1$. Then the first choice is between lottery $A = \begin{pmatrix} z_1 & z_3 \\ 0.5 & 0.5 \end{pmatrix}$ and lottery $B = \begin{pmatrix} z_2 \\ 1 \end{pmatrix}$. Since Anton prefers $B$ to $A$, $\mathbb{E}[U(B)] = a > \mathbb{E}[U(A)] = 1(0.5) + 0(0.5) = 0.5$, that is, $a > 0.5$. The second choice is a choice between lotteries $C = \begin{pmatrix} z_1 & z_3 \\ 0.05 & 0.95 \end{pmatrix}$

and $D = \begin{pmatrix} z_2 & z_3 \\ 0.10 & 0.90 \end{pmatrix}$. $\mathbb{E}[U(C)] = 1(0.05) + 0(0.95) = 0.05$ and $\mathbb{E}[U(D)] = a(0.10) + 0(0.90) = 0.10a$. Since $a > 0.5$ (because he prefers $B$ to $A$), it follows that $\mathbb{E}[U(D)] > 0.05 = \mathbb{E}[U(C)]$. Thus, according to Expected Utility Theory, $D$ ought to be preferred to $C$. Hence Anton's preferences are not compatible with Expected Utility Theory.

Exercise 5.10 Ann has the opportunity to invest her entire wealth of \$450,000 in a project that with probability ½ will be successful and with probability ½ will fail. If the project is successful, after one year she will get back \$1 Million (thus making a profit of \$550,000), while if the project fails she will get back nothing (that is, she will lose her initial investment of \$450,000). The alternative is to put her \$450,000 into a savings account that pays interest at the annual rate of 5%. Ann only cares about her wealth one year from now (and prefers more money to less) and her von Neumann-Morgenstern utility-of-money function is given by $U(\$x) = \sqrt{\frac{x}{1{,}000}}$ (thus, for example, $U(\$400,000) = \sqrt{\frac{400{,}000}{1{,}000}} = \sqrt{400} = 20$).

**(a)** What will Ann choose?

**(b)** What would a risk-neutral person choose?

**Solution to Exercise 5.10.**

**(a)** Let $I$ stand for 'investment' and $S$ for 'savings account'. The two choices correspond to the following lotteries: $I = \begin{pmatrix} \$1,000,000 & \$0 \\ \frac{1}{2} & \frac{1}{2} \end{pmatrix}$ and $S = \begin{pmatrix} \$472,500 \\ 1 \end{pmatrix}$ (if she puts her money in the savings account she will get a net return of $\frac{5}{100} 450,000 = 22,500$ increasing her wealth to \$472,500). $\mathbb{E}[U(I)] = \frac{1}{2}\sqrt{1,000} + \frac{1}{2}\sqrt{0} = 15.81$ and $\mathbb{E}[U(S)] = \sqrt{472.5} = 21.74$. Thus Ann will choose to put her money into the savings account.

**(b)** A risk neutral person would choose the investment since $\mathbb{E}[I] = 500,000 > \mathbb{E}[S] = 472,500$.

Exercise 5.11 Emily's preferences satisfy the von Neumann-Morgenstern axioms. There are three possible states of health for Emily: $a$, $b$ and $c$. Emily prefers $a$ to $b$ and $b$ to $c$. Her current state of health is $b$. She has two options:

- Option 1: do nothing, in which case her state of health remains $b$ for sure.
- Option 2: undergo an operation, in which case there is a 25% chance that her state of health will be $a$ and a 75% chance that her state of health will be $c$.

Emily cannot make up her mind as to what to do, because she is indifferent between the two options. Her doctor then offers her a third option:

- Option 3: a drug treatment that, with probability $p$, will have no effect (so that her state of health will remain $b$), with probability ⅟16 will put her in state of health $a$ and with the remaining probability will put her in state of health $c$.

For what values of $p$ does she prefer Option 3 to the other two options?

**Solution to Exercise 5.11.**

We can construct the normalized utility function so that $U(a) = 1$ and $U(c) = 0$. It follows that the expected utility of Option 2 is ¼. Hence (since she is indifferent between Options 1 and 2), $U(b) = \frac{1}{4}$. Then the expected utility of of Option 3 is

$$\frac{1}{16}U(a) + pU(b) + \left(1 - p - \frac{1}{16}\right)U(c) = \frac{1}{16} + p\frac{1}{4}.$$

Thus, Emily prefers Option 3 to the other two options if and only if

$$\frac{1}{16} + p\frac{1}{4} > \frac{1}{4} \quad \text{that is,} \quad p > \frac{3}{4}.$$

Exercise 5.12 Alice's von Neumann-Morgenstern utility-of-money function is $U(\$m) = \sqrt{m}$. Consider the following lotteries:

- Lottery A: she gets \$16 with probability ¼, \$36 with probability ⅓ and \$144 with probability 5⁄12.
- Lottery B: she gets $\$x$ for sure.
- Lottery C: she gets \$100 with probability ½ and $\$y$ with probability ½.

For what values of $x$ and $y$ is Alice is indifferent among these three lotteries? ■

**Solution to Exercise 5.12.** $\mathbb{E}[U(A)] = \frac{1}{4}\sqrt{16} + \frac{1}{3}\sqrt{36} + \frac{5}{12}\sqrt{144} = 8$, $\mathbb{E}[U(B)] = \sqrt{x}$ and $\mathbb{E}[U(C)] = \frac{1}{2}\sqrt{100} + \frac{1}{2}\sqrt{y}$. The three are equal if anf only if $x = 64$ and $y = 36$.

Exercise 5.13 Laila is risk neutral and prefers more money to less. Write her normalized utility function for the following sums of money: \$225, \$175, \$100, \$25. ■

**Solution to Exercise 5.13.** For a risk-neutral person one can take as utility function the identity function: $U(\$m) = m$. To normalize this utility function proceed as follows:

| outcome | utility | | outcome | utility | | outcome | utility |
|---|---|---|---|---|---|---|---|
| \$225 | 225 | | \$225 | 200 | | \$225 | 1 |
| \$175 | 175 | $\xrightarrow[\text{subtract 25}]{}$ | \$175 | 150 | $\xrightarrow[\text{divide by 200}]{}$ | \$175 | $\frac{3}{4}$ |
| \$100 | 100 | | \$100 | 75 | | \$100 | $\frac{3}{8}$ |
| \$25 | 25 | | \$25 | 0 | | \$25 | 0 |

Exercise 5.14 Consider again the situation described in Exercise 5.7: Robert has developed a knee injury and is consulting his doctor on what to do. The doctor tells him that he has two options: (1) rest for a few months, or (2) have knee surgery. In the doctor's experience with similar injuries, resting leads to complete recovery (outcome $a$) in 30 out of a 100 cases, while in the remaining 70 cases the knee remains in bad shape (outcome $b$). As for Option 2, the doctor can report success in 65 out of 100 surgeries; success means that the knee will return to normal after a lengthy period of physical therapy (outcome $c$), while failure of surgery leads to the knee being in a

worse state than it is now (outcome $d$). Robert ranks the possible outcomes as follows ($x \succ y$ means that outcome x is preferred to outcome y): $a \succ c \succ b \succ d$. Suppose that Robert's preferences satisfy the axioms of expected utility theory. Robert comes to you for advice on what to do. You ask him how he feels about the following lotteries:

$$E = \begin{pmatrix} a & b \\ \frac{1}{8} & \frac{7}{8} \end{pmatrix}, \quad F = \begin{pmatrix} b & c \\ \frac{4}{5} & \frac{1}{5} \end{pmatrix}, \quad G = \begin{pmatrix} a & d \\ \frac{3}{8} & \frac{5}{8} \end{pmatrix}, \quad H = \begin{pmatrix} c & d \\ \frac{1}{2} & \frac{1}{2} \end{pmatrix}$$

He tells you that he is indifferent between $E$ and $F$ and he is also indifferent between $G$ and $H$.

**(a)** What is the his expected utility if he decides to rest and his expected utility if he decides to have surgery?

**(b)** Suppose that Robert decides not to have surgery. A while later the doctor informs Robert that there is a new type of surgery that is better than the older type: instead of being successful in 65 out of 100 cases, the new surgery has been reported to be successful in $n$ out of 100 cases ($n > 65$). Based on this information, Robert decides to have surgery. What can you say about the value of $n$?

**Solution to Exercise 5.14.**

**(a)** The answer, of course, depends on what utility function we use to represent Robert's preferences. Let us use the normalized utility function: $U(a) = 1, U(b) = x, U(c) = y, U(d) = 0$ (thus $x < y$). From the indifference between $E$ and $F$ we get that $\frac{1}{8} + \frac{7}{8}x = \frac{4}{5}x + \frac{1}{5}y$ and from the second indifference we get that $\frac{3}{8} = \frac{1}{2}y$. Solving we get $U(b) = \frac{1}{3}$ and $U(c) = \frac{3}{4}$. Thus the expected utility of resting is $\frac{3}{10}U(a) + \frac{7}{10}U(b) = \frac{3}{10} + \frac{7}{30} = \frac{8}{15} = 0.5333$ and the expected utility of surgery is $\frac{65}{100}U(c) + \frac{35}{100}U(d) = \frac{65}{100} \times \frac{3}{4} + \frac{35}{100} \times 0 = \frac{39}{80} = 0.4875$. Thus the best decision for Robert is to rest.

**(b)** It must be that $\frac{n}{100} \times \frac{3}{4} \geq \frac{8}{15}$, that is, $n \geq \frac{640}{9} = 71.11$.

Exercise 5.15 Bill is thinking of investing in a Silicon Valley start-up. His von Neumann-Morgenstern utility-of-money function is

$$U(\$w) = 9 + w - \frac{w^2}{100}$$

where $w$ denotes wealth measured in thousands of dollars (thus $w = 1$ means \$1,000). Bill's total wealth is \$20,000 and he would have to invest it all. With probability ⅖ the start-up will be a failure and Bill will lose all his investment; with probability ⅗ it will succeed and he will get his initial investment back plus a profit of \$30,000.

**(a)** Will Bill make the investment?

**(b)** Would a risk-neutral person make the investment?

**Solution to Exercise 5.15.**

**(a)** If he does not invest, his utility is $9+20-\frac{400}{100}=25$. If he invests, his expected utility from the investment is: $\frac{2}{5}\left(9+0-\frac{0}{100}\right)+\frac{3}{5}\left(9+50-\frac{2{,}500}{100}\right)=24$. Thus he will not make the investment.

**(b)** A risk neutral person ranks lotteries according to their expected value. The lottery corresponding to no investment is $\begin{pmatrix} \$20,000 \\ 1 \end{pmatrix}$, whose expected value is 20,000, and the lottery corresponding to investing is $\begin{pmatrix} 0 & \$50,000 \\ \frac{2}{5} & \frac{3}{5} \end{pmatrix}$, whose expected value is $\frac{2}{5}0+\frac{3}{5}(50,000)=30,000$. Thus a risk neutral person would make the investment.

Exercise 5.16 Gwen, who obeys the axioms of Expected Utility Theory, is faced with four possible outcomes: $A, B, C$ and $D$. Her ranking of these outcomes is $A \succ B \succ C \succ D$. Gwen is indifferent between the certainty of B and a lottery where there is a 25% probability of $A$ and a 75% probability of $D$. She is also indifferent between the certainty of $C$ and a lottery where there is a 20% probability of $B$ and a 80% probability of $D$.

**(a)** Construct a von Neumann-Morgenstern utility function that reflects these preferences and is such that the largest utility is 80 and the smallest utility is 20.

**(b)** How does Gwen rank the lotteries $L=\begin{pmatrix} A & C & D \\ \frac{1}{10} & \frac{2}{5} & \frac{1}{2} \end{pmatrix}$ and $M=\begin{pmatrix} B & C \\ \frac{2}{5} & \frac{3}{5} \end{pmatrix}$?

**(c)** For what value of $p$ is Gwen indifferent between $L=\begin{pmatrix} A & C & D \\ \frac{1}{10} & \frac{2}{5} & \frac{1}{2} \end{pmatrix}$ and $N=\begin{pmatrix} B & C \\ p & 1-p \end{pmatrix}$?

**(d)** Normalize the utility function of Part (a).

■

**Solution to Exercise 5.16.**

**(a)** $U(A)=80$ and $U(D)=20$. Then the expected utility of lottery $\begin{pmatrix} A & D \\ \frac{1}{4} & \frac{3}{4} \end{pmatrix}$ is $\frac{1}{4}80+\frac{3}{4}20=35$. Hence $U(B)=35$. Thus the expected utility of lottery $\begin{pmatrix} B & D \\ \frac{1}{5} & \frac{4}{5} \end{pmatrix}$ is $\frac{1}{5}35+\frac{4}{5}20=23$, so that $U(C)=23$.

**(b)** $\mathbb{E}[U(L)]=\frac{1}{10}80+\frac{2}{5}23+\frac{1}{2}20=27.2$ and $\mathbb{E}[U(M)]=\frac{2}{5}35+\frac{3}{5}23=27.8$ thus Gwen prefers $M$ to $L$.

**(c)** We need $35p+23(1-p)=27.2$. Thus $p=\frac{35}{100}=35\%$.

**(d)** Start from $\begin{array}{cccc} A & B & C & D \\ 80 & 35 & 23 & 20 \end{array}$, subtract 20: $\begin{array}{cccc} A & B & C & D \\ 60 & 15 & 3 & 0 \end{array}$ and finally divide by 60 to get $\begin{array}{cccc} A & B & C & D \\ 1 & \frac{15}{60}=\frac{1}{4} & \frac{3}{60}=\frac{1}{20} & 0 \end{array}$.

Exercise 5.17 Karl is planning a trip. Let $m$ be the amount of money that Karl spends on his trip (measured in dollars). The utility from his upcoming vacation is primarily a function of how much money he spends on it and given by: $U(\$m)=\sqrt{m+625}$. Karl has set aside \$5,000 to spend on this trip.

**(a)** If there is a 10% probability that Karl will lose all his cash on his vacation, what is Karl's expected utility from the trip?

**(b)** Karl is considering converting his entire \$5,000 into travelers checks. Travelers checks cost 1% of face value (that is, if you give $\$x$ to the bank to be converted into travelers checks, the bank deducts a fee of $\$0.01x$ and gives you travelers checks with a nominal or face value of $\$0.99x$). Will Karl's utility be higher if he converts his \$5,000 into travelers checks or if he keeps his \$5,000 cash? [Recall that there is a 10% probability that he will lose his cash; on the other hand, travelers checks involve no risk, because, if you lose them, they will be replaced at no charge.]

**(c)** What is the maximum amount (the maximum fee) that Karl would be willing to pay to convert his \$5,000 cash into travelers checks? ■

**Solution to Exercise 5.17.**

**(a)** Karl's expected utility is $0.10\times\sqrt{625}+0.90\times\sqrt{625+5000}=70$.

**(b)** If Karl buys travelers cheques he will spend $0.01(5000)=\$50$ and he will have a guaranteed sum of \$4,950 to spend with a corresponding utility of $\sqrt{625+4950}=74.67$. Thus he should buy travelers cheques.

**(c)** The relevant equation is $\sqrt{625+5000-x}=70$. Its solution is $x=725$. Thus Karl would be willing to pay up to \$725 to insure his travel money.

Exercise 5.18 Ann's wealth is \$310 and her von Neumann-Morgenstern utility-of-money function is $U(\$m)=\sqrt{m}$. She is offered a chance to pay \$100 to enter a room where Bob is going to toss a fair coin (in her presence) and if the coin comes up Heads, Bob will give her \$300, while if it comes up Tails then he will give her nothing.

**(a)** Represent, in terms of wealth levels, the lottery that Ann faces after she has paid \$100 and entered the room.

**(b)** Should Ann have paid \$100 to enter the room?

**(c)** Ann paid \$100 to enter the room and was unlucky, because the coin came up Tails. She has now left the room in tears and is about to leave the building, when

Bob calls her and offers her the same deal: "pay \$100 to come back to the room, I will toss a fair coin and give you \$300 if the outcome is Heads and nothing if the outcome is Tails". Should she accept?

**(d)** Suppose that Ann decided to pay \$100 for the second time, entered the room for the second time and was unlucky for the second time, because the coin came up Tails again. She has now left the room, again in tears, and is about to leave the building, when Bob calls her and offers her the same deal once more: "pay \$100 to come back to the room, I will toss a fair coin and give you \$300 if the outcome is Heads and nothing if the outcome is Tails". Should she accept?

**(e)** What can you can you learn from this exercise?

**Solution to Exercise 5.18.**

**(a)** $\begin{pmatrix} \$(310-100) & \$(310-100+300) \\ \frac{1}{2} & \frac{1}{2} \end{pmatrix} = \begin{pmatrix} \$210 & \$510 \\ \frac{1}{2} & \frac{1}{2} \end{pmatrix}$.

**(b)** If she does not pay \$100 to enter the room, her utility is $\sqrt{310} = 17.61$. If she pays \$100 to enter the room, her expected utility is $\frac{1}{2}\sqrt{210} + \frac{1}{2}\sqrt{510} = 18.54$. Thus it was a rational decision to pay \$100 and enter the room.

**(c)** Her wealth is now \$210. If she says No, then her expected utility is $\sqrt{210} = 14.49$. If she says Yes then her expected utility is $\frac{1}{2}\sqrt{110} + \frac{1}{2}\sqrt{410} = 15.37$. Thus she should pay \$100 again and enter the room again.

**(d)** Her wealth is now \$110. If she says No, then her utility is $\sqrt{110} = 10.49$. If she says Yes then her expected utility is $\frac{1}{2}\sqrt{10} + \frac{1}{2}\sqrt{310} = 10.39$. Thus this time she should say No.

**(e)** The lesson here is that the same lottery is judged differently depending on the initial wealth: typically, a richer person is more wiling to take risks than a poorer person.

Exercise 5.19 Anna and Bella are roommates. They have the same initial wealth of \$5,000, but different von Neumann-Morgenstern utility-of-money functions: Anna's is $V(\$w) = 10\sqrt{w}$ while Bellas is $U(\$w) = \ln(w)$ (ln denotes the natural logarithm). All of this is common knowledge between Anna and Bella. Bella's uncle gave her as birthday gift a lottery ticket which with probability 10% will yield a prize of \$16,000 and with the remaining probability no prize. Bella makes the following proposal to Anna:

> "If you like, you can choose one of these two options.
> Option 1: you give me \$500 now, for me to keep, and if I win the prize then I will give you 30% of the prize.
> Option 2: you give me \$300 now, for me to keep, and if I win the prize then I will give you 25% of the prize".

**(a)** Will Anna accept one of the deals offered by Bella and, if so, which one?

**(b)** Is it in Bella's interest to make the above offer? [You can assume that Bella is able to figure out Part (a)].

**Solution to Exercise 5.19.**

**(a)** If Anna refuses, her utility is $U(5,000) = 10\sqrt{5,000} = 707.107$. If Anna chooses Option 1 then her expected utility is

$$\frac{9}{10}10\sqrt{5,000-500}+\frac{1}{10}10\sqrt{5,000-500+\frac{30}{100}16,000} = 700.175.$$

If Anna chooses Option 2 then her expected utility is

$$\frac{9}{10}10\sqrt{5,000-300}+\frac{1}{10}10\sqrt{5,000-300+\frac{25}{100}16,000} = 710.283.$$

Thus the best choice for her is to accept Option 2.

**(b)** Bella can figure out, as we did, that Anna will choose Option 2. If Bella had not made the offer her expected utility would have been

$$\frac{9}{10}\ln(5,000)+\frac{1}{10}\ln(5,000+16,000) = 8.661.$$

With Option 2 her expected utility is

$$\frac{9}{10}\ln(5,000+300)+\frac{1}{10}\ln\left(5,000+300+\frac{75}{100}16,000\right) = 8.694.$$

Thus it is in her interest to make that offer.

Exercise 5.20 John is a dishonest bank clerk. His von Neumann Morgenstern utility-of-money function is $U(\$m) = \sqrt{m}$. His initial wealth is \$40,000. By stealing small amounts at a time he can steal \$4,100 without attracting too much attention. However there is routine monitoring of the accounts and there is a small chance that this monitoring will uncover the missing funds and that he will be caught. Let $p$ be the probability of being caught stealing. If he is caught, he will have to reimburse the \$4,100 that he has stolen and, in addition, he will have to pay a fine of \$3,900 for stealing. (He will also be fired and will have to find another job but for the purpose of this exercise we will ignore this side of the story.) We want to see if John, who has no moral principles, will choose to steal, or will be deterred by the risk of being caught and having to pay the fine.

**(a)** Will John steal if $p = 0.1$?

**(b)** The bank where John is working is losing money and the management realizes that funds are disappearing because some employees are stealing. Thus they decide to better monitor their employees. This stricter monitoring increases the probability of an employee being caught when stealing money. What is the minimum probability $p$ which will discourage John from stealing?(Assume that he has not stolen yet, so that his initial wealth is still \$40,000.)

**(c)** Monitoring is costly to the bank. As an alternative to increased monitoring (thus we are back to assuming that $p = 0.1$), the bank is considering increasing the fine for stealing. What is the smallest fine that would discourage John from stealing?

**(d)** Amy is in the same situation as John, but at a different bank. Her utility-of-money function is $V(\$m) = \left(\frac{m}{10{,}000}\right)^2$.

d.1 Answer question (a) for Amy.

d.2 Answer question (b) for Amy.

■

**Solution to Exercise 5.20.**

**(a)** If John does not steal, his utility is $\sqrt{40{,}000} = 200$. If he steals his utility is $0.9\sqrt{44,100} + 0.1\sqrt{36,100} = 208$. Thus he will steal.

**(b)** It is the solution to $(1-p)\sqrt{44,100} + p\sqrt{36,100} = 200$ which is 0.5.

**(c)** It is the solution to $0.9\sqrt{44,100} + 0.1\sqrt{40,000 - x} = 200$ which is \$27,900.

**(d)** Calculations for Amy:

d.1 If Amy does not steal, her utility is 16. If she steals her utility is $0.9\,(4.41)^2 + 0.1\,(3.61)^2 = 18.8065$. Thus she will steal.

d.2 It is the solution to $(1-p)\,(4.41)^2 + p\,(3.61)^2 = 16$ which is 0.53742.

Exercise 5.21 Jim is a member of a club that requires participation in an annual party. Jim derives no pleasure from attending the annual party. Everybody who attends has to pay \$1,000 to be admitted to the party. At the end of the party there is a lottery, in which all the names of the participants are entered and half the names are drawn at random. Those whose names are drawn are given their \$1,000 back.

**(a)** What is the maximum that Jim would be willing to pay his brother Mitch to replace him at the party if his von Neumann-Morgenstern utility-of-money function is $U(\$m) = \ln(m)$ (ln denotes the natural logarithm) and his current wealth is \$50,000?

**(b)** What is the maximum that Jim would be willing to pay his brother Mitch to replace him at the party if his von Neumann-Morgenstern utility-of-money function is $U(\$m) = \ln(m)$ (ln denotes the natural logarithm) and his current wealth is \$100,000?

**(c)** What is the maximum that Jim would be willing to pay his brother Mitch to replace him at the party if his von Neumann-Morgenstern utility-of-money function is $U(\$m) = \sqrt{m}$ and his current wealth is \$50,000?

**(d)** What is the maximum that Jim would be willing to pay his brother Mitch to replace him at the party if his von Neumann-Morgenstern utility-of-money function is $U(\$m) = \sqrt{m}$ and his current wealth is \$100,000?

■

**Solution to Exercise 5.21.**

**(a)** It is given by the solution to $\ln(50,000 - x) = 0.5\ln(50,000) + 0.5\ln(49,000)$ which is \$502.53.

**(b)** It is given by the solution to $\ln(100,000 - x) = 0.5\ln(100,000) + 0.5\ln(99,000)$ which is \$501.25.

**(c)** It is given by the solution to $\sqrt{50,000-x} = 0.5\sqrt{50,000} + 0.5\sqrt{49,000}$ which is \$501.26.

**(d)** It is given by the solution to $\sqrt{100,000-x} = 0.5\sqrt{100,000} + 0.5\sqrt{99,000}$ which is \$500.63.

Exercise 5.22 Adele (who prefers more money to less), when faced with the choice between lotteries $A$ and $B$ below, declared that she preferred $A$ to $B$:

$$A = \begin{pmatrix} \$500 \\ 1 \end{pmatrix} \qquad B = \begin{pmatrix} \$1,000 & \$500 & \$100 \\ 0.85 & 0.12 & 0.03 \end{pmatrix}$$

She also said that she was indifferent between $C$ and $D$ below:

$$C = \begin{pmatrix} \$500 & \$100 \\ 0.15 & 0.85 \end{pmatrix} \qquad D = \begin{pmatrix} \$1,000 & \$100 \\ 0.12 & 0.88 \end{pmatrix}$$

Is she rational according to the theory of expected utility? ■

**Solution to Exercise 5.22.** Let $U$ be Adele's von Neumann-Morgenstern utility function, normalized so that $U(\$1,000) = 1$, $U(\$500) = a$ and $U(\$100) = 0$. The expected utility of $A$ is $a$, $\mathbb{E}[U(B)] = 0.85 + 0.12a$, $\mathbb{E}[U(C)] = 0.15a$ and $\mathbb{E}[U(D)] = 0.12$. Since she prefers $A$ to $B$, $a > 0.85 + 0.12a$, that is, $0.88a > 0.85$ or $a > \frac{85}{88} = 0.9659$. Hence $\mathbb{E}[U(C)] = 0.15a > 0.15(0.9659) = 0.1449$. Since $\mathbb{E}[U(C)] > 0.1449 > 0.12 = \mathbb{E}[U(D)]$, she ought to prefer $C$ to $D$. Hence she is not rational according to Expected Utility Theory.

Exercise 5.23 Your current wealth is \$60. I am going to go to another room and toss a coin. If it comes up heads (which happens with probability ½), I will put \$2 in an opaque envelope and seal it; if it comes up tails (which happens with probability ½), I will put \$100 in an opaque envelope and seal it (your friend, who is a karate black belt, is going to come with me to make sure that I do not cheat). I then come back with the sealed envelope, which, as far as you know, might contain \$2 or \$100. I give you a choice between (1) passing and (2) buying the envelope from me for $\$P$.

**(a)** Suppose you are risk neutral and that $P = 50$. Which option will you choose?

**(b)** Now suppose that I do not know what your von Neumann-Morgenstern utility function is, but you tell me that if $P = 40$ then you are indifferent between the two options above. Having learned this, I change my offer and ask you to choose between the following options.

OPTION 1: I roll a die; if it comes up 1, I'll give you \$60; if it comes up either 2 or 4 or 6, no money will exchange hands; if it comes up 3 or 5, you will give me \$38.

OPTION 2: I roll a die; if it comes up 1 or 6, I'll give you \$60; if it comes up 2, 3, 4 or 5, you will give me \$38. Which option should I expect you to choose?

**(c)** Suppose now that your von Neumann-Morgenstern utility-of-money function is $U(\$m) = \frac{400m - m^2}{100}$ and that $P = 41$. If your choice is between (1) passing and (2) buying the envelope (which contains either \$2 or \$100, with equal probability) from me for \$41, which option will you choose?

**Solution to Exercise 5.23.**

**(a)** Getting the envelope is playing a lottery with expected value $\frac{1}{2}(2)+\frac{1}{2}(100) = \$51$. Hence, if risk neutral, you will choose to buy the envelope for \$50.

**(b)** The possible outcomes in terms of wealth are: $60+60 = 120$, 60 and $60-38 = 22$. Let $U$ be your normalized von Neumann-Morgenstern utility function. Then $U(\$120) = 1, U(\$60) = x$ and $U(\$22) = 0$ with $0 < x < 1$. Since you are indifferent between the two original options, $U(\$60) = \frac{1}{2}U(\$120)+\frac{1}{2}U(\$22) = \frac{1}{2}$. Now consider the new options.

Option 1 is the lottery $L_1 = \begin{pmatrix} \$22 & \$60 & \$120 \\ \frac{1}{3} & \frac{1}{2} & \frac{1}{6} \end{pmatrix}$ while Option 2 is the lottery $L_2 = \begin{pmatrix} \$22 & \$120 \\ \frac{2}{3} & \frac{1}{3} \end{pmatrix}$. $\mathbb{E}[U(L_1)] = \frac{1}{3}\times 0+\frac{1}{2}\times\frac{1}{2}+\frac{1}{6}\times 1+ = 0.417$ and $\mathbb{E}[U(L_2)] = \frac{2}{3}\times 0+\frac{1}{3}\times 1 = 0.33$. Thus you should choose Option 1.

**(c)** If you do not exchange, your wealth is \$60 and your utility is $U(\$60) = \frac{400\times 60-60^2}{100} = 204$. If you buy the envelope for \$41 you face the following wealth lottery $\begin{pmatrix} \$119 & \$21 \\ \frac{1}{2} & \frac{1}{2} \end{pmatrix}$, whose expected utility is $\frac{1}{2}U(21)+\frac{1}{2}U(119) = 206.99$. Hence you will choose to buy the envelope for \$41.

Exercise 5.24 Let the set of basic outcomes be $Z = \{z_1, z_2, z_3, z_4\}$. Let $\Delta(Z)$ denote the set of lotteries over $Z$. Sue has von Neumann-Morgenstern preferences over $\Delta(Z)$, which are represented by the following von Neumann-Morgenstern utility function:

| outcome | $z_1$ | $z_2$ | $z_3$ | $z_4$ |
|---|---|---|---|---|
| $U$ : | 2 | 0 | 8 | 6 |

**(a)** Which of the following two lotteries will Sue choose?

$$L_1 = \begin{pmatrix} z_1 & z_4 \\ \frac{3}{4} & \frac{1}{4} \end{pmatrix} \qquad L_2 = \begin{pmatrix} z_2 & z_3 \\ \frac{1}{2} & \frac{1}{2} \end{pmatrix}$$

**(b)** Which of the following utility functions are also a representation of Sue's preferences?

| outcome | $z_1$ | $z_2$ | $z_3$ | $z_4$ |
|---|---|---|---|---|
| $U_1$ : | 6 | 4 | 12 | 10 |
| $U_2$ : | 2 | 1 | 5 | 4 |
| $U_3$ : | 8 | 6 | 14 | 12 |

**(c)** Write two lotteries that, according to expected utility, are ranked differently by the following von Neumann-Morgenstern utility functions:

| outcome | $z_1$ | $z_2$ | $z_3$ | $z_4$ |
|---|---|---|---|---|
| $V_1$ : | $\frac{1}{4}$ | 0 | $\frac{3}{4}$ | 1 |
| $V_2$ : | 4 | 2 | 5 | 6 |

**(d)** Tom satisfies the axioms of expected utility. His ranking of Z is as follows:

$$\begin{array}{ll} \text{best} & z_3 \\ & z_1 \\ & z_2 \\ \text{worst} & z_4 \end{array}$$

Tom is indifferent between getting $z_1$ for sure and the lottery $\begin{pmatrix} z_3 & z_4 \\ \frac{2}{5} & \frac{3}{5} \end{pmatrix}$; furthermore, he is indifferent between getting $z_2$ for sure and the lottery

$$\begin{pmatrix} z_1 & z_2 & z_3 & z_4 \\ \frac{1}{10} & \frac{1}{10} & \frac{1}{10} & \frac{7}{10} \end{pmatrix}.$$

Find his normalized von Neumann-Morgenstern utility function.

**Solution to Exercise 5.24.**

**(a)** $\mathbb{E}[U(L_1)] = 2 \times \frac{3}{4} + 6 \times \frac{1}{4} = 3$ and $\mathbb{E}[U(L_2)] = 0 \times \frac{1}{2} + 8 \times \frac{1}{2} = 4$. Thus she will choose $L_2$.

**(b)** All of them. In fact, each of the utility functions can be obtained from $U$ by multiplying by a positive number $b$ and adding a constant $a$. To go from $U$ to $U_1$ we take $b = 1$ and $a = 4$. To go from $U$ to $U_2$ we take $b = \frac{1}{2}$ and $a = 1$. To go from $U$ to $U_3$ we take $b = 1$ and $a = 4$. To go from $U$ to $U_3$ we take $b = 1$ and $a = 6$.
How can one find the values of $a$ and $b$? Take, for example $U$ and $U_2$. We want to see if there are real numbers $a$ and $b$, with $b > 0$, such that $U_2(z) = a + bU(z)$, for every $z$. We need

$$\begin{aligned} a + 2b &= 2 \\ a + 0b &= 1 \\ a + 8b &= 5 \\ a + 6b &= 4 \end{aligned}$$

From the second equation we get $a = 1$. Replacing in the others we get

$$\begin{aligned} 1 + 2b &= 2 \\ 1 + 8b &= 5 \\ 1 + 6b &= 4 \end{aligned}$$

which all give $b = \frac{1}{2}$.

**(c)** According to $V_1$ the individual would be indifferent between $L_1 = \begin{pmatrix} z_1 \\ 1 \end{pmatrix}$ and

$L_2 = \begin{pmatrix} z_2 & z_4 \\ \frac{3}{4} & \frac{1}{4} \end{pmatrix}$, while according to V2 the individual would prefer $L_1$ to $L_2$, because $\mathbb{E}[V_2(L_1)] = 4$ and $\mathbb{E}[V_2(L_2)] = \frac{3}{4} \times 2 + \frac{1}{4} \times 6 = 3$. [Note that this pair of lotteries is just one of many possibilities.]

**(d)** We can set $U(z_3) = 1$ and $U(z_4) = 0$. Since he is indifferent between $\begin{pmatrix} z_1 \\ 1 \end{pmatrix}$ and $\begin{pmatrix} z_3 & z_4 \\ \frac{2}{5} & \frac{3}{5} \end{pmatrix}$, it follows that $U(z_1) = 1 \times \frac{2}{5} + 0 \times \frac{3}{5} = \frac{2}{5}$. Since he is indifferent between $\begin{pmatrix} z_2 \\ 1 \end{pmatrix}$ and $\begin{pmatrix} z_1 & z_2 & z_3 & z_4 \\ \frac{1}{10} & \frac{1}{10} & \frac{1}{10} & \frac{7}{10} \end{pmatrix}$ it follows that $U(z_2) = \frac{1}{10} \times \frac{2}{5} + \frac{1}{10} \times U(z_2) + \frac{1}{10} \times 1 + \frac{7}{10} \times 0$. Multiplying both sides by 50 we get $50U(z_2) = 2 + 5U(z_2) + 5$. Thus $U(z_2) = \frac{7}{45}$.

Exercise 5.25 Susan's wealth (measued in dollars) is $W > 0$. She is told the following:

> "When you went to the concert last week, you were inadvertently exposed to a rare and fatal virus. The probability of actually contracting the disease is 1 in 100, but once you develop the disease there is no known cure. However, you can be given an injection now that stops the development of the disease (with no side effects). How much would you be willing to pay for the injection?"

Suppose that Susan's preferences are very natural: conditional on being alive, she prefers to have more wealth than less wealth, while conditional on being dead she does not care what wealth she leaves behind (she has no next of kin to leave her money to); furthermore, quite naturally, she prefers to be alive than dead. Suppose that Susan's reply is that if she were to pay an amount equal to half her wealth, she would be indifferent between getting the injection and not getting the injection. Assume that Susan's preferences satisfy the von Neumann-Morgenstern axioms.

**(a)** Identify the basic outcomes and calculate Susan's normalized utility function.

**(b)** Suppose now that the information she was given is revised as follows.

> "The probability of actually contracting the disease is 4 in 100 and the injection would not guarantee that the you do not develop the disease: it only reduces the probability of getting the disease from 4 in 100 to 3 in 100."

Would Susan still be indifferent between not getting the injection and paying half of her wealth to get the injection? ■

**Solution to Exercise 5.25.**

**(a)** There are four possible outcomes: (1) Susan is alive and has her entire wealth; denote this outcome by W, (2) Susan is alive and has half of her wealth; denote this outcome by $\frac{W}{2}$, (3) Susan dies and leaves behind her entire wealth: denote this outcome by D, (4) Susan dies and leaves behind half of her wealth: denote this outcome by $\frac{D}{2}$. Her preferences are: $W \succ \frac{W}{2} \succ \frac{D}{2} \sim D$. Assign utility 1 to $W$ and utility 0 to each of $D$ and $\frac{D}{2}$. She said that she is indifferent between the following two lotteries;

$$\begin{pmatrix} W & D \\ \frac{99}{100} & \frac{1}{100} \end{pmatrix} \quad \text{and} \quad \begin{pmatrix} \frac{W}{2} \\ 1 \end{pmatrix}$$

Since the expected utility of the lottery on the left is $\frac{99}{100} \times 1 + \frac{1}{100} \times 0 = \frac{99}{100}$, it follows that $U\left(\frac{W}{2}\right) = \frac{99}{100} \times 1 + \frac{1}{100} \times 0 = \frac{99}{100}$.

**(b)** Let

$$L = \begin{pmatrix} W & D \\ \frac{96}{100} & \frac{4}{100} \end{pmatrix} \text{ and } M = \begin{pmatrix} \frac{W}{2} & \frac{D}{2} \\ \frac{97}{100} & \frac{3}{100} \end{pmatrix}.$$

Then the expected utility of $L$ is $\frac{96}{100} \times 1 = 0.96$ and the expected utility of $M$ is $\frac{99}{100} \times \frac{97}{100} = 0.9603$. Thus she is better off paying half of her wealth to get the injection (whereas before she was indifferent). Hence she would be willing to pay more than half of her wealth to get the injection in this case.

## 5.3 Expected utility: the axioms

The exercises in this section deal with the axioms of Expected Utility Theory (Volume 1, Chapter 5, Section 5.2, pp. 175-181).

Exercise 5.26 Sam has a debilitating disease. Call his current state of health $z_1$. His doctor presents him with a list of possible treatments, which involve some of the following basic outcomes:

$z_1$ current state of health
$z_2$ complete recovery
$z_3$ paralysis of the lower extremities
$z_4$ partial recovery with minor symptoms

For example, one possible "treatment" is to do nothing, which would correspond to the lottery $\begin{pmatrix} z_1 \\ 1 \end{pmatrix}$, another possible treatment is to take a drug called "Nihilfacet" that can be viewed as the lottery $\begin{pmatrix} z_1 & z_2 \\ 0.95 & 0.05 \end{pmatrix}$, etc. Sam's ranking of the basic outcomes is:

$$\begin{array}{ll} \text{best} & z_2 \\ & z_4 \\ & z_1 \\ \text{worst} & z_3 \end{array}$$

Sam tells his doctor that he would not consider any treatments that involve paralysis (outcome $z_3$) with *any* positive probability: he would view any such treatment as worse than his current state of health (outcome $z_1$). Does Sam satisfy the axioms of Expected Utility Theory? ■

**Solution to Exercise 5.26.**
No, Sam's preferences imply that, for every $p < 1$,

$$\begin{pmatrix} z_1 \\ 1 \end{pmatrix} \succ \begin{pmatrix} z_2 & z_3 \\ p & 1-p \end{pmatrix}$$

which violates the Continuity Axiom (Axiom 3, Volume 1, Chapter 5, p. 175). The Continuity Axiom requires tha there be a probability $p \in (0,1)$ such that

$$\begin{pmatrix} z_1 \\ 1 \end{pmatrix} \sim \begin{pmatrix} z_2 & z_3 \\ p & 1-p \end{pmatrix}.$$

Exercise 5.27 Consider the following set of basic axioms: $Z = \{z_1, z_2, z_3, z_4, z_5\}$ and the following compound lottery:

$$C = \begin{pmatrix} \begin{pmatrix} z_1 & z_3 & z_4 \\ \frac{1}{3} & \frac{1}{3} & \frac{1}{3} \end{pmatrix} & \begin{pmatrix} z_2 & z_3 \\ \frac{1}{2} & \frac{1}{2} \end{pmatrix} & z_3 & \begin{pmatrix} z_3 & z_4 \\ \frac{1}{4} & \frac{3}{4} \end{pmatrix} & z_1 \\ \frac{1}{12} & \frac{1}{4} & \frac{1}{6} & \frac{5}{12} & \frac{1}{12} \end{pmatrix}$$

Find the simple lottery $L$ corresponding to $C$ (Definition 5.3.2, Volume 1, Section 5.3, p. 176). ■

**Solution to Exercise 5.27.**
The simple lottery is as follows:

$$L = \begin{pmatrix} z_1 & z_2 & z_3 & z_4 & z_5 \\ \frac{1}{9} & \frac{1}{8} & \frac{61}{144} & \frac{49}{144} & 0 \end{pmatrix}$$

The probabilities are calculated as follows:

$$\begin{aligned} z_1 &: \ \tfrac{1}{3} \times \tfrac{1}{12} + \tfrac{1}{12} = \tfrac{1}{9} \\ z_2 &: \ \tfrac{1}{2} \times \tfrac{1}{4} = \tfrac{1}{8} \\ z_3 &: \ \tfrac{1}{3} \times \tfrac{1}{12} + \tfrac{1}{2} \times \tfrac{1}{4} + \tfrac{1}{6} + \tfrac{1}{4} \times \tfrac{5}{12} = \tfrac{61}{144} \\ z_4 &: \ \tfrac{1}{3} \times \tfrac{1}{12} + \tfrac{3}{4} \times \tfrac{5}{12} = \tfrac{49}{144} \end{aligned}$$

# 6. Strategic-form Games

## 6.1 Strategic-form games with cardinal payoffs

The exercises in this section deal with the notion of *game in strategic form with cardinal payoffs* (Definition 6.1.2, Volume 1, Chapter 6, Section 6.1).

Exercise 6.1 Consider the following situation. There are two players, Ann and Brad. Ann has \$4 in her purse (her entire wealth), while Brad has \$12 (his entire wealth). They are both given a red card and a black card. Each player selects one of the two cards and puts it face down on the table, while hiding the other card. The two cards on the table are then turned. If they are the same color, a fair coin is tossed and if it comes up Heads, Ann has to give her \$4 to Brad, while if it comes up Tails, Brad has to give his \$12 to Ann. If one of the cards is black and the other is red then a symmetric 6-face die is rolled and if it comes up 1 or 2, Ann gives her \$4 to Brad, otherwise (that is, if it comes up 3, 4, 5 or 6) Brad gives his \$12 to Ann.

**(a)** Represent this situation as a game-frame in strategic form by identifying basic outcomes with pairs $(x,y)$, where $x$ is Ann's final wealth and $y$ is Brad's final wealth.

**(b)** Convert the game-frame of Part (a) into a game assuming that each player is selfish and greedy (that is, cares only about her/his own wealth and prefers more money to less) and that Ann's von Neumann-Morgenstern utility-of-money function is $U(\$m) = \sqrt{m}$, while Brad's von Neumann-Morgenstern utility-of-money function is $V(\$m) = 4m - \frac{1}{16}m^2$.

**(c)** Rewrite the game of Part (b) under the assumption that Ann and Brad are still selfish and greedy, but risk neutral.

■

**Solution to Exercise 6.1.**

**(a)** The game-frame is as follows, where 'B' stands for 'Black' and 'R' for 'Red':

| | | Brad | |
|---|---|---|---|
| | | $B$ | $R$ |
| **Ann** | $B$ | $\begin{pmatrix} (\$16,\$0) & (\$0,\$16) \\ \frac{1}{2} & \frac{1}{2} \end{pmatrix}$ | $\begin{pmatrix} (\$16,\$0) & (\$0,\$16) \\ \frac{2}{3} & \frac{1}{3} \end{pmatrix}$ |
| | $R$ | $\begin{pmatrix} (\$16,\$0) & (\$0,\$16) \\ \frac{2}{3} & \frac{1}{3} \end{pmatrix}$ | $\begin{pmatrix} (\$16,\$0) & (\$0,\$16) \\ \frac{1}{2} & \frac{1}{2} \end{pmatrix}$ |

**(b)** Ann's utility function is $U(\$x,\$y) = \sqrt{x}$ and Brad's utility function is $V(\$x,\$y) = 4y - \frac{1}{16}y^2$. Thus the expected utility of the lottery corresponding to the coin toss is: for Ann $\frac{1}{2}\sqrt{16} = 2$ and for Brad $\frac{1}{2}\left[4(16) - \frac{1}{16}(16^2)\right] = \frac{1}{2}48 = 24$ and the expected utility of the lottery corresponding to the roll of the die is: for Ann $\frac{2}{3}\sqrt{16} = \frac{8}{3}$ and for Brad $\frac{1}{3}\left[4(16) - \frac{1}{16}(16^2)\right] = \frac{1}{3}48 = 16$. Hence the game is as follows:

| | | Brad | |
|---|---|---|---|
| | | $B$ | $R$ |
| **Ann** | $B$ | 2   24 | 8/3   16 |
| | $R$ | 8/3   16 | 2   24 |

**(c)** If both Ann and Brad are risk neutral then we can take, for each player, the identity function as the utility-of-money function so that Ann's utility function is is $U(\$x,\$y) = x$ and Brad's utility function is $V(\$x,\$y) = y$. Thus the expected utility of the lottery corresponding to the coin toss is: for Ann $\frac{1}{2}16 = 8$ and for Brad $\frac{1}{2}16 = 8$ and the expected utility of the lottery corresponding to the roll of the die is: for Ann $\frac{2}{3}16 = \frac{32}{16}$ and for Brad $\frac{1}{3}16 = \frac{16}{3}$. Hence the game is as follows:

| | | Brad | |
|---|---|---|---|
| | | $B$ | $R$ |
| **Ann** | $B$ | 8   8 | 32/3   16/3 |
| | $R$ | 32/3   16/3 | 8   8 |

**Exercise 6.2** Consider the game-frame shown in Figure 6.1. Player 1's ranking of the outcomes is as follows (as usual if outcome $x$ is above outcome $y$ then $x$ is preferred to $y$, while if they are listed next to each other then one is considered to be just as good as the other):

$$\begin{array}{ll} \text{best} & z_2, z_5 \\ & z_3, z_4 \\ \text{worst} & z_1, z_6 \end{array}$$

Player 1 has von Neumann-Morgenstern preferences over lotteries over $\{z_1, z_2, \ldots, z_6\}$. He is indifferent between the following two lotteries:

$$L_1 = \begin{pmatrix} z_1 & z_3 & z_4 & z_6 \\ \frac{1}{6} & \frac{1}{6} & \frac{1}{3} & \frac{1}{3} \end{pmatrix} \quad \text{and} \quad L_2 = \begin{pmatrix} z_2 & z_5 & z_6 \\ \frac{1}{8} & \frac{1}{4} & \frac{5}{8} \end{pmatrix}.$$

Player 2's ranking of the outcomes is as follows:

$$\begin{array}{ll} \text{best} & z_1 \\ & z_4 \\ & z_5 \\ \text{worst} & z_2, z_3, z_6 \end{array}$$

Also Player 2 has von Neumann-Morgenstern preferences. She is indifferent between the following two lotteries:

$$L_3 = \begin{pmatrix} z_1 & z_2 & z_3 & z_4 \\ \frac{16}{25} & \frac{3}{25} & \frac{1}{25} & \frac{1}{5} \end{pmatrix} \quad \text{and} \quad L_4 = \begin{pmatrix} z_4 \\ 1 \end{pmatrix}$$

and she is also indifferent between getting $z_5$ for sure and the lottery $\begin{pmatrix} z_1 & z_2 \\ \frac{1}{5} & \frac{4}{5} \end{pmatrix}$.

Find von Neumann-Morgenstern utility functions that represent the players' preferences and use them to construct a game based on the given game-frame ■

| | | Player 2 | |
|---|---|---|---|
| | | $D$ | $E$ |
| Player 1 | $A$ | $z_1$ | $z_2$ |
| | $B$ | $z_3$ | $z_4$ |
| | $C$ | $z_5$ | $z_6$ |

Figure 6.1: The game-frame for Exercise 6.2.

**Solution to Exercise 6.2.**

Let us start with Player 1. We can construct the normalized utility function by setting $U_1(z_2) = U_1(z_5) = 1$ and $U_1(z_1) = U_1(z_6) = 0$. Let $U_1(z_3) = U_1(z_4) = a$. Then $\mathbb{E}[U_1(L_1)] = \frac{1}{2}a$ and $\mathbb{E}[U_1(L_2)] = \frac{3}{8}$; hence from $\frac{1}{2}a = \frac{3}{8}$ we get that $U_1(z_3) = U_1(z_4) = \frac{3}{4}$. For convenience, let us multiply all the payoffs by 100 (this is allowed by the second theorem of Expected Utility: Theorem 5.2.2, Volume 1, Chapter 5, Section 5.2,). Then Player 1's utility function is given by

| outcome | $U_1$ |
|---|---|
| $z_2, z_5$ | 100 |
| $z_3, z_4$ | 75 |
| $z_1, z_6$ | 0 |

Now consider Player 2. Set $U_2(z_1) = 1$ and $U_2(z_2) = U_2(z_3) = U_2(z_6) = 0$. Then, since $\mathbb{E}[U_2(L_3)] = \frac{16}{25} + \frac{1}{5}U_2(z_4)$, from $\frac{16}{25} + \frac{1}{5}U_2(z_4) = U_2(z_4)$ we get that $U_2(z_4) = \frac{4}{5}$. Similarly, we get that $U_2(z_5) = \frac{1}{5}$. Multiplying all the payoffs by 100 we get the following utility function:

| outcome | $U_2$ |
|---|---|
| $z_1$ | 100 |
| $z_4$ | 80 |
| $z_5$ | 20 |
| $z_2, z_3, z_6$ | 0 |

Thus the game is as shown in Figure 6.2.

| Player 1 \ Player 2 | $D$ | $E$ |
|---|---|---|
| $A$ | 0, 100 | 100, 0 |
| $B$ | 75, 0 | 75, 80 |
| $C$ | 100, 20 | 0, 0 |

Figure 6.2: The game based on the game-frame of Figure 6.1.

Exercise 6.3 A bank has only two depositors. Each has deposited \$1,024. The bank has invested the total amount in a long-term project which will yield a return, if it is allowed to reach maturity. However, if the bank is forced to liquidate the project before maturity, then it will only be able to recover \$1,568 in total. There have been rumors that the banking sector is going to fail and the two depositors have to decide, independently of each other, whether to go to the bank now to withdraw their funds. If both decide to withdraw, then the bank has to liquidate the investment and can only give $\$(\frac{1}{2}1,568) = \$784$ to each depositor. If both decide not to withdraw, then the investment will reach maturity and each depositor will get \$1,600 (her initial deposit of \$1,024 plus interest in the amount of \$576). If one decides to withdraw and the other does not, then the bank has to liquidate the investment and will be able to return the initial deposit (namely, \$1,024) to the person who withdraws, while the other will be left with only \$529 (that is, the remaining amount $\$(1,568 - 1,024) = 544$ minus a fee of \$15).

**(a)** Represent this situation as a strategic-form game, assuming that each depositor is selfish, greedy and risk-neutral.

**(b)** Does any player have a dominant strategy?

**(c)** Represent the above situation as a strategic-form game, this time assuming that both players are risk averse and their von Neumann-Morgenstern utility-of-money function is $U(\$m) = \sqrt{m}$. Continue to assume that they are selfish and greedy.

**(d)** Does any player have a dominant strategy?

**Solution to Exercise 6.3.**

**(a)** For each player we can take as utility-of-money function the identity function: $U(\$m) = m$. Thus the game is as follows, where 'W' stands for 'Withdraw' and 'N' for 'Not withdraw':

| | | Player 2 | | | |
|---|---|---|---|---|---|
| | | $W$ | | $N$ | |
| Player 1 | $W$ | 784 | 784 | 1,024 | 529 |
| | $N$ | 529 | 1,024 | 1,600 | 1,600 |

Figure 6.3: The game for Part (a) of Exercise 6.3.

**(b)** Neither player has a dominant strategy: $W$ is best against $W$ but $N$ is best against $N$.

**(c)** The game is as follows:

| | | Player 2 | |
|---|---|---|---|
| | | $W$ | $N$ |
| Player 1 | $W$ | 28 28 | 32 23 |
| | $N$ | 23 32 | 40 40 |

Figure 6.4: The game for Part (c) of Exercise 6.3.

**(d)** Neither player has a dominant strategy: $W$ is best against $W$ but $N$ is best against $N$.

## 6.2 Mixed strategies

Exercise 6.4 Ken suggests the following game to Gail:

> You choose either the letter $i$ or the letter $o$, while I, independently, will choose one of the following letters: $t$ or $x$. If the letter you choose followed by the letter I choose form a word in the English language (no abbreviations allowed) then I will pay you \$1 if it is a pronoun and \$3 if it is a noun, while in every other case you pay me \$2.

[If you do not know whether a combination of letters forms a word in the English language, ask Google or Alexa or Siri or ChatGPT any of the other wonders of today's world! In the old days one would have used a dictionary!]

**(a)** Represent the strategic-form game assuming that both players are selfish, greedy and risk-neutral.

**(b)** Explain why the following is not a Nash equilibrium:

$$\left[\begin{pmatrix} i & o \\ \frac{1}{2} & \frac{1}{2} \end{pmatrix}, \begin{pmatrix} t & x \\ \frac{5}{8} & \frac{3}{8} \end{pmatrix}\right]$$

**(c)** Verify that the following *is* a Nash equilibrium:

$$\left[\begin{pmatrix} i & o \\ \frac{5}{8} & \frac{3}{8} \end{pmatrix}, \begin{pmatrix} t & x \\ \frac{5}{8} & \frac{3}{8} \end{pmatrix}\right]$$

■

**Solution to Exercise 6.4.**

**(a)** For each player we can take as utility-of-money function the identity function: $U(\$m) = m$. Thus the game is as follows:

| | | Ken | |
|---|---|---|---|
| | | $t$ | $x$ |
| Gail | $i$ | 1, −1 | −2, 2 |
| | $o$ | −2, 2 | 3, −3 |

**(b)** It is not a Nash equilibrium because Ken's payoff is

$$\frac{1}{2}\times\frac{5}{8}\times(-1)+\frac{1}{2}\times\frac{3}{8}\times 2+\frac{1}{2}\times\frac{5}{8}\times 2+\frac{1}{2}\times\frac{3}{8}\times(-3)=\frac{1}{8}=0.125$$

while if Ken, instead of playing the mixed strategy $\begin{pmatrix} t & x \\ \frac{5}{8} & \frac{3}{8} \end{pmatrix}$, were to play the pure strategy $t$, then his payoff would be larger: $\frac{1}{2}\times(-1)+\frac{1}{2}\times 2=\frac{1}{2}=0.5$. That is, the mixed strategy $\begin{pmatrix} t & x \\ \frac{5}{8} & \frac{3}{8} \end{pmatrix}$ is not a best reply to Gail's mixed strategy $\begin{pmatrix} i & o \\ \frac{1}{2} & \frac{1}{2} \end{pmatrix}$.

**(c)** Given Ken's mixed strategy $\begin{pmatrix} t & x \\ \frac{5}{8} & \frac{3}{8} \end{pmatrix}$, if Gail were to play the pure strategy $i$ her payoff would be $\frac{5}{8}\times 1+\frac{3}{8}\times(-2)=-\frac{1}{8}$ and if she were to play the pure strategy $o$ then her payoff would be the same: $\frac{5}{8}\times(-2)+\frac{3}{8}\times 3=-\frac{1}{8}$. Hence Gail's payoff is $-\frac{1}{8}$, no matter what strategy (pure or mixed) she uses. Thus any strategy of Gail's is a best reply to Ken's mixed strtategy; in particular Gail's mixed strategy $\begin{pmatrix} i & o \\ \frac{5}{8} & \frac{3}{8} \end{pmatrix}$ is a best reply to Ken's mixed strategy.

Similar calculations show that, given Gail's mixed strategy $\begin{pmatrix} i & o \\ \frac{5}{8} & \frac{3}{8} \end{pmatrix}$, Ken gets the same payoff, namely $\frac{1}{8}$, no matter what strategy (pure or mixed) he uses, so that any strategy of his is a best reply to Gail's mixed strategy, in particular the mixed strategy $\begin{pmatrix} t & x \\ \frac{5}{8} & \frac{3}{8} \end{pmatrix}$.

Thus $\left[\begin{pmatrix} i & o \\ \frac{5}{8} & \frac{3}{8} \end{pmatrix},\begin{pmatrix} t & x \\ \frac{5}{8} & \frac{3}{8} \end{pmatrix}\right]$ is a Nash equilibrium.

## 6.3 Computing the mixed-strategy Nash equilibria

**Exercise 6.5** Find all the Nash equilibria, including the mixed-strategy ones, of the game shown in Figure 6.5. ■

Player 2

| Player 1 | D | E | F |
|---|---|---|---|
| A | 2   1 | 1   −1 | −1   0 |
| B | 1   1 | 4   −1 | 0   2 |
| C | 1   −2 | 0   0 | −2   4 |

Figure 6.5: The game for Exercise 6.5.

**Solution to Exercise 6.5.**
As a first step let us apply the IDSDS procedure (iterated deletion of strictly dominated strategies) to reduce the game. For Player 1, $C$ is strictly dominated by $A$ and, for Player 2, $E$ is strictly dominated by $F$. Deleting $C$ and $E$ we are left with the following game:

Player 2

| Player 1 | D | F |
|---|---|---|
| A | 2   1 | −1   0 |
| B | 1   1 | 0   2 |

This game has two pure-strategy Nash equilibria: $(A,D)$ and $(B,F)$. Generically, the number of equilibria is odd, hence there ought to be also a mixed-strategy equilibrium. To find it, let $p$ be the probability with which Player 1 plays $A$ and $q$ the probability with which Player 2 plays $D$. Then, given $q$, Player 1 must be indifferent between playing $A$ for sure and playing $B$ for sure, that is, the following equation must be satisfied:

$$\underbrace{2q-1(1-q)}_{\text{if } A \text{ for sure}} = \underbrace{q+(1-q)0}_{\text{if } B \text{ for sure}}.$$

Solving this equation we get $q=\frac{1}{2}$. Similarly, given $p$, Player 2 must be indifferent between playing $D$ for sure and playing $F$ for sure, that is, the following equation must be satisfied:

$$\underbrace{p+(1-p)}_{\text{if } D \text{ for sure}} = \underbrace{p(0)+(1-p)2}_{\text{if } F \text{ for sure}}.$$

Solving this equation we get $p=\frac{1}{2}$. Thus the mixed-strategy Nash equilibrium is $\left[\begin{pmatrix} A & B \\ \frac{1}{2} & \frac{1}{2} \end{pmatrix}, \begin{pmatrix} D & F \\ \frac{1}{2} & \frac{1}{2} \end{pmatrix}\right]$.

Exercise 6.6 Consider again the situation described in Exercise 6.1.

**(a)** Find the Nash equilibria (including the mixed-strategy ones) of the game of Part (b) of Exercise 6.1.

**(b)** Find the Nash equilibria (including the mixed-strategy ones) of the game of Part (c) of Exercise 6.1.

**Solution to Exercise 6.6.**

**(a)** The game is as follows:

| | | Brad: $B$ | | Brad: $R$ | |
|---|---|---|---|---|---|
| Ann | $B$ | 2 | 24 | 8/3 | 16 |
| | $R$ | 8/3 | 16 | 2 | 24 |

There are no pure-strategy Nash equilibria. To find the mixed-strategy equilibrium, let $p$ be the probability with which Ann play $B$ and $q$ the probability with which Brad play $B$. Then $p$ and $q$ must satisfy the following equations:

- For Ann: $2q + \frac{8}{3}(1-q) = \frac{8}{3}q + 2(1-q)$.
- For Brad: $24p + 16(1-p) = 16p + 24(1-p)$.

The solution is $p = q = \frac{1}{2}$. Thus the Nash equilibrium is: $\left[\begin{pmatrix} B & R \\ \frac{1}{2} & \frac{1}{2} \end{pmatrix}, \begin{pmatrix} B & R \\ \frac{1}{2} & \frac{1}{2} \end{pmatrix}\right]$.

**(b)** The game is as follows:

| | | Brad: $B$ | | Brad: $R$ | |
|---|---|---|---|---|---|
| Ann | $B$ | 8 | 8 | 32/3 | 16/3 |
| | $R$ | 32/3 | 16/3 | 8 | 8 |

Again, there are no pure-strategy Nash equilibria. To find the mixed-strategy equilibrium, let $p$ be the probability with which Ann play $B$ and $q$ the probability with which Brad play $B$. Then $p$ and $q$ must satisfy the following equations:

- For Ann: $8q + \frac{32}{3}(1-q) = \frac{32}{3}q + 8(1-q)$.
- For Brad: $8p + \frac{16}{3}(1-p) = \frac{16}{3}p + 8(1-p)$.

The solution is $p = q = \frac{1}{2}$. Thus the Nash equilibrium is: $\left[\begin{pmatrix} B & R \\ \frac{1}{2} & \frac{1}{2} \end{pmatrix}, \begin{pmatrix} B & R \\ \frac{1}{2} & \frac{1}{2} \end{pmatrix}\right]$.

Exercise 6.7 Let $G_1$ and $G_2$ be two strategic-form, two-player games with the same strategy sets: $S_1 = \{A,B\}$ and $S_2 = \{C,D\}$. Player 1's payoff function is the same in $G_1$ and $G_2$, while Player 2's payoff function is different in the two games. Suppose that

$$\left[\begin{pmatrix} A & B \\ p_1 & 1-p_1 \end{pmatrix}, \begin{pmatrix} C & D \\ q_1 & 1-q_1 \end{pmatrix}\right] \quad \text{with } 0 < p_1 < 1 \text{ and } 0 < q_1 < 1$$

is a Nash equilibrium of game $G_1$ and

$$\left[\begin{pmatrix} A & B \\ p_2 & 1-p_2 \end{pmatrix}, \begin{pmatrix} C & D \\ q_2 & 1-q_2 \end{pmatrix}\right] \quad \text{with } 0 < p_2 < 1 \text{ and } 0 < q_2 < 1$$

is a Nash equilibrium of game $G_2$.

**(a)** Explain why it must be that $q_1 = q_2$.

**(b)** Must it also be that $p_1 = p_2$?

**Solution to Exercise 6.7.**

**(a)** The probability with which Player 2 plays each strategy must be such that Player 1 is indifferent between his two strategies. Since Player 1's payoffs are the same in the two games, the equation yielding Player 2's equilibrium strategy is the same for both games.

**(b)** No. Since Player 2's payoffs are different in the two games, the equation yielding Player 1's equilibrium strategy in game $G_1$ will, in general, be different from the equation yielding Player 1's equilibrium strategy in game $G_2$.

Exercise 6.8 Consider the game shown in Figure 6.6, where the payoffs are von Neumann-Morgenstern payoffs.

**(a)** Find all the pure-strategy Nash equilibria.

**(b)** Explain why, if a strategy profile is a Nash equilibrium, then Player 1's strategy must be a pure strategy.

**(c)** Find all the Nash equilibria where Player 2 does not play a pure strategy.

| | | Player 2 | | | | | |
|---|---|---|---|---|---|---|---|
| | | *A* | | *B* | | *C* | |
| Player 1 | *N* | 30 | 2 | 30 | 2 | 10 | 1 |
| | *S* | 60 | 2 | 25 | 10 | 25 | 10 |

Figure 6.6: The game for Exercise 6.8.

**Solution to Exercise 6.8.**

**(a)** The pure-strategy Nash equilibria are: $(N,B)$ and $(S,C)$.

**(b)** Suppose that Player 1 plays a mixed strategy that assigns positive probability to both $N$ and $S$. Let $p$ be the probability assigned to $N$ (thus $0 < p < 1$). Then Player 2's payoff is:
- 2, if she plays $A$,
- $2p + 10(1-p) = 10 - 8p$, if she plays $B$, and
- $p + 10(1-p) = 10 - 9p$, if she plays $C$.

Since $0 < p < 1$, $10 - 8p > 2$ and $10 - 8p > 10 - 9p$, so that Player 2's unique best reply would be the pure strategy $B$. But then Player 1's unique best reply to $B$ is the pure strategy $N$. In other words, if Player 1 mixes then Player 2 wants to play $B$, but then Player 1 does not want to mix against $B$.

**(c)** Consider first the case where Player 1 plays $N$ with probability 1. Then Player 2 would assign positive probability only to $A$ and $B$ (because $C$ gives a lower payoff against $N$ than $A$ and $B$). Let $q$ be the probability of $A$ and $(1-q)$ the probability of $B$. Then, for Player 1, $N$ gives a payoff of 30 while $S$ gives a payoff of $25 + 35q$. Thus it must be that $30 \geq 25 + 35q$, that is, it must be that $q \leq \frac{1}{7}$. Hence, for every $0 < q \leq \frac{1}{7}$, the following is a Nash equilibrium:

$$\left[\begin{pmatrix} N & S \\ 1 & 0 \end{pmatrix}, \begin{pmatrix} A & B & C \\ q & 1-q & 0 \end{pmatrix}\right]$$

Now consider the case where Player 1 plays $S$ with probability 1. Then Player 2 would assign positive probability only to $B$ and $C$ (because $A$ gives a lower payoff against $S$ than $B$ and $C$). Let $q$ be the probability of $B$ and $(1-q)$ the probability of $C$. Then, for Player 1, $S$ gives a payoff of 25 while $N$ gives a payoff of $10 + 20q$. Thus it must be that $25 \geq 10 + 20q$, that is, it must be that $q \leq \frac{3}{4}$. Hence, for every $0 < q \leq \frac{3}{4}$, the following is a Nash equilibrium:

$$\left[\begin{pmatrix} N & S \\ 0 & 1 \end{pmatrix}, \begin{pmatrix} A & B & C \\ 0 & q & 1-q \end{pmatrix}\right]$$

Exercise 6.9 Consider the game shown in Figure 6.7, where the payoffs are von Neumann-Morgenstern payoffs.

**(a)** For what values of $x$ and $y$ is the following a Nash equilibrium?

$$\left[\begin{pmatrix} A & B \\ \frac{1}{5} & \frac{4}{5} \end{pmatrix}, \begin{pmatrix} C & D \\ \frac{3}{4} & \frac{1}{4} \end{pmatrix}\right]$$

**(b)** Suppose that $x = y = 2$. Find the mixed-strategy Nash equilibrium and calculate the payoffs of both players at the Nash equilibrium.

■

| | | Player 2 | |
|---|---|---|---|
| | | $C$ | $D$ |
| Player 1 | $A$ | $x$ $y$ | 3 0 |
| | $B$ | 6 2 | 0 4 |

Figure 6.7: The game for Exercise 6.9.

**Solution to Exercise 6.9.**

**(a)** Player 1 must be indifferent between playing $A$ for sure and playing $B$ for sure: $x\frac{3}{4}+3\frac{1}{4}=6\frac{3}{4}$. Thus $x=5$. Similarly, Player 2 must be indifferent between playing $C$ for sure and playing $D$ for sure: $y\frac{1}{5}+2\frac{4}{5}=4\frac{4}{5}$. Thus $y=8$.

**(b)** Let $p$ be the probability of $A$ and $q$ the probability of $B$. Then, Player 1 must be indifferent between playing $A$ for sure and playing $B$ for sure: $2q+3(1-q)=6q$. This gives $q=\frac{3}{7}$. Similarly, Player 2 must be indifferent between playing $C$ for sure and playing $D$ for sure: $2=4(1-p)$. This gives $p=\frac{1}{2}$. Thus the Nash equilibrium is

$$\left[\begin{pmatrix} A & B \\ \frac{1}{2} & \frac{1}{2} \end{pmatrix}, \begin{pmatrix} C & D \\ \frac{3}{7} & \frac{4}{7} \end{pmatrix}\right]$$

The equilibrium payoffs are $\frac{18}{7}=2.57$ for Player 1 and 2 for Player 2.

Exercise 6.10 Consider the game-frame shown in Figure 6.8, where $w_1, w_2, \ldots, w_6$ are basic outcomes. The players rank the outcomes as indicated below (if outcome $x$ is above outcome $y$ then $x$ is preferred to $y$, and if $x$ and $y$ are written next to each other then the player is indifferent between the two):

$$\text{Player 1's ranking: } \begin{pmatrix} \text{best} & w_2, w_5 \\ & w_4 \\ & w_1, w_3 \\ \text{worst} & w_6 \end{pmatrix} \qquad \text{Player 2's ranking: } \begin{pmatrix} \text{best} & w_1 \\ & w_6 \\ & w_4, w_5 \\ & w_3 \\ \text{worst} & w_2 \end{pmatrix}$$

**(a)** One player has a strategy that is strictly dominated. Identify the player and the strategy.

[Hint: in order to answer questions (b), (c) and (d), you can make your life a lot easier if you simplify the game on the basis of your answer to Part (a).]

Player 1 satisfies the von Neumann-Morgenstern axioms and is indifferent between $w_3$ for sure and the lottery $\begin{pmatrix} w_4 & w_6 \\ \frac{1}{4} & \frac{3}{4} \end{pmatrix}$.

**(b)** Suppose that Player 1 believes that Player 2 is going to play $L$ with probability ¼ and $R$ with probability ¾. Which strategy should Player 1 choose?

Player 2 satisfies the von Neumann-Morgenstern axioms and is indifferent between $w_6$ for sure and the lottery $\begin{pmatrix} w_1 & w_3 \\ \frac{4}{5} & \frac{1}{5} \end{pmatrix}$ and is also indifferent between $w_4$ for sure and the lottery $\begin{pmatrix} w_3 & w_6 \\ \frac{1}{2} & \frac{1}{2} \end{pmatrix}$.

**(c)** Suppose that Player 2 believes that Player 1 is going to play $T$ with probability ½ and $B$ with probability ½. Which strategy should Player 2 choose?

**(d)** Find all the (pure- and mixed-strategy) Nash equilibria of this game.

| | | Player 2 | | |
|---|---|---|---|---|
| | | $L$ | $M$ | $R$ |
| Player 1 | $T$ | $w_1$ | $w_2$ | $w_3$ |
| | $B$ | $w_4$ | $w_5$ | $w_6$ |

Figure 6.8: The game for Exercise 6.10.

**Solution to Exercise 6.10.**

**(a)** Since Player 2 prefers $w_3$ to $w_2$ and prefers $w_6$ to $w_5$, strategy $M$ is strictly dominated by strategy $R$. Thus Player 2 will not choose (or assign positive probability to) $M$ and from now on we can restrict attention to the following reduced game-frame:

| | | Player 2 | |
|---|---|---|---|
| | | $L$ | $R$ |
| Player 1 | $T$ | $w_1$ | $w_3$ |
| | $B$ | $w_4$ | $w_6$ |

**(b)** Of the remaining outcomes, for Player 1 $w_4$ is the best (we can assign utility 1 to it) and $w_6$ is the worst (we can assign utility 0 to it). Thus, since she is indifferent between $w_3$ for sure and the lottery $\begin{pmatrix} w_4 & w_6 \\ \frac{1}{4} & \frac{3}{4} \end{pmatrix}$, the utility of $w_3$ is ¼ and so is the utility of $w_1$. Thus playing $T$ gives Player 1 an expected payoff of $\frac{1}{4} \times \frac{1}{4} + \frac{3}{4} \times \frac{1}{4} = \frac{1}{4}$, while playing $B$ gives Player 1 an expected utility of $\frac{1}{4} \times 1 + \frac{3}{4} \times 0 = \frac{1}{4}$. Hence she is indifferent between playing $T$ and playing $B$ (and also any mixed strategy).

**(c)** In the reduced game, for Player 2 $w_1$ is the best outcome (we can assign utility 1 to it) and $w_3$ is the worst outcome (we can assign utility 0 to it). Thus, since Player 2 is indifferent between $w_6$ and the lottery $\begin{pmatrix} w_1 & w_3 \\ \frac{4}{5} & \frac{1}{5} \end{pmatrix}$, the utility of $w_6$

is ⅘. Hence the expected utility of the lottery $\begin{pmatrix} w_3 & w_6 \\ \frac{1}{2} & \frac{1}{2} \end{pmatrix}$ is $\frac{1}{2} \times 0 + \frac{1}{2} \times \frac{4}{5} = \frac{2}{5}$ and thus the utility of $w_4$ is ⅖. So playing $L$ gives Player 2 an expected payoff of $\frac{1}{2} \times 1 + \frac{1}{2} \times \frac{2}{5} = \frac{7}{10}$, while playing $R$ gives an expected payoff of $\frac{1}{2} \times 0 + \frac{1}{2} \times \frac{4}{5} = \frac{2}{5}$. Hence Player 2 should play $L$.

**(d)** Using the calculations of Parts (b) and (c) the game is as follows:

| | | Player 2 | | | |
|---|---|---|---|---|---|
| | | $L$ | | $R$ | |
| Player 1 | $T$ | ¼ | 1 | ¼ | 0 |
| | $B$ | 1 | ⅖ | 0 | ⅘ |

There are no pure-strategy Nash equilibria. At a mixed-strategy Nash equilibrium, each player must be indifferent between his/her two pure strategies. From Part (b) we already know that Player 1 is indifferent if Player 2 plays $L$ with probability ¼ and $R$ with probability ¾. Let $p$ be the probability with which Player 1 plays $T$. Then for Player 2 to be indifferent between $L$ and $R$ we need $p + \frac{2}{5}(1-p) = \frac{4}{5}(1-p)$, that is $p = \frac{2}{7}$. Thus the Nash equilibrium is

$$\left[ \begin{pmatrix} T & B \\ \frac{2}{7} & \frac{5}{7} \end{pmatrix}, \begin{pmatrix} L & R \\ \frac{1}{4} & \frac{3}{4} \end{pmatrix} \right]$$

Exercise 6.11 Find the Nash equilibrium of the game shown in Figure 6.9, where the payoffs are von Neumann-Morgenstern payoffs.

Player 3: $E$

| | | Player 2 | | | | | |
|---|---|---|---|---|---|---|---|
| | | $C$ | | | $D$ | | |
| Player 1 | $A$ | 5 | 1 | 6 | 0 | 4 | 1 |
| | $B$ | 4 | 2 | 2 | 6 | 0 | 3 |

Player 3: $F$

| | | Player 2 | | | | | |
|---|---|---|---|---|---|---|---|
| | | $C$ | | | $D$ | | |
| Player 1 | $A$ | 8 | 0 | 5 | 0 | 8 | 0 |
| | $B$ | 0 | 2 | 1 | 9 | 1 | 2 |

Figure 6.9: The game for Exercise 6.11.

**Solution to Exercise 6.11.**
For Player 3, $F$ is strictly dominated by $E$. Thus Player 3 will play $E$ with probability 1. There are no pure-strategy Nash equilibria. Let $p$ be the probability with which Player 1 plays $A$ and $q$ the probability with which Player 2 plays $C$. Then $q$ is given by the solution to the following equation (the left-hand side is what Player 1 gets if he plays $A$ and the right-hand side is what he gets if he plays $B$): $5q = 4q + (1-q)6$. Thus $q = \frac{6}{7}$. Similarly, $p$ is given by the solution to the following equation (the left-hand side is what Player 2 gets if she plays $C$ and the right-hand side is what she gets if she plays $D$): $p + 2(1-p) = 4p$. Thus $p = \frac{2}{5}$. Hence the Nash equilibrium is:

$$\left[ \begin{pmatrix} A & B \\ \frac{2}{5} & \frac{3}{5} \end{pmatrix}, \begin{pmatrix} C & D \\ \frac{6}{7} & \frac{1}{7} \end{pmatrix}, \begin{pmatrix} E & F \\ 1 & 0 \end{pmatrix} \right]$$

Exercise 6.12 Find all the Nash equilibria of the game of Exercise 6.2 of Section 1, shown in Figure 6.2. ■

**Solution to Exercise 6.12.**
For convenience, the game of Figure 6.2 is reproduced below:

| | | Player 2: D | | Player 2: E | |
|---|---|---|---|---|---|
| Player 1 | A | 0 | 100 | 100 | 0 |
| | B | 75 | 0 | 75 | 80 |
| | C | 100 | 20 | 0 | 0 |

There is only one pure-strategy Nash equilibrium, namely $(C, D)$. To see if there are any mixed-strategy Nash equilibria, let $q$ be probability with which Player 2 plays $D$ and let us see what Player 1's best reply to $q$ is. Figure 6.10 shows the graphs of the following functions:

$$\begin{array}{ll} \text{Player 1's payoff from playing } A: & A(q) = 100(1-q) \\ \text{Player 1's payoff from playing } B: & B(q) = 75 \\ \text{Player 1's payoff from playing } C: & C(q) = 100q \end{array}$$

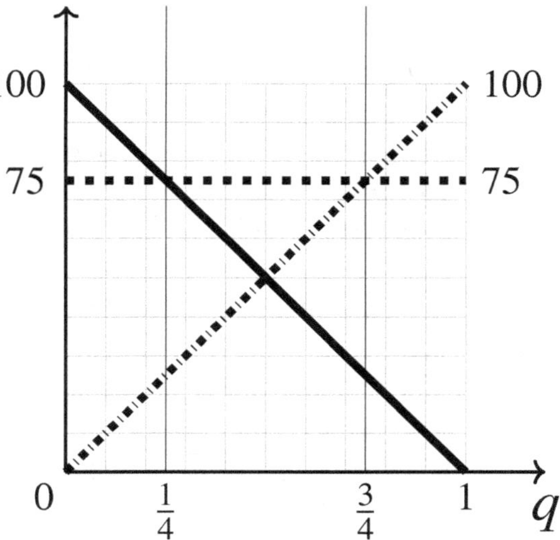


The continuous line from (0,100) to (1,0) is $A(q)$.
The dotted line from (0,75) to (1,75) is $B(q)$.
The dash-dotted line from (0,0) to (1,100) is $C(q)$.

Figure 6.10: The graph of $A(q)$, $B(q)$ and $C(q)$.

From Figure 6.10 one can see that Player 1's best reply to $q \in [0,1]$, denoted by $BR_1(q)$, is as follows:

$$BR_1(q) = \begin{cases} A & \text{if } q < \frac{1}{4} \\ A \text{ or } B \text{ or any mixture of } A \text{ and } B & \text{if } q = \frac{1}{4} \\ B & \text{if } \frac{1}{4} < q < \frac{3}{4} \\ B \text{ or } C \text{ or any mixture of } B \text{ and } C & \text{if } q = \frac{3}{4} \\ C & \text{if } q > \frac{3}{4} \end{cases}$$

Let us consider all possible values of $q \in [0,1]$.

- There is no Nash equilibrium with $q < \frac{1}{4}$: Player 1's best reply to $q < \frac{1}{4}$ is to play the pure strategy $A$ but then Player 2's best reply to $A$ is $q = 1$.
- There is no Nash equilibrium with $\frac{1}{4} < q < \frac{3}{4}$: Player 1's best reply to $q \in \left(\frac{1}{4}, \frac{3}{4}\right)$ is to play the pure strategy $B$ but then Player 2's best reply to $B$ is $q = 0$.
- The only Nash equilibrium with $q > \frac{3}{4}$: is the pure-strategy equilibrium $(C, D)$ where $q = 1$.
- Now consider the case where $q = \frac{1}{4}$. For Player 1 any mixed strategy over $\{A, B\}$ is a best reply to $q = \frac{1}{4}$. In order for Player 2 to rationally choose $q = \frac{1}{4}$, she must be indifferent between playing $D$ and playing $E$. Let $p$ be the probability that Player 1 assigns to $A$. Then, for Player 2, $D$ gives an expected payoff of $100p$ and $E$ an expected payoff of $80(1-p)$. Thus it must be that $100p = 80(1-p)$, that is, $p = \frac{4}{9}$.

Thus the following is a Nash equilibrium:

$$\left[\begin{pmatrix} A & B & C \\ \frac{4}{9} & \frac{5}{9} & 0 \end{pmatrix}, \begin{pmatrix} D & E \\ \frac{1}{4} & \frac{3}{4} \end{pmatrix}\right].$$

- Finally, consider the case where $q = \frac{3}{4}$. For Player 1 any mixed strategy over $\{B, C\}$ is a best reply to $q = \frac{3}{4}$. In order for the mixed strategy $\begin{pmatrix} D & E \\ \frac{3}{4} & \frac{1}{4} \end{pmatrix}$ to be optimal for Player 2 it must be the case that $D$ and $E$ yield the same payoff. Let $\begin{pmatrix} B & C \\ p & 1-p \end{pmatrix}$ be Player 1's strategy. Then, by playing $D$, Player 2 gets a payoff of $20(1-p)$ and, by playing $E$, Player 2 gets a payoff of $80p$. Thus we need $20(1-p) = 80p$, that is, $p = \frac{1}{5}$. Thus the following is a Nash equilibrium:

$$\left[\begin{pmatrix} A & B & C \\ 0 & \frac{1}{5} & \frac{4}{5} \end{pmatrix}, \begin{pmatrix} D & E \\ \frac{3}{4} & \frac{1}{4} \end{pmatrix}\right].$$

Exercise 6.13 Consider the following two-player game: $S_1 = S_2 = [0,1]$ and the payoff functions $\pi_1 : [0,1]^2 \to \mathbb{R}$ and $\pi_2 : [0,1]^2 \to \mathbb{R}$ are as follows:

$$\pi_1(x_1,x_2) = \begin{cases} x_1 & \text{if } (x_1,x_2) \neq (1,1) \\ 0 & \text{if } (x_1,x_2) = (1,1) \end{cases} \quad \text{and} \quad \pi_2(x_1,x_2) = \begin{cases} x_2 & \text{if } (x_1,x_2) \neq (1,1) \\ 0 & \text{if } (x_1,x_2) = (1,1) \end{cases}$$

**(a)** Prove hat this game has no pure-strategy Nash equilibria.

**(b)** Prove that this game has no mixed-strategy Nash equilibria if the support of the mixed strategies is a *finite* number of pure strategies (that is, if each mixed strategy assigns positive probability to only a finite number of pure strategies).

■

**Solution to Exercise 6.13.**

**(a)** Suppose that Player 1's strategy is $x_1 = 1$. Then Player 2 has no best reply, because if she chooses $x_2 = 1$ her payoff is 0 whereas if she chooses $x_2 < 1$ her payoff is $x_2$ (thus she would want to choose the largest $x_2$ which is strictly less than 1 and, of course, there is no such number because of the open interval problem).
Suppose that Player 1's pure strategy is $\hat{x}_1 < 1$. Then Player 2's best response is $x_2 = 1$ but then, by the above argument, $\hat{x}_1$ is not a best reply to $x_2 = 1$.

**(b)** If Player 1 chooses a mixed strategy with finite support that assigns zero probability to $x_1 = 1$ (that is, $\sigma_1(1) = 0$) then Player 2's best reply is the pure strategy $x_2 = 1$ but then (by the argument in Part (a)) Player 1's mixed strategy is not a best reply to $x_2 = 1$. Suppose that Player 1 chooses a mixed strategy $\sigma_1$ with finite support that assigns positive probability to $x_1$ (that is, $\sigma_1(1) = \alpha$ with $0 < \alpha < 1$; the case where $\alpha = 1$ was considered in Part (a)). Then Player 2 has no best reply, because if she chooses $x_2 = 1$ then her payoff is $0 \times \alpha + 1 \times (1-\alpha) = 1-\alpha$ whereas if she chooses $x_2$ arbitrarily close to 1 then her payoff is $x_2$ (thus she would want to choose the largest $x_2$ which is strictly less than 1 and, of course, there is no such number because of the open interval problem).

Exercise 6.14 Consider again the situation, involving a bank and two depositors, described in Exercise 6.3.

**(a)** Find all the Nash equilibria (in pure as well as mixed strategies) of the game of Part (a) of Exercise 6.3, shown in Figure 6.3.

**(b)** Find all the Nash equilibria (in pure as well as mixed strategies) of the game of Part (c) of Exercise 6.3, shown in Figure 6.4.

**(c)** Define a *bank run* as the event that at least one of the two depositors withdraws her money before the investment matures. Using the Nash equilibria found in Parts (a) and (b) above, is a bank run more likely when the two depositors are risk neutral or when they are both risk averse?

**Solution to Exercise 6.14.**

**(a)** The game is reproduced below:

| | | Player 2 | |
|---|---|---|---|
| | | $W$ | $N$ |
| Player 1 | $W$ | 784   784 | 1,024   529 |
| | $N$ | 529   1,024 | 1,600   1,600 |

There are two pure-strategy Nash equilibria: $(W,W)$ and $(N,N)$. To find the mixed-strategy equilibrium, let $p$ be the probability with which Player 1 chooses $W$ and $q$ the probability with which Player 2 chooses $W$. Then $p$ and $q$ must be the solutions to the following equations:

$$784q + 1024(1-q) = 529q + 1600(1-q)$$

and

$$784p + 1024(1-p) = 529p + 1600(1-p).$$

The solutions are $p = q = \frac{192}{277}$. Thus the mixed-strategy Nash equilibrium is:

$$\left[\begin{pmatrix} W & N \\ \frac{192}{277} & \frac{85}{277} \end{pmatrix}, \begin{pmatrix} W & N \\ \frac{192}{277} & \frac{85}{277} \end{pmatrix}\right].$$

**(b)** The game is reproduced below:

| | | Player 2 | |
|---|---|---|---|
| | | $W$ | $N$ |
| Player 1 | $W$ | 28   28 | 32   23 |
| | $N$ | 23   32 | 40   40 |

Again, there are two pure-strategy Nash equilibria: $(W,W)$ and $(N,N)$. To find the mixed-strategy equilibrium, let $p$ be the probability with which Player 1 chooses $W$ and $q$ the probability with which Player 2 chooses $W$. Then $p$ and $q$ must be the solutions to the following equations:

$$28q + 32(1-q) = 23q + 40(1-q)$$

and

$$28p + 32(1-p) = 23p + 40(1-p).$$

The solutions are $p = q = \frac{8}{13}$. Thus the mixed-strategy Nash equilibrium is:

$$\left[ \begin{pmatrix} W & N \\ \frac{8}{13} & \frac{5}{13} \end{pmatrix}, \begin{pmatrix} W & N \\ \frac{8}{13} & \frac{5}{13} \end{pmatrix} \right].$$

**(c)** The probability of **no** bank run is equal to the probability that both players choose $N$, which is as follows:

When both risk neutral $\frac{85}{277} \times \frac{85}{277} = 0.0942$

When both risk averse $\frac{5}{13} \times \frac{5}{13} = 0.1479.$

The probability of a bank run is equal to 1 minus the probability of no bank run. Thus it is as follows:

When both are risk neutral $1 - 0.0942 = 0.9058 \approx 90\%$

When both are risk averse $1 - 0.1479 = 0.8521 \approx 85\%.$

Hence the probability of a bank run is higher when the two players are risk neutral.

Exercise 6.15 Consider the strategic-form game shown in Figure 6.11 where the numbers are von Neumann-Morgenstern payoffs. For what values of $x$ and $y$ is the following a Nash equilibrium?

$$\left[ \begin{pmatrix} A & B & C \\ \frac{5}{11} & \frac{6}{11} & 0 \end{pmatrix}, \begin{pmatrix} D & E \\ \frac{5}{8} & \frac{3}{8} \end{pmatrix} \right]$$

■

Player 2

| | | D | | E | |
|---|---|---|---|---|---|
| Player 1 | A | 5 | $y$ | 1 | 7 |
| | B | 2 | 8 | 6 | 3 |
| | C | $x$ | 1 | $x$ | 9 |

Figure 6.11: The game for Exercise 6.15.

**Solution to Exercise 6.15.**

Since $5 \times \frac{5}{8} + \frac{3}{8} = 2 \times \frac{5}{8} + 6 \times \frac{3}{8} = 3.5$, Player 1 is indifferent between playing $A$ and playing $B$, with an expected payoff of 3.5. Thus it is optimal for Player 1 to play the mixed strategy $\begin{pmatrix} A & B & C \\ \frac{5}{11} & \frac{6}{11} & 0 \end{pmatrix}$, as long as playing $C$ does not yield a higher payoff. Thus it must be that $x \leq 3.5$. For Player 2 it is optimal to play (any) mixed strategy if and only if she expects the same payoff from playing $D$ and from playing $E$, that is, if and only if $y \times \frac{5}{11} + 8 \times \frac{6}{11} = 7 \times \frac{5}{11} + 3 \times \frac{6}{11}$, which is true if and only if $y = 1$. Thus the answer is: $x \leq 3.5$ and $y = 1$.

Exercise 6.16 Consider the game-frame in strategic form shown in Figure 6.12, where $z_1, z_2, \ldots, z_6$ are the possible outcomes.

Player 1 ranks the outcomes as follows: $z_3 \succ z_2 \sim z_5 \succ z_1 \sim z_4 \succ z_6$ and Player 2 ranks them as follows: $z_5 \sim z_6 \succ z_1 \succ z_2 \sim z_4 \succ z_3$. Both players have von Neumann Morgenstern preferences over lotteries involving these outcomes.

**(a)** Does any player have a strictly dominated strategy?

**(b)** Explain why $\left(\begin{array}{ccc|cc} A & B & C & D & E \\ \frac{1}{3} & \frac{1}{3} & \frac{1}{3} & \frac{1}{2} & \frac{1}{2} \end{array}\right)$ is not a Nash equilibrium.

**(c)** Is $\left(\begin{array}{ccc|cc} A & B & C & D & E \\ 0 & 1 & 0 & \frac{1}{4} & \frac{3}{4} \end{array}\right)$ a Nash equilibrium?

**(d)** Is $\left(\begin{array}{ccc|cc} A & B & C & D & E \\ \frac{3}{5} & \frac{2}{5} & 0 & 0 & 1 \end{array}\right)$ a Nash equilibrium?

**(e)** Player 1 is indifferent between $z_2$ for sure and the lottery $\begin{pmatrix} z_1 & z_3 & z_4 \\ \frac{1}{4} & \frac{1}{2} & \frac{1}{4} \end{pmatrix}$. Construct Player 1's normalized von Neumann Morgenstern utility function over the **relevant** outcomes. [To determine what outcomes are relevant, make use of your answer to Part (a).]

**(f)** When Player 1 plays the mixed strategy $\begin{pmatrix} A & B & C \\ \frac{3}{5} & \frac{2}{5} & 0 \end{pmatrix}$, Player 2 is indifferent between playing $D$ and playing $E$. Construct Player 2's normalized von Neumann Morgenstern utility function over the relevant outcomes.

**(g)** Find the Nash equilibrium of the original game.

| Player 1 \ Player 2 | $D$ | $E$ |
|---|---|---|
| $A$ | $z_1$ | $z_2$ |
| $B$ | $z_3$ | $z_4$ |
| $C$ | $z_5$ | $z_6$ |

Figure 6.12: The game-frame for Exercise 6.16.

**Solution to Exercise 6.16.**

**(a)** Yes, for Player 1, $C$ is strictly dominated by $B$ (because $z_3 \succ z_5$ and $z_4 \succ z_6$).

**(b)** Because at a Nash equilibrium a strictly dominated strategy (in this case $C$) must be assigned zero probability.

**(c)** No, because when Player 1 plays $B$, Player 2 strictly prefers $E$ to $D$ (since for Player 2 $z_4 \succ z_3$) while mixing between $D$ and $E$ requires indifference.

**(d)** No, because when Player 2 plays $E$, Player 1 strictly prefers $A$ to $B$ while mixing between $A$ and $B$ requires indifference.

For parts (e) and (f) proceed as follows. Since $C$ is strictly dominated, we can reduce the game as follows:

| Player 1 \ Player 2 | $D$ | $E$ |
|---|---|---|
| $A$ | $z_1$ | $z_2$ |
| $B$ | $z_3$ | $z_4$ |

**(e)** Player 1's ranking of the relevant outcomes is $z_3 \succ z_2 \succ z_1 \sim z_4$. We can assign utility 1 to $z_3$ and utility 0 to $z_1$ and $z_4$. Since Player 1 is indifferent between $z_2$ for sure and the lottery $\begin{pmatrix} z_1 & z_3 & z_4 \\ \frac{1}{4} & \frac{1}{2} & \frac{1}{4} \end{pmatrix}$, the utility of $z_2$ is $\frac{1}{2}$. Thus Player 1's utility function is: $\begin{matrix} z_1 & z_2 & z_3 & z_4 \\ 0 & \frac{1}{2} & 1 & 0 \end{matrix}$.

**(f)** Player 2's ranking of the relevant outcomes is $z_1 \succ z_2 \sim z_4 \succ z_3$. We can assign utility 1 to $z_1$ and utility 0 to $z_3$. When Player 1 plays the mixed strategy $\begin{pmatrix} A & B & C \\ \frac{3}{5} & \frac{2}{5} & 0 \end{pmatrix}$, Player 2 with $E$ gets utility $U(z_2) = U(z_4)$ and with $D$ she gets expected utility of $\frac{3}{5}U(z_1) + \frac{2}{5}U(z_3) = \frac{3}{5} \times 1 + \frac{2}{5} \times 0 = \frac{3}{5}$. Thus $U(z_2) = U(z_4) = \frac{3}{5}$. Hence Player 2's normalized utility function is $\begin{matrix} z_1 & z_2 & z_3 & z_4 \\ 1 & \frac{3}{5} & 0 & \frac{3}{5} \end{matrix}$. Hence the game reduces to the following:

| | | Player 2 | |
|---|---|---|---|
| | | $D$ | $E$ |
| Player 1 | $A$ | 0 1 | $\frac{1}{2}$ $\frac{3}{5}$ |
| | $B$ | 1 0 | 0 $\frac{3}{5}$ |

Let $p$ be the probability of $A$ and $q$ the probability of $D$. Indifference for Player 2 requires $p = \frac{3}{5}$ and indifference for Player 1 requires $\frac{1}{2}(1-q) = q$, that is, $q = \frac{1}{3}$. Thus the Nash equilibrium of the original game is: $\left(\begin{array}{ccc|cc} A & B & C & D & E \\ \frac{3}{5} & \frac{2}{5} & 0 & \frac{1}{3} & \frac{2}{3} \end{array}\right)$.

## 6.4 Strict dominance and rationalizability

The exercises in this section deal with the notion of a pure strategy being strictly dominated by a mixed strategy and the notion of cardinal IDSDS procedure (Definition 6.4.1, Volume 1, Section 4, p. 205).

Exercise 6.17 Consider the strategic-form game shown in Figure 6.13 where the numbers are von Neumann-Morgenstern payoffs. For every value of the parameter $a$, find all the Nash equilibria, including the mixed-strategy ones. ■

| | | Player 2 | |
|---|---|---|---|
| | | $L$ | $R$ |
| Player 1 | $u$ | 5 4 | 0 $a$ |
| | $m$ | 3 8 | 0 2 |
| | $c$ | 1 1 | 3 5 |
| | $d$ | 1 0 | 2 9 |

Figure 6.13: The game for Exercise 6.17.

**Solution to Exercise 6.17.**

Let us start with the pure-strategy Nash equilibria.

- $(c,R)$ is a Nash equilibrium for every value of $a$ and the only Nash equilibrium if $a > 4$.
- $(u,L)$ is a Nash equilibrium for every $a \leq 4$.

Now let us see if there are any mixed-strategy equilibria. Player 1 has two strictly dominated strategies: $m$ and $d$. Strategy $m$ is strictly dominated by any mixed strategy of the form $\begin{pmatrix} u & c \\ p & 1-p \end{pmatrix}$ with $\frac{1}{2} < p < 1$, while strategy $d$ is strictly dominated by any mixed strategy of the form $\begin{pmatrix} u & c \\ p & 1-p \end{pmatrix}$ with $0 < p < \frac{1}{3}$. Deleting $m$ and $d$ we are left with the following game:

| | | Player 2 | |
|---|---|---|---|
| | | $L$ | $R$ |
| Player 1 | $u$ | 5 4 | 0 $a$ |
| | $c$ | 1 1 | 3 5 |

In the reduced game, we have the following:

- If $a > 4$ then, for Player 2, $R$ strictly dominates $L$ and thus at a Nash equilibrium Player 2 must play $R$ with probability 1 and then for Player 1 the best reply is to play $c$ with probability 1. Thus the existence of Nash equilibria that are not pure-strategy equilibria requires $a \leq 4$.

- If $a = 4$, $\left[\begin{pmatrix} u & c \\ 1 & 0 \end{pmatrix}, \begin{pmatrix} L & R \\ q & 1-q \end{pmatrix}\right]$ is a Nash equilibrium if and only if $5q \geq q + 3(1-q)$, that is, if and only if $\frac{3}{7} \leq q \leq 1$.

- If $a < 4$ then $\left[\begin{pmatrix} u & c \\ \frac{4}{8-a} & \frac{4-a}{8-a} \end{pmatrix}, \begin{pmatrix} L & R \\ \frac{3}{7} & \frac{4}{7} \end{pmatrix}\right]$ is a Nash equilibrium.

Thus the Nash equilibria of the original game of Figure **??** are:

- If $a > 4$, $(c, R)$.

- If $a = 4$, $(c, R)$ and $(u, L)$ and, for every $q \in \left[\frac{3}{7}, 1\right]$, $\left[\begin{pmatrix} u & m & c & d \\ 1 & 0 & 0 & 0 \end{pmatrix}, \begin{pmatrix} L & R \\ q & 1-q \end{pmatrix}\right]$.

- If $a < 4$ $(c, R)$ and $(u, L)$ and $\left[\begin{pmatrix} u & m & c & d \\ \frac{4}{8-a} & 0 & \frac{4-a}{8-a} & 0 \end{pmatrix}, \begin{pmatrix} L & R \\ \frac{3}{7} & \frac{4}{7} \end{pmatrix}\right]$.

Exercise 6.18 The famous British spy 008 has to choose one of four routes to ski down a mountain: $a, b, c$ or $d$ (listed in order of length). Shorter routes are more susceptible to be struck by an avalanche. Simultaneously and independently, the notorious Russian rival 009 has to choose whether to use (strategy $E$) or not use (strategy $N$) his irreplaceable explosive device to cause an avalanche. The von Neumann-Morgenstern payoffs of this game are shown in Figure 6.15.

**(a)** There is one pure strategy that 008 should not choose. Which one is it and why?

**(b)** Let $q$ the probability that 008 attaches to 009 selecting strategy $E$. Explain what 008 should do if $q < \frac{1}{4}$ and if $q > \frac{2}{5}$.

**(c)** Construct the best reply function of 008 to any possible mixed strategy of 009.

**(d)** Find all the Nash equilibria of this game.

■

| | | 009 | |
|---|---|---|---|
| | | $E$ | $N$ |
| 008 | $a$ | 0   6 | 12   0 |
| | $b$ | 3   5 | 11   1 |
| | $c$ | 4   2 | 10   2 |
| | $d$ | 6   0 | 9   3 |

Figure 6.14: The game for Exercise 6.18.

**Solution to Exercise 6.18.**

**(a)** Strategy $c$ is strictly dominated by an appropriate mixture of $b$ and $d$, for example by the mixed strategy $m = \begin{pmatrix} b & d \\ \frac{3}{5} & \frac{2}{5} \end{pmatrix}$. In fact, against $E$, $m$ gives an expected payoff of $\frac{3}{5} \times 3 + \frac{2}{5} \times 6 = \frac{21}{5} = 4.2 > 4$ and, against $F$, $m$ gives an expected payoff of $\frac{3}{5} \times 11 + \frac{2}{5} \times 9 = \frac{51}{5} = 10.2 > 10$. Thus 008 should not choose $c$ (equivalently, she should play $c$ with probability 0). [Yes, you were wrong to assume that 008 is a man, just because 007 is!] Thus we can restrict attention to the following reduced game:

| | | 009 | |
|---|---|---|---|
| | | $E$ | $N$ |
| 008 | $a$ | 0   6 | 12   0 |
| | $b$ | 3   5 | 11   1 |
| | $d$ | 6   0 | 9   3 |

**(b)** The reduced game has no pure-strategy Nash equilibria. Let $q \in [0,1]$ be the probability with which 009 plays $E$ and let us see what the best reply to $q$ is for 008. Figure 6.15 shows the graphs of the following functions:

Payoff from playing $a$: $a(q) = 12(1-q)$
Payoff from playing $b$: $b(q) = 3q + 11(1-q) = 11 - 8q$
Payoff from playing $d$: $d(q) = 6q + 9(1-q) = 9 - 3q$

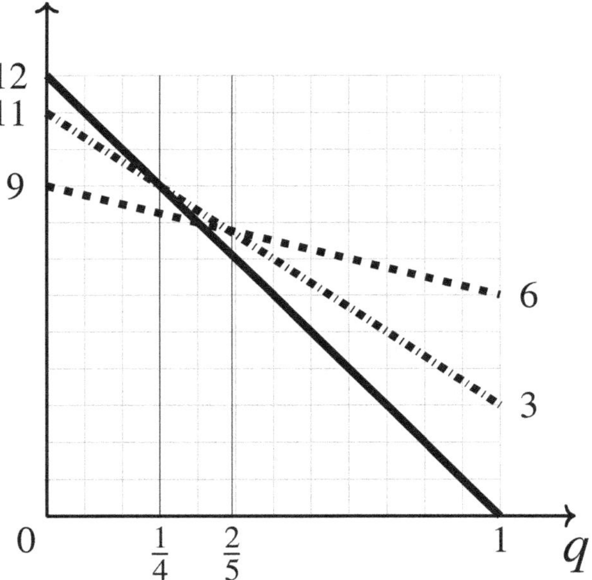


The continuous line from (0,12) to (1,0) is $a(q)$
The dash-dotted line from (0,11) to (1,3) is $b(q)$
The dotted line from (0,9) to (1,6) is $d(q)$

Figure 6.15: The graph of $a(q)$, $b(q)$ and $d(q)$.

From Figure 6.15 one can see that if $q < \frac{1}{4}$ then 008 should choose $a$ and if $q > \frac{2}{5}$ then 008 should choose $d$.

**(c)** From Figure 6.15 one can see that the best reply of 008 to $q \in [0,1]$, denoted by $BR_{008}(q)$ is as follows:

$$BR_{008}(q) = \begin{cases} a & \text{if } q < \frac{1}{4} \\ a \text{ or } b \text{ or any mixture of } a \text{ and } b & \text{if } q = \frac{1}{4} \\ b & \text{if } \frac{1}{4} < q < \frac{2}{5} \\ b \text{ or } d \text{ or any mixture of } b \text{ and } d & \text{if } q = \frac{2}{5} \\ d & \text{if } q > \frac{2}{5} \end{cases}$$

**(d)** We consider all possible values of $q$.

- There is no Nash equilibrium with $q < \frac{1}{4}$: 008's best reply to $q < \frac{1}{4}$ is to play the pure strategy $a$ but then 009's best reply to $a$ is $q = 1$.
- There is no Nash equilibrium with $\frac{1}{4} < q < \frac{2}{5}$: 008's best reply to $\frac{1}{4} < q < \frac{2}{5}$ is to play the pure strategy $b$ but then 009's best reply to $b$ is $q = 1$.
- There is no Nash equilibrium with $q > \frac{2}{5}$: 008's best reply to $q > \frac{2}{5}$ is to play the pure strategy $d$ but then 009's best reply to $b$ is $q = 0$.

- Now consider the case where $q = \frac{1}{4}$. For 008 any mixed strategy over $\{a, b\}$ is a best reply to $q = \frac{1}{4}$. However, when 008 plays $d$ with zero probability, for 009 the expected payoff from $E$ is larger than the expected payoff from $N$, so that 009 would want to choose $q = 1$, not $q = \frac{1}{4}$.

- Finally, consider the case where $q = \frac{2}{5}$. For 008 any mixed strategy over $\{b, d\}$ is a best reply to $q = \frac{2}{5}$. In order for the mixed strategy $\begin{pmatrix} E & N \\ \frac{2}{5} & \frac{3}{5} \end{pmatrix}$ to be optimal for 009 it must be the case that $E$ and $N$ yield the same payoff. Let $\begin{pmatrix} b & d \\ p & 1-p \end{pmatrix}$ be 008's strategy. Then, by playing $E$, 009 gets a payoff of $5p$ and, by playing $N$, 009 gets a payoff of $p + 3(1-p) = 3 - 2p$. Thus we need $5p = 3 - 2p$, that is, $p = \frac{3}{7}$. Thus the Nash equilibrium of the original game is

$$\left[ \begin{pmatrix} a & b & c & d \\ 0 & \frac{3}{7} & 0 & \frac{4}{7} \end{pmatrix}, \begin{pmatrix} E & N \\ \frac{2}{5} & \frac{3}{5} \end{pmatrix} \right].$$

Exercise 6.19 Find all the Nash equilibria of the game shown in Figure 6.16. ■

| | | Player 2: $D$ | | $E$ | | $F$ | |
|---|---|---|---|---|---|---|---|
| Player 1 | $A$ | 8 | 0 | 0 | 4 | 2 | 3 |
| | $B$ | 4 | 2 | 2 | 6 | 8 | 4 |
| | $C$ | 0 | 2 | 6 | 1 | 4 | 1 |

Figure 6.16: The game for Exercise 6.19.

**Solution to Exercise 6.19.**

For Player 2, $F$ is strictly dominated by an appropriate mixture of $D$ and $E$, for example by the mixed strategy $\begin{pmatrix} D & E \\ \frac{1}{5} & \frac{4}{5} \end{pmatrix}$. Deleting $F$ we are left with the game shown in Figure 6.17.

| | | Player 2 | |
|---|---|---|---|
| | | $D$ | $E$ |
| Player 1 | $A$ | 8 0 | 0 4 |
| | $B$ | 4 2 | 2 6 |
| | $C$ | 0 2 | 6 1 |

Figure 6.17: The reduced game after deleting Player 2's strategy $F$.

In the reduced game, Player 1's strategy $B$ is strictly dominated by an appropriate mixture of $A$ and $C$, for example by the mixed strategy $\begin{pmatrix} A & C \\ \frac{5}{8} & \frac{3}{8} \end{pmatrix}$. Deleting $B$ we are left with the game shown in Figure 6.18.

| | | Player 2 | |
|---|---|---|---|
| | | $D$ | $E$ |
| Player 1 | $A$ | 8 0 | 0 4 |
| | $C$ | 0 2 | 6 1 |

Figure 6.18: The reduced game after deleting Player 1's strategy $B$ from the game of Figure 6.17.

To find a Nash equilibrium of this game, let $p$ the probability that Player 1's strategy assigns to $A$ and $q$ the probability that Player 2's strategy assigns to $D$. Then $p$ and $q$ must solve the following equations: $8q = 6(1-q)$ and $2(1-p) = 4p + (1-p)$. The solution is: $p = \frac{1}{5}$ and $q = \frac{3}{7}$. Thus the Nash equilibrium of the original game of Figure 6.16 is:

$$\left[ \begin{pmatrix} A & B & C \\ \frac{1}{5} & 0 & \frac{4}{5} \end{pmatrix}, \begin{pmatrix} D & E & F \\ \frac{3}{7} & \frac{4}{7} & 0 \end{pmatrix} \right].$$

**Exercise 6.20** Apply the cardinal IDSDS procedure to the game shown in Figure 6.19. ■

| | | Player 2 | | | | | | | |
|---|---|---|---|---|---|---|---|---|---|
| | | E | | F | | G | | H | |
| Player 1 | A | 2 | 2 | 5 | 3 | 3 | 3 | 5 | 1 |
| | B | 3 | 8 | 7 | 1 | 2 | 0 | 2 | 6 |
| | C | 5 | 4 | 8 | 5 | 3 | 6 | 1 | 3 |
| | D | 0 | 1 | 2 | 2 | 4 | 0 | 2 | 0 |

Figure 6.19: The game for Exercise 6.20.

**Solution to Exercise 6.20.**

For Player 1, $B$ is strictly dominated by any mixed strategy $\begin{pmatrix} A & C \\ p & 1-p \end{pmatrix}$ with $\frac{1}{4} < p < \frac{1}{3}$ (e.g. $p = \frac{7}{24}$). For Player 2, $H$ is strictly dominated by $E$. Deleting $B$ and $H$ yields the following reduced game:

| | | Player 2 | | | | | |
|---|---|---|---|---|---|---|---|
| | | E | | F | | G | |
| Player 1 | A | 2 | 2 | 5 | 3 | 3 | 3 |
| | C | 5 | 4 | 8 | 5 | 3 | 6 |
| | D | 0 | 1 | 2 | 2 | 4 | 0 |

In the reduced game, For Player 2 $E$ is strictly dominated by $F$. After deleting $E$ we obtain the following game:

| | | Player 2 | | | |
|---|---|---|---|---|---|
| | | F | | G | |
| Player 1 | A | 5 | 3 | 3 | 3 |
| | C | 8 | 5 | 3 | 6 |
| | D | 2 | 2 | 4 | 0 |

In this game, for Player 1, $A$ is strictly dominated by any mixed strategy $\begin{pmatrix} C & D \\ p & 1-p \end{pmatrix}$ with $\frac{1}{2} < p < 1$. After deleting $A$ there are no further strategies that are strictly dominated and thus the deletion procedure stops. Thus, the output of the cardinal IDSDS procedure is the follwing set of strategy profiles: $\{C,F), (C,G), (D,F), (D,G)\}$.

**Exercise 6.21** Apply the cardinal IDSDS procedure to the game shown in Figure 6.20. ■

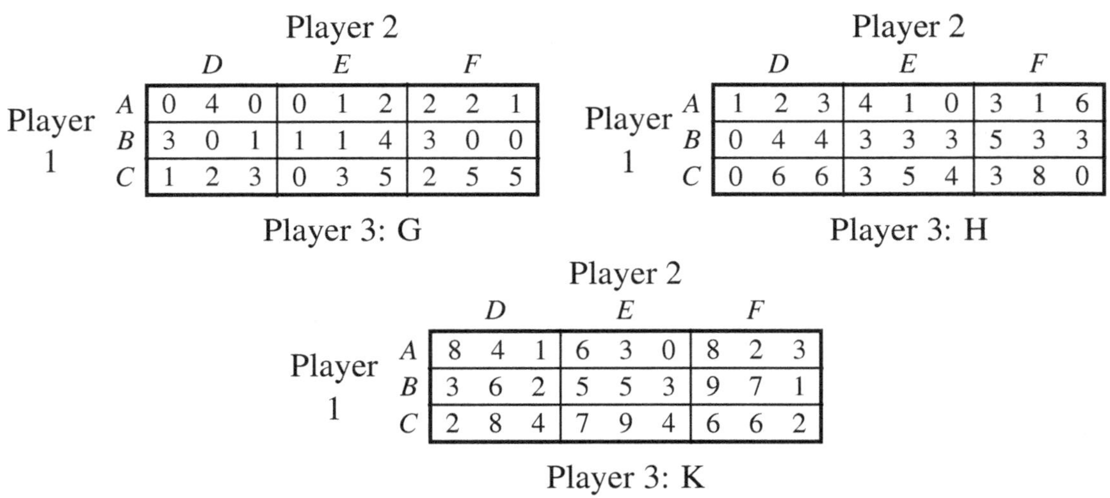

Player 3: G

| Player 1 \ Player 2 | D | E | F |
|---|---|---|---|
| A | 0 4 0 | 0 1 2 | 2 2 1 |
| B | 3 0 1 | 1 1 4 | 3 0 0 |
| C | 1 2 3 | 0 3 5 | 2 5 5 |

Player 3: H

| Player 1 \ Player 2 | D | E | F |
|---|---|---|---|
| A | 1 2 3 | 4 1 0 | 3 1 6 |
| B | 0 4 4 | 3 3 3 | 5 3 3 |
| C | 0 6 6 | 3 5 4 | 3 8 0 |

Player 3: K

| Player 1 \ Player 2 | D | E | F |
|---|---|---|---|
| A | 8 4 1 | 6 3 0 | 8 2 3 |
| B | 3 6 2 | 5 5 3 | 9 7 1 |
| C | 2 8 4 | 7 9 4 | 6 6 2 |

Figure 6.20: The game for Exercise 6.20.

**Solution to Exercise 6.21.**

For Player 3, strategy $K$ is strictly dominated by the mixed strategy $\begin{pmatrix} G & H \\ \frac{1}{2} & \frac{1}{2} \end{pmatrix}$. After deleting $K$ we are left with the following game:

Player 3: G

| Player 1 \ Player 2 | D | E | F |
|---|---|---|---|
| A | 0 4 0 | 0 1 2 | 2 2 1 |
| B | 3 0 1 | 1 1 4 | 3 0 0 |
| C | 1 2 3 | 0 3 5 | 2 5 5 |

Player 3: H

| Player 1 \ Player 2 | D | E | F |
|---|---|---|---|
| A | 1 2 3 | 4 1 0 | 3 1 6 |
| B | 0 4 4 | 3 3 3 | 5 3 3 |
| C | 0 6 6 | 3 5 4 | 3 8 0 |

Figure 6.21: The reduced game after deleting strategy $K$ for Player 3.

In the game of Figure 6.21, for Player 1 strategy $C$ is strictly dominated by the mixed strategy $\begin{pmatrix} A & B \\ \frac{1}{2} & \frac{1}{2} \end{pmatrix}$. After deleting $C$ we are left with the following game:

Player 3: G

| Player 1 \ Player 2 | $D$ | $E$ | $F$ |
|---|---|---|---|
| $A$ | 0 4 0 | 0 1 2 | 2 2 1 |
| $B$ | 3 0 1 | 1 1 4 | 3 0 0 |

Player 3: H

| Player 1 \ Player 2 | $D$ | $E$ | $F$ |
|---|---|---|---|
| $A$ | 1 2 3 | 4 1 0 | 3 1 6 |
| $B$ | 0 4 4 | 3 3 3 | 5 3 3 |

Figure 6.22: The reduced game after deleting strategy $C$ from the game of Figure 6.21.

In the game of Figure 6.22, for Player 2 strategy $F$ is strictly dominated by the mixed strategy $\begin{pmatrix} D & E \\ \frac{1}{2} & \frac{1}{2} \end{pmatrix}$. After deleting $F$ we are left with the following game, which constitutes the output of the cardinal IDSDS procedure, since there are no more strategies that are strictly dominated.

Player 3: $G$

| Player 1 \ Player 2 | $D$ | $E$ |
|---|---|---|
| $A$ | 0 4 0 | 0 1 2 |
| $B$ | 3 0 1 | 1 1 4 |

Player 3: $H$

| Player 1 \ Player 2 | $D$ | $E$ |
|---|---|---|
| $A$ | 1 2 3 | 4 1 0 |
| $B$ | 0 4 4 | 3 3 3 |

More precisely, the output of the IDSDS procedure is the folowing set of strategy profiles:

$$\{(A,D,G),(A,E,G),(B,D,G),(B,E,G),(A,D,H),(A,E,H),(B,D,H),(B,E,H)\}.$$

## 6.5 More difficult exercises

The exercises in this section are more difficult and challenging than the previous ones.

**Exercise 6.22** Consider the following claim:

> "In a two-player strategic-form game with cardinal payoffs, if the pure-strategy pair $(s_1, s_2)$ survives the cardinal version of the IDSDS procedure (iterated deletion of strictly dominated strategies) then there is a mixed-strategy Nash equilibrium where Player 1 plays $s_1$ with positive probability (possibly equal to 1) and Player 2 plays $s_2$ with positive probability (possibly equal to 1)."

Is this claim true? If your answer is 'Yes', then provide a proof, and if your answer is 'No', then provide a counterexample. ■

**Solution to Exercise 6.22.**
The claim is not true. Consider, for example, the following game:

| | | Player 2 | | | | | |
|---|---|---|---|---|---|---|---|
| | | *D* | | *E* | | *F* | |
| Player 1 | *A* | 2 | 0 | 0 | 2 | 1 | 2 |
| | *B* | 0 | 2 | 2 | 0 | 1 | 0 |
| | *C* | 2 | 0 | 0 | 2 | 0 | 0 |

Since there are no strictly dominated strategies, the cardinal IDSDS procedure leaves the game unchanged, so that every pure-strategy pair survives it. Consider $(C, F)$. Is there a Nash equilibrium where Player 1 plays $C$ with positive probability and Player 2 plays $F$ with positive probability? The answer is No, because $C$ cannot be played with positive probability when $F$ is played with positive probability and, similarly, $F$ cannot be played with positive probability when $C$ is played with positive probability, for the following reasons:

- For Player 1, $C$ is weakly dominated by $A$ and gives a lower payoff than $A$ whenever Player 2 plays $F$ with positive probability (transfer any probability from $C$ to $A$ and Player 1's payoff increases).
- For Player 2, $F$ is weakly dominated by $E$ and gives a lower payoff than $E$ whenever Player 1 plays $C$ with positive probability (transfer any probability from $F$ to $E$ and Player 2's payoff increases).

Exercise 6.23 Consider a strategic-form game with von Neumann-Morgenstern payoffs where there is a pure strategy of Player 1, call it $D$, that is strictly dominated by a mixed strategy of Player 1. Must it be the case that if $\sigma_1$ is a mixed strategy of Player 1 that is part of a Nash equilibrium, then $\sigma_1$ dominates $D$? If your answer is 'Yes', prove it; if it is 'No' then provide an example. ■

**Solution to Exercise 6.23.**

The answer is No. Consider, for example, the following game:

| | | Player 2 | |
|---|---|---|---|
| | | $L$ | $R$ |
| Player 1 | $A$ | 4   1 | 0   1 |
| | $B$ | 0   1 | 4   1 |
| | $C$ | 2   1 | 2   1 |
| | $D$ | 1   0 | 2.5   2 |

For Player 1 pure strategy $D$ is strictly dominated by the mixed strategy $\begin{pmatrix} B & C \\ \frac{9}{20} & \frac{11}{20} \end{pmatrix}$.

In this game the following is a Nash equilibrium:

$$\left[\begin{pmatrix} A & B & C & D \\ \frac{1}{3} & \frac{1}{3} & \frac{1}{3} & 0 \end{pmatrix}, \begin{pmatrix} L & R \\ \frac{1}{2} & \frac{1}{2} \end{pmatrix}\right]$$

and yet $\begin{pmatrix} A & B & C & D \\ \frac{1}{3} & \frac{1}{3} & \frac{1}{3} & 0 \end{pmatrix}$ does not dominate (not even weakly) $D$ (indeed it is worse than $D$ against $R$).

Exercise 6.24 [This exercise requires the use of calculus.]

A bank has only two depositors. Each has deposited $\$X$. The bank has invested the total amount in a long-term project which will yield a return, if it is allowed to reach maturity. However, if the bank is forced to liquidate the project before maturity then it will only be able to recover $\$Y$ in total, with $X < Y < 2X$. There have been rumors that the banking sector is going to fail and the two depositors have to decide, independently of each other, whether to go to the bank now to withdraw their funds. If both decide to withdraw, then the bank has to liquidate the investment and can only give $\$\frac{1}{2}Y$ to each depositor. If both decide not to withdraw, then the investment will reach maturity and each depositor will get $\$(1+r)X$: her initial deposit plus interest ($r > 0$). If one decides to withdraw and the other does not, then the bank has to liquidate the investment and will be able to return the initial deposit (namely, $\$X$) to the person who withdraws, while the other will be left with $\$(Y-X)$.

**(a)** Assuming that each depositor is selfish, greedy and risk-neutral, represent this situation as a strategic-form game.

**(b)** Does any player have a dominant strategy? If yes, identify it and state whether it is weakly or strictly dominant.

**(c)** Find all the Nash equilibria of this game, including the mixed-strategy ones.

**(d)** Define a bank run as the event that at least one of the two depositors withdraws her money before the investment matures. Focusing on the mixed-strategy equilibrium, how does the probability of a bank run vary with the parameters $r$, $X$ and $Y$?

In what follows assume that $X = 1,024$ and $Y = 1,800$.

**(e)** Find the mixed-strategy Nash equilibrium.

**(f)** Calculate the probability of a bank run at the mixed-strategy equilibrium. Call this probability $\rho$ and calculate the derivative of $\rho$ with respect to $r$.

In what follows continue to assume that $X = 1,024, Y = 1,800$ and add the assumption that $r = \frac{33}{256}$.

**(g)** Assume now that both players are still selfish and greedy but risk averse and their von Neumann-Morgenstern utility-of-money function is $U(\$m) = \sqrt{m}$. Find the mixed-strategy Nash equilibrium and calculate the probability of a bank run at the mixed-strategy equilibrium.

**(h)** Is a bank run more likely when the two depositors are risk neutral or when they are both risk averse with utility-of-money function $U(\$m) = \sqrt{m}$ ? [Use the mixed-strategy Nash equilibrium to answer this question.]

**Solution to Exercise 6.24.**

**(a)** The game is as follows (given the fact that players are selfish, greedy and risk neutral, we can take for each player as utility-of-money function the identity function: $U(\$m) = m$), where W means 'withdraw' and N 'not withdraw'.

| | | Player 2 | | | |
|---|---|---|---|---|---|
| | | $W$ | | $N$ | |
| Player 1 | $W$ | $\frac{Y}{2}$ | $\frac{Y}{2}$ | $X$ | $Y-X$ |
| | $N$ | $Y-X$ | $X$ | $(1+r)X$ | $(1+r)X$ |

**(b)** Since $\frac{Y}{2} > Y - X$ (because $Y < 2X$), W is better than N if the other player chooses W. Since $r > 0$, N is better than W if the other player chooses N. Thus neither player has a dominant strategy.

**(c)** There are two pure-strategy Nash equilibria: $(W,W)$ and $(N,N)$. To find the symmetric mixed-strategy equilibrium, let p be the probability with which each player play $W$. Then p must be such that $\frac{Y}{2}p + X(1-p) = (Y-X)p + (1+r)X(1-p)$. Thus $p = \frac{2rX}{2(1+r)X-Y}$.

**(d)** The probability that there is no bank run is $(1-p)^2$ so that the probability that there is a bank run is $1-(1-p)^2 = p(2-p) = \frac{4rX[(2+r)X-Y]}{[2(1+r)X-Y]^2}$. Call this expression $f(r,X,Y)$. Then $\frac{\partial f(r,X,Y)}{\partial r} = \frac{4X(2X-Y)^2}{[2(1+r)X-Y]^3} > 0$; thus the higher the interest rate the higher the probability of a bank run.

**(e)** $\frac{\partial f(r,X,Y)}{\partial X} = -\frac{4rY(2X-Y)}{[2(1+r)X-Y]^3} < 0$; thus the higher the initial investment, the lower the probability of a bank run.
$\frac{\partial f(r,X,Y)}{\partial Y} = \frac{4rX(2X-Y)}{[2(1+r)X-Y]^3} > 0$; thus the higher the recovery amount $Y$, the higher the probability of a bank run.

**(f)** When X = 1,024 and Y = 1,800 the game becomes:

| | | Player 2 | | | |
|---|---|---|---|---|---|
| | | $W$ | | $N$ | |
| Player 1 | $W$ | 900 | 900 | 1024 | 776 |
| | $N$ | 776 | 1024 | $(1+r)1024$ | $(1+r)1024$ |

Plugging these values in the expression of Part (c) we get that at a Nash equilibrium the probability of $W$ is $p = \frac{256r}{256r+31}$.

**(g)** The probability that there is a bank run at the mixed-strategy Nash equilibrium is $\rho = 1-(1-p)^2 = 1-\left(1-\frac{256r}{256r+31}\right)^2 = 1-\frac{961}{(256r+31)^2}$. Thus $\frac{d\rho}{dr} = \frac{492032}{(256r+31)^3} > 0$.

**(h)** With the added assumption that $r = \frac{33}{256}$, the monetary outcomes are:

| | | Player 2 | |
|---|---|---|---|
| | | $W$ | $N$ |
| Player 1 | $W$ | \$900 \$900 | \$1024 \$776 |
| | $N$ | \$776 \$1024 | \$1156 \$1156 |

So that the payoffs are:

| | | Player 2 | |
|---|---|---|---|
| | | $W$ | $N$ |
| Player 1 | $W$ | 30 30 | 32 27.857 |
| | $N$ | 27.857 32 | 34 34 |

If $p$ is the probability of $W$ at the mixed-strategy equilibrium, then it must be the solution to $30p+32(1-p) = 27.857p+34(1-p)$. Thus $p = 0.4827$, so that the probability of a bank run is $1-(1-0.4827)^2 = 0.7324$. On the other hand, when the players are risk neutral the probability of a bank run is 0.7654 (obtained by replacing $r$ with $\frac{33}{256}$ in the expression obtained in Part (f)) . Thus the probability is higher if the players are risk neutral.

**Exercise 6.25** Consider the strategic-form game with von Neumann-Morgenstern payoffs shown in Figure 6.23, call it $G$, where, for every $i \in \{1,2\}$, $a_i - b_i - c_i + d_i \neq 0$. Assume that the parameters are such that $G$ has a completely mixed Nash equilibrium $\sigma$ (that is, every pure strategy in $\sigma$ is played with positive probability) and there is no other completely mixed Nash equilibrium (although there may be other Nash equilibria where some pure strategies are played with zero probability).

**(a)** Prove that Player 1 does not have a weakly dominant strategy.

**(b)** Suppose that we increase payoff $a_1$ in such a way that the sign of $(a_1 - c_1)$ does not change. Call the resulting game $G'$.

  **(1)** Explain why $G'$ must also have one and only one completely mixed Nash equilibrium. Denote this equilibrium by $\sigma'$.

  **(2)** Specify which components of $\sigma'$ differ from those of $\sigma$. For those that differ, specify the directions of the changes in values as functions of the game's payoff parameters.

■

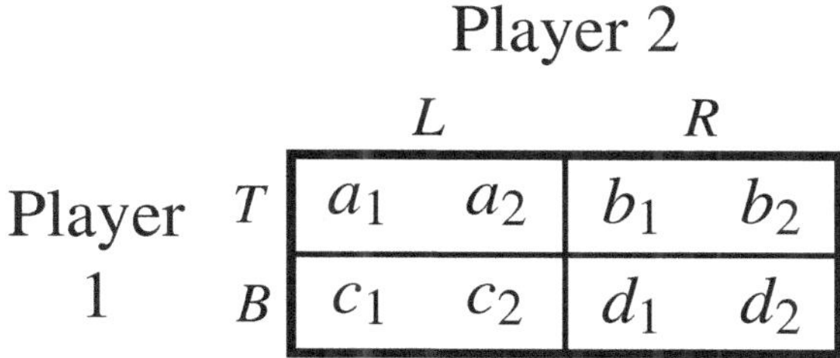

| | | L | | R | |
|---|---|---|---|---|---|
| Player 1 | T | $a_1$ | $a_2$ | $b_1$ | $b_2$ |
| | B | $c_1$ | $c_2$ | $d_1$ | $d_2$ |

Figure 6.23: The game for Exercise 6.25.

**Solution to Exercise 6.25.**

**(a)** There are two ways of proving this.

Method 1. Suppose that $T$ weakly dominates $B$; then either

- Case 1: $a_1 > c_1$ and $b_1 \geq d_1$, or
- Case 2: $a_1 \geq c_1$ and $b_1 > d_1$.

Let $q$ (with $0 < q < 1$) be the probability with which Player 2 plays $L$ at the completely mixed-strategy Nash equilibrium. Then, in both Cases 1 and 2, $a_1 q + b_1(1 - q) > c_1 q + d_1(1 - q)$, so that $T$ is strictly better than $B$ (and thus Player 1 is not indifferent between $T$ and $B$, which is a requirement for Player 1 to mix between $T$ and $B$ at a Nash equilibrium). The proof for the case where $B$ weakly dominates $T$ is similar (reverse the inequalities).

Method 2. As shown below, $\text{sgn}(a_1 - c_1) = \text{sgn}(d_1 - b_1)$ so that if $a_1 > c_1$ then $d_1 > b_1$ and if $a_1 < c_1$ then $d_1 < b_1$.

**(b)** Let $q = \sigma_2(L)$ be the probability with which Player 2 plays L at the completely mixed equilibrium.

**(1)** Player 1 must be indifferent between playing $T$ and playing $B$, that is, it must be that $a_1 q + b_1(1 - q) = c_1 q + d_1(1 - q)$ . Solving for $q$ we get $\sigma_2(L) = \frac{d_1 - b_1}{(a_1 - c_1) + (d_1 - b_1)}$ . Since $0 < \sigma_2(L) < 1$ , $\frac{1}{\sigma_2(L)} > 1$ , that is, $\frac{(a_1 - c_1) + (d_1 - b_1)}{d_1 - b_1} = \frac{a_1 - c_1}{d_1 - b_1} + 1 > 1$. Hence it must be that $\frac{a_1 - c_1}{d_1 - b_1} > 0$, implying that $\text{sgn}(a_1 - c_1) = \text{sgn}(d_1 - b_1)$.

**(2)** When $a_1$ is increased to $a_1'$ there is no effect on Player 1's equilibrium mixed strategy (that depends only on Player 2's payoffs), that is $\sigma_1 = \sigma_1'$. Player 2's mixed equilibrium strategy changes to

$$\sigma_2'(L) = \frac{d_1 - b_1}{(a_1' - c_1) + (d_1 - b_1)} \tag{6.1}$$

Since $\text{sgn}(a_1' - c_1) = \text{sgn}(a_1 - c_1) = \text{sgn}(d_1 - b_1)$ we have that $\sigma_2'(L) \in (0, 1)$; since (6.1) uniquely determines $\sigma_2'(L)$, the completely mixed equilibrium of $G'$ is unique. The numerator and denominator of (6.1) have the same sign; thus $\sigma_2'(L)$ is less than $\sigma_2(L)$ if the numerator and denominator are positive, and is greater than $\sigma_2(L)$ if they are negative. (Of course $\sigma_2(R)$ is affected in the opposite way.)

Exercise 6.26 The Department of Homeland Security (DHS) has been reliably informed that an arms shipment is being smuggled into the country in a vessel scheduled to arrive on August 20 either in the port of San Francisco or in the port of San Diego. The DHS has available a total of 15 coast-guard boats, each of which can stop and thoroughly inspect 10 vessels in one day. It is expected that, on August 20, 200 vessels will be approaching San Francisco and 100 vessels will be approaching San Diego. Nothing is known about the characteristics of the vessel carrying the illegal arms, but it is known that, if the vessel is directed to San Francisco, it is carrying a larger shipment than if it is directed to San Diego. All of this is common knowledge between the DHS and the terrorist organization (TO) behind the shipment. This situation can be seen as a simultaneous game, where the DHS must decide how many of the 15 coast-guard boats should be posted in San Diego (possibly none, possibly all; the remaining boats will be posted in San Francisco), the TO must decide which port to send the shipment to and there are four possible outcomes: (1) a large shipment is intercepted by the DHS ($I_L$), (2) a small shipment is intercepted by the DHS ($I_S$), (3) a large shipment arrives undetected ($U_L$) and (4) a small shipment arrives undetected ($U_S$). The ranking of the outcomes is as follows. For the the DHS: $I_L \succ I_S \succ U_L \sim U_S$ and for the TO: $U_L \succ U_S \succ I_L \sim I_S$. Both the DHS and the TO have von Neumann-Morgenstern preferences. For the DHS the normalized utility of $I_L$ is twice the normalized utility of $I_S$. For the TO the normalized utility of $U_L$ is twice the normalized utility of $U_S$. Note that the size of the shipment is tied to the destination: large if directed to San Francisco and small if directed to San Diego.

**(a)** Find the normalized von Neumann-Morgenstern utility functions of the two players.

**(b)** If the DHS assigns 8 coast-guard boats to San Francisco and 7 to San Diego, what is the probability that the arms shipment will be intercepted if it is directed to San Diego? What is the probability that the arms shipment will be intercepted if it is directed to San Francisco?

**(c)** Does the DHS have any strategies that are dominated? If your answer is No then prove it and if your answer is Yes then list the dominated strategies and state whether they are weakly or strictly dominated.

**(d)** Prove that there are no pure strategy Nash equilibria.

**(e)** Prove that there is no mixed-strategy Nash equilibrium where the DHS uses an undominated pure strategy and the TO uses a completely mixed strategy.

**(f)** Using the insight you got from the calculations for part (e), find a Nash equilibrium of the game and prove that it is a Nash equilibrium. [Note: remember that necessary conditions are not always sufficient.]

■

**Solution to Exercise 6.26.**

**(a)** For the DHS: $U(I_L) = 1$, $U(I_S) = \frac{1}{2}$, $U(U_L) = U(U_S) = 0$.

For the TO: $U(U_L) = 1$, $U(U_S) = \frac{1}{2}$, $U(I_L) = U(I_S) = 0$.

**(b)** If the arms are shipped to San Diego then, since there are 100 boats arriving in

San Diego and 7 boats can inspect a total of 70 boats, the probability of finding the arms is $\frac{7}{10}$. If the arms are shipped to San Francisco then, since there are 200 boats arriving in San Francisco and the 8 boats can inspect a total of 80 boats, the probability of finding the arms is $\frac{80}{200} = \frac{2}{5}$.

**(c)** Assigning 10 boats to San Diego guarantees that, if the arms shipment is directed there, it will be intercepted, because all the approaching boats will be inspected. Assigning more than 10 boats to San Diego does not increase the probability of intercepting the arms shipment if it is directed to San Diego (that probability is already 1 with 10 boats) but decreases the probability of intercepting the arms shipment if it is directed to San Francisco. Thus, letting $n$ be the number of boats assigned to San Diego, every $n$ such that $10 < n \leq 15$ is weakly dominated by $n = 10$.

**(d)** Let $n$ be the number of boats assigned to San Diego, let $F$ be TO's pure strategy of sending the shipment to San Francisco and $D$ be TO's pure strategy of sending the shipment to San Diego. The unique best response to $F$ is $n = 0$, but the unique best response to $n = 0$ is $D$. Similarly, a best response to $D$ is any $n \geq 10$ (and no other) but the unique best response to any $n \geq 10$ is $F$. Thus there is no pure-strategy Nash equilibrium.

**(e)** Fix an arbitrary $n$ with $0 \leq n \leq 10$ (these are the undominated pure strategies). Then the expected payoffs are as follows:

| | TO: $F$ | | TO: $D$ | |
|---|---|---|---|---|
| DHS $n$ | $\frac{10(15-n)}{200}$ | $1 - \frac{10(15-n)}{200}$ | $\frac{10n}{100}\left(\frac{1}{2}\right)$ | $\left(1 - \frac{10n}{100}\right)\left(\frac{1}{2}\right)$ |

which simplifies to:

| | TO: $F$ | | TO: $D$ | |
|---|---|---|---|---|
| DHS $n$ | $\frac{75-5n}{100}$ | $\frac{25+5n}{100}$ | $\frac{5n}{100}$ | $\frac{50-5n}{100}$ |

In order for the strategy profile $\left(n, \begin{pmatrix} F & D \\ p & 1-p \end{pmatrix}\right)$ with $0 < p < 1$ to be a Nash equilibrium it is necessary that TO be indifferent between playing $F$ for sure and playing $D$ for sure, that is, that $\frac{25+n}{100} = \frac{50-5n}{100}$ which gives $n = 2.5$, not an integer.

**(f)** While $n = 2.5$ is not a possible choice, it is possible to obtain it as an average, that is, the DHS could play the following mixed strategy: $\begin{pmatrix} n=2 & n=3 \\ \frac{1}{2} & \frac{1}{2} \end{pmatrix}$. Can this be part of a Nash equilibrium? Against $\begin{pmatrix} n=2 & n=3 \\ \frac{1}{2} & \frac{1}{2} \end{pmatrix}$ $F$ yields an expected payoff of $\frac{1}{2}\left(\frac{35}{100}\right) + \frac{1}{2}\left(\frac{40}{100}\right) = \frac{75}{200}$, while $D$ yields an expected payoff of $\frac{1}{2}\left(\frac{40}{100}\right) + \frac{1}{2}\left(\frac{35}{100}\right) = \frac{75}{200}$. Thus any mixture of $F$ and $D$ is a best reply to $\begin{pmatrix} n=2 & n=3 \\ \frac{1}{2} & \frac{1}{2} \end{pmatrix}$. Now let us see what value of $p$ in the mixed strategy $\left(n, \begin{pmatrix} F & D \\ p & 1-p \end{pmatrix}\right)$ makes the DHS indifferent between playing $n = 2$ and playing $n = 3$. The expected payoff from playing $n = 2$ is $\frac{65}{100}p + \frac{10}{100}(1-p) = \frac{10+55p}{100}$, while the expected payoff from playing $n = 3$ is $\frac{60}{100}p + \frac{15}{100}(1-p) = \frac{15+45p}{100}$. The two are equal if and only if $p = \frac{1}{2}$, in which case DHS's expected payoff is $\frac{75}{200}$. Thus we have found a candidate for a Nash equilibrium, namely $\left(\begin{pmatrix} 2 & 3 \\ \frac{1}{2} & \frac{1}{2} \end{pmatrix}, \begin{pmatrix} F & D \\ \frac{1}{2} & \frac{1}{2} \end{pmatrix}\right)$. To complete the proof that it is indeed a Nash equilibrium we need to show that the DHS cannot get a payoff higher than $\frac{75}{200}$ against $\begin{pmatrix} F & D \\ \frac{1}{2} & \frac{1}{2} \end{pmatrix}$ with a value of $n$ different from 2 and 3. Playing $n \leq 10$ against $\begin{pmatrix} F & D \\ \frac{1}{2} & \frac{1}{2} \end{pmatrix}$ yields $\frac{1}{2}\left(\frac{75-5n}{100}\right) + \frac{1}{2}\left(\frac{5n}{100}\right) = \frac{75}{200}$; thus the DHS is indifferent among all its un-dominated strategies and therefore $\begin{pmatrix} 2 & 3 \\ \frac{1}{2} & \frac{1}{2} \end{pmatrix}$ is indeed a best reply to $\begin{pmatrix} F & D \\ \frac{1}{2} & \frac{1}{2} \end{pmatrix}$ (as would be any other mixed strategy).

Exercise 6.27 [Parts (c) and (d) are more easily answered using calculus, but the use of calculus is not necessary.]

Suppose that one party is able to inflict harm on another party through dishonest behavior, but it is possible to detect such dishonest behavior and punish the cheater. It seems reasonable to believe that there is an inverse relationship between the magnitude of the punishment suffered by those caught cheating and the frequency of cheating. Test this intuition in the following buyer-seller situation. The seller knows the quality of his product and can either be "honest" (H) or "cheat" (C), i.e. claim that the quality is higher than it actually is. The buyer does not know the quality and chooses between "trusting" (T), i.e. buying without inspection, and "inspecting" (I), i.e. paying an expert to examine the good. The von Neumann-Morgenstern payoffs are as shown in Figure 6.24, with $c > a$, $b > d$, $\alpha > \beta$, $\delta > \gamma$.

**(a)** Find the Nash equilibrium of this game.

**(b)** An increase in the fine for cheating can be thought of as a reduction in the value of $\beta$. Would an increase in the fine for cheating reduce the probability that the Seller cheats? Explain why.

**(c)** Assume that, conditional on being honest, the Seller prefers being trusted to being inspected (either because being inspected involves a cost or because of psychological reasons), that is, assume that $\gamma > \alpha$. Would an increase in the fine for cheating increase or reduce the Seller's expected payoff?

■

| | | Seller | |
|---|---|---|---|
| | | $H$ | $C$ |
| Buyer | $I$ | $a \quad \alpha$ | $b \quad \beta$ |
| | $T$ | $c \quad \gamma$ | $d \quad \delta$ |

Figure 6.24: The game for Exercise 6.27 with $c > a$, $b > d$, $\alpha > \beta$, $\delta > \gamma$.

**Solution to Exercise 6.27.**

**(a)** There is only one Nash equilibrium which is in mixed strategies, as follows:

$$\left(\begin{array}{cc|cc} Inspect & Trust & Honest & Cheat \\ \frac{\delta-\gamma}{(\alpha-\beta)+(\delta-\gamma)} & \frac{\alpha-\beta}{(\alpha-\beta)+(\delta-\gamma)} & \frac{b-d}{(c-a)+(b-d)} & \frac{c-a}{(c-a)+(b-d)} \end{array}\right)$$

**(b)** The probability of cheating is independent of $\beta$. The reason for this is that the probabilities for the Seller must be such that the Buyer is indifferent between his two strategies. Thus they depend not on the Seller's payoff, but on the Buyer's payoffs.

**(c)** An increase in $\beta$ will lead to a reduction in the probability of inspecting:

$$\frac{\partial}{\partial \beta}\left(\frac{\delta-\gamma}{(\alpha-\beta)+(\delta-\gamma)}\right) = \frac{\delta-\gamma}{(\alpha-\beta+\delta-\gamma)^2} > 0$$

**(d)** The expected payoffs at the Nash equilibrium are:

$$\text{For the Buyer: } \frac{bc-ad}{(c-a)+(b-d)} \qquad \text{For the Seller: } \frac{\alpha\delta-\beta\gamma}{(\alpha-\beta)+(\delta-\gamma)}.$$

Since $\gamma > \alpha$, that is, $\alpha-\gamma < 0$, the Seller's expected payoff is decreasing in $\beta$:

$$\frac{\partial}{\partial\beta}\left(\frac{\alpha\delta-\beta\gamma}{(\alpha-\beta)+(\delta-\gamma)}\right) = \frac{(\alpha-\gamma)(\delta-\gamma)}{(\alpha-\beta+\delta-\gamma)^2} < 0$$

so that a higher fine (i.e. a lower $\beta$) makes the Seller better off! The reason is that, the higher the fine, the lower the probability of being inspected and caught!

# 7. Extensive-form Games

## 7.1 Cardinal payoffs in extensive-form games

The exercises in this section deal with the notions of *behavioral strategy* and *extensive-form game with cardinal payoffs* (Definitions 7.1.1 and 7.1.2, Volume 1, Chapter 7, Section 1).

Exercise 7.1 In answering this question, consider only extensive-form game-frames that are non-trivial, in the sense that, at every information set, the corresponding player has at least two choices.

**(a)** If an extensive-form game-frame is such that, for every player, the set of mixed strategies coincides with the set of behavioral strategies (that is, an object is a mixed strategy for a player if and only if it is also a behavioral strategy), what property must this game-frame satisfy?

**(b)** Draw an extensive-form game-frame with only two players, where Player 1 has only one information set, while Player 2 has two information sets, each consisting of two nodes.

**(1)** Write one completely mixed strategy for Player 2 (that is, a mixed strategy that assigns positive probability to each pure strategy).

**(2)** Write one completely mixed behavioral strategy for Player 2.

■

**Solution to Exercise 7.1.**

**(a)** The game-frame must be such that every player has only one information set.

**(b)** For example the following game:

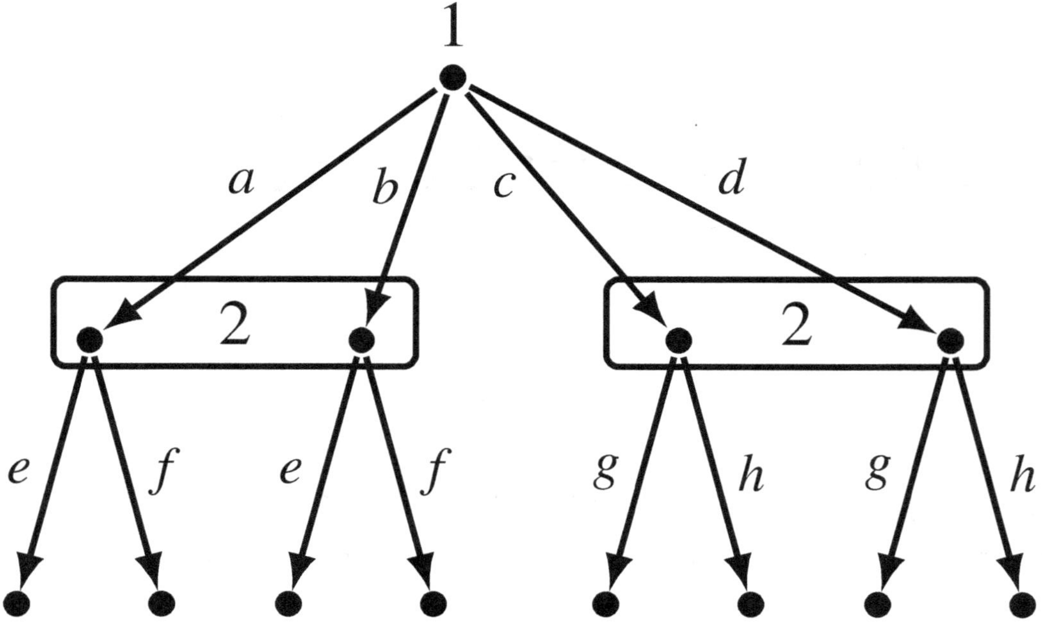


**(1)** The following is an example of a completely mixed strategy of Player 2:

$$\begin{pmatrix} eg & eh & fg & fh \\ \frac{1}{10} & \frac{3}{10} & \frac{2}{10} & \frac{4}{10} \end{pmatrix}.$$

**(2)** The following is an example of a completely mixed behavioral strategy of Player 2:

$$\left(\begin{array}{cc|cc} e & f & g & h \\ \frac{4}{5} & \frac{1}{5} & \frac{2}{3} & \frac{1}{3} \end{array}\right).$$

**Exercise 7.2** Consider the following interaction between a police officer (Player 1) and a motorist (Player 2). At the start of the interaction, the police officer observes the motorist speeding. The officer has two choices: pull the motorist over to give her a ticket (T) or pull her over to give her a warning (W). If the officer pulls her over, then the motorist decides whether to stay put (s) or drive away (d). When the motorist makes this decision, she only knows that the officer has pulled her over, but cannot tell whether the officer has decided to give her a ticket or a warning. If the motorist drives away then she is equally likely to get caught (c) and to escape (e). Thus the possible outcomes are as follows:

$z_1$ Player 2 drives away and is caught.
$z_2$ Player 2 drives away and escapes.
$z_3$ Player 2 stays put and gets a ticket.
$z_4$ Player 2 stays put and gets a warning.

**(a)** Draw an extensive-form game-frame that represents this interaction.

**(b)** Suppose that both players have von Neumann-Morgenstern preferences over the set of lotteries involving the basic outcomes $z_1, \ldots, z_4$.
The police officer (Player 1) ranks the basic outcomes as follows: $z_4 \succ z_3 \succ z_1 \succ z_2$ and is indifferent between $z_3$ for sure and the lottery $\begin{pmatrix} z_2 & z_4 \\ \frac{3}{16} & \frac{13}{16} \end{pmatrix}$.
Furthermore, he is indifferent between $z_1$ for sure and the lottery $\begin{pmatrix} z_2 & z_3 \\ \frac{5}{13} & \frac{8}{13} \end{pmatrix}$.
Construct a von Neumann-Morgenstern utility function for Player 1 with 20 as the lowest value and 100 as the largest value.

**(c)** The motorist (Player 2) ranks the basic outcomes as follows: $z_2 \succ z_4 \succ z_3 \succ z_1$ and is indifferent between the following two lotteries:

$$\begin{pmatrix} z_4 \\ 1 \end{pmatrix} \quad \text{and} \quad \begin{pmatrix} z_1 & z_2 \\ \frac{1}{5} & \frac{4}{5} \end{pmatrix}.$$

Furthermore, she is indifferent between the following two lotteries:

$$\begin{pmatrix} z_1 & z_4 \\ \frac{3}{8} & \frac{5}{8} \end{pmatrix} \quad \text{and} \quad \begin{pmatrix} z_3 \\ 1 \end{pmatrix}.$$

Construct a von Neumann-Morgenstern utility function for Player 2 with 0 as the lowest value and 10 as the largest value.

**(d)** Using the von Neumann-Morgenstern utility functions of Parts (b) and (c) turn the extensive-form game-frame of part (a) into a game and write the corresponding strategic-form game.

**(e)** Find all the Nash equilibria (including the mixed-strategy ones) of the game of Part (d).

■

**Solution to Exercise 7.2.**

**(a)** There are two ways of representing the interaction. In one representation the random event is explicitly modeled as a move by Nature, as follows:

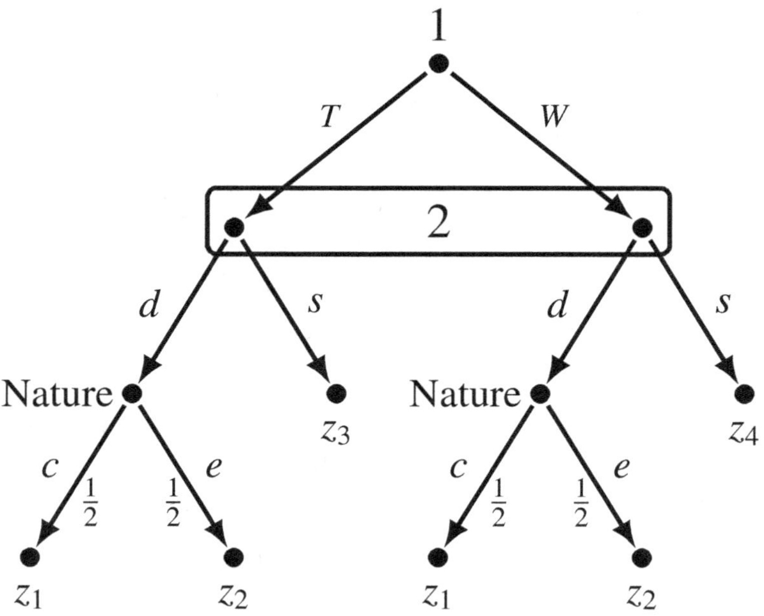


In the other representation the random event is shown as a lottery associated with a terminal node, as follows:

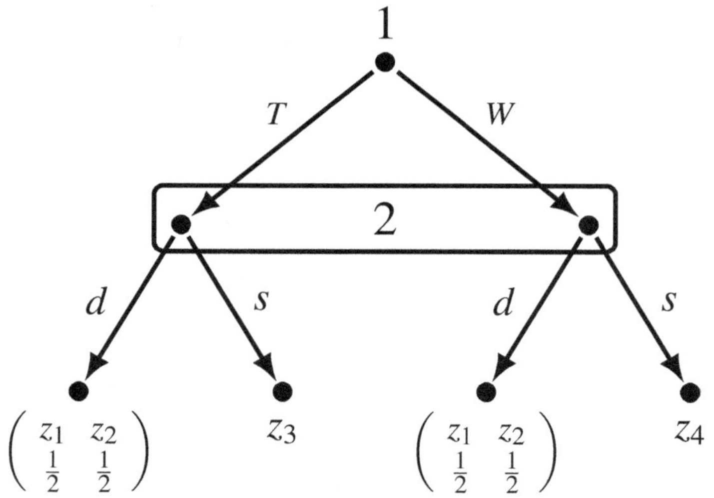


**(b)** $U_1(z_4) = 100$ and $U_1(z_2) = 20$. From the first indifference we get that $U_1(z_3) = \frac{13}{16} \times 100 + \frac{3}{16} \times 20 = 85$ and from the second indifference we get that $U_1(z_1) = \frac{8}{13} \times 85 + \frac{5}{13} \times 20 = 60$. Thus the police officer's von Neumann-Morgenstern utility function is:

| $z_1$ | $z_2$ | $z_3$ | $z_4$ |
|---|---|---|---|
| 60 | 20 | 85 | 100 |

**(c)** $U_2(z_2) = 10$ and $U_2(z_1 = 0$. From the first indifference we get that $U_2(z_4) = \frac{1}{5} \times 0 + \frac{4}{5} \times 10 = 8$ and from the second indifference we get that $U_2(z_3) = \frac{5}{8} \times 0 + \frac{3}{8} \times 8 = 3$. Thus the motorist's von Neumann-Morgenstern utility function is:

| $z_1$ | $z_2$ | $z_3$ | $z_4$ |
|---|---|---|---|
| 0 | 10 | 3 | 8 |

**(d)** The expected utility of lottery $\begin{pmatrix} z_1 & z_2 \\ \frac{1}{2} & \frac{1}{2} \end{pmatrix}$ is 40 for Player 1 and 5 for Player 2. Thus the extensive-form game and corresponding strategic-form game are as follows:

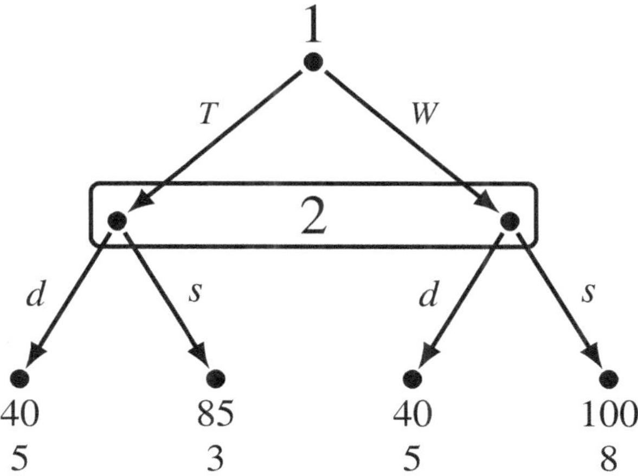


| | | Player 2: $d$ | | Player 2: $s$ | |
|---|---|---|---|---|---|
| Player 1 | $T$ | 40 | 5 | 85 | 3 |
| | $W$ | 40 | 5 | 100 | 8 |

**(e)** The game has two pure-strategy Nash equilibria: $(T,d)$ and $(W,s)$. If Player 2 plays $s$ with positive probability, then, for Player 1, $W$ yields a higher expected payoff than $T$ and thus Player 1's best response is to play the pure strategy $W$, in which case Player 2 wants to play $s$ with probability 1, so that we are back to the pure-strategy Nash equilibrium $(W,s)$. On the other hand, if Player 2 plays the pure strategy $d$ then Player 1 is indifferent between any of his mixed strategies. Let $p$ be the probability with which Player 1 plays $T$; then, for Player 2, $d$ is a best response to the mixed strategy $\begin{pmatrix} T & W \\ p & 1-p \end{pmatrix}$ if and only if $5 \geq 3p + 8(1-p)$, that is, if $p \geq \frac{3}{5}$. Thus, for every $p \geq \frac{3}{5}$, the following is a mixed-strategy Nash equilibrium: $\left(\begin{array}{cc|cc} T & W & d & s \\ p & 1-p & 1 & 0 \end{array}\right)$. Hence there is an infinite number of mixed-strategy equilibria.

Exercise 7.3 This exercise asks you to compare the Nash equilibria of various extensive-form games with the same payoffs and different information structures.

**GAME 1**. This is a game with perfect information. Player 1 moves first choosing between U and D. Player 2 sees Player 1's move and chooses between L and R. The payoffs are as follows (as usual, the left number is the payoff of Player 1 and the right number the payoff of Player 2):

| | $L$ | $R$ |
|---|---|---|
| $U$ | 3 2 | 1 1 |
| $D$ | 5 0 | 2 2 |

**(a)** Draw the extensive-form game and write the corresponding strategic-form game.

**(b)** Find all the pure-strategy Nash equilibria.

**(c)** Find the output of the iterated deletion of strictly dominated strategies (IDSDS).

**(d)** Find the output of the iterated deletion of weakly dominated strategies (IDWDS).

**(e)** Find the backward-induction solution.

**GAME 2**. In this game again Player 1 moves first choosing between U and D. If Player 1 chooses U then Nature chooses either $a$ with probability $1-\varepsilon$ or $b$ with probability $\varepsilon$, where $0<\varepsilon<1$. If Player 1 chooses D then Nature chooses either $a$ with probability $\varepsilon$ or $b$ with probability $1-\varepsilon$. Player 2 moves after Nature, observing Nature's choice but without observing Player 1's choice. Player 2 chooses between L and R. The payoffs are the same as above; in particular, they depend only on the choices of Player 1 and 2 and not on Nature's choices.

**(f)** Draw the extensive-form game and write the corresponding strategic-form game. [Recall that the entries of the strategic form are *expected* payoffs.]

**(g)** Find the pure-strategy Nash equilibria of this game, for every $\varepsilon \in (0,1)$.

**(h)** What happens as $\varepsilon \to 0$?

■

**Solution to Exercise 7.3.**

**(a)** The extensive-form game and corresponding strategic-form game are as follows (in the strategic form, a strategy $xy$ of Player 2 is to be interpreted as "play $x$ if Player 1 plays $U$ and play $y$ if Player 1 plays $D$"):

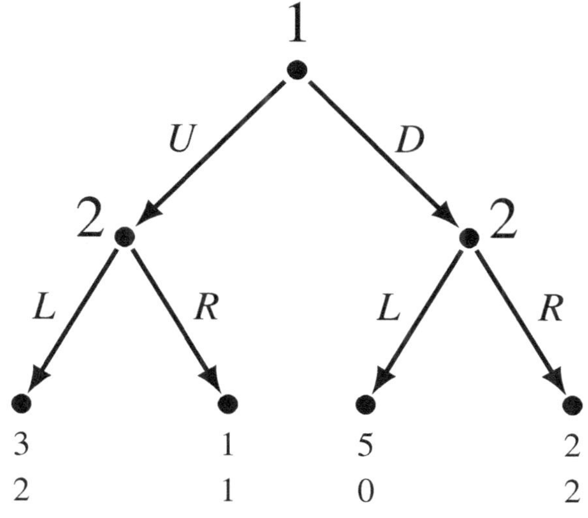


Player 2

| | | LL | LR | RL | RR |
|---|---|---|---|---|---|
| Player 1 | U | 3   2 | 3   2 | 1   1 | 1   1 |
| | D | 5   0 | 2   2 | 5   0 | 2   2 |

**(b)** The pure-strategy Nash equilibria are: $(U, LR)$ and $(D, RR)$.
**(c)** The output of IDSDS is as follows:

Player 2

| | | LL | LR | RR |
|---|---|---|---|---|
| Player 1 | U | 3   2 | 3   2 | 1   1 |
| | D | 5   0 | 2   2 | 2   2 |

**(d)** The output of IDWDS is $(U,LR)$.
**(e)** The backward-induction solution is $(U,LR)$.
**(f)** The extensive-form game and corresponding strategic-form game are as follows (in the strategic form, a strategy $xy$ of Player 2 is to be interpreted as "play $x$ if Nature selects $a$ and play $y$ if Nature selects $b$"):

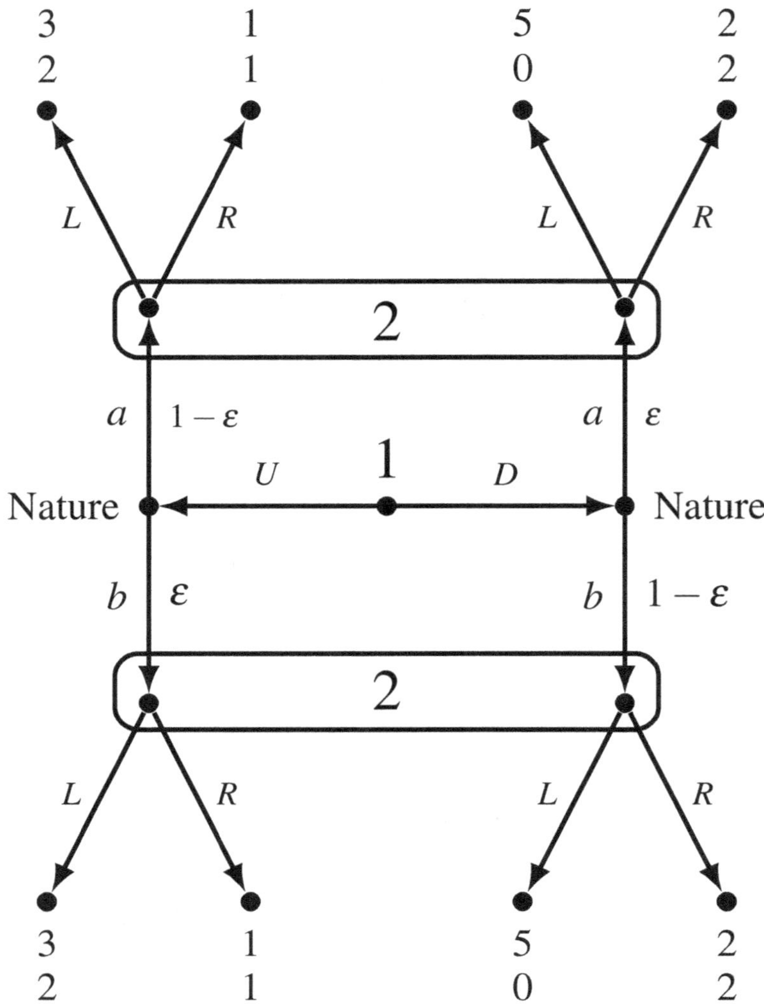


| | | Player 2 | | | |
|---|---|---|---|---|---|
| | | $LL$ | $RL$ | $LR$ | $RR$ |
| Player | $U$ | 3 2 | $1+2\varepsilon$ $1+\varepsilon$ | $3-2\varepsilon$ $2-\varepsilon$ | 1 1 |
| 1 | $D$ | 5 0 | $5-3\varepsilon$ $2\varepsilon$ | $2+3\varepsilon$ $2-2\varepsilon$ | 2 2 |

**(g)** For every value of $\varepsilon$, $(D, RR)$ is a Nash equilibrium. There is no other pure-strategy Nash equilibrium where Player 1 plays $D$, no matter what the value of $\varepsilon$ is: Player 2's best reply to $D$ is $RR$ (since, for every $0 < \varepsilon < 1$, $2 > 2\varepsilon > 0$ and $2 > 2 - 2\varepsilon$). Furthermore, there is no pure-strategy Nash equilibrium where Player 1 plays $U$, because Player 2's best reply to $U$ is $LL$ and Player 1's best reply to $LL$ is $D$ and not $U$.

**(h)** There is a discontinuity at $\varepsilon = 0$: along the sequence, as long as $\varepsilon > 0$, there is only one Nash equilibrium, namely $(D, RR)$, but at $\varepsilon = 0$ a second one appears, namely $(U, LR)$.

Exercise 7.4 Consider the extensive-form game-frame shown in Figure 7.1.

**(a)** Construct the associated strategic-form for the game based on that frame where the players' von Neumann-Morgenstern utility functions are as follows:

| Player 1: | $z_1$ | $z_2$ | $z_3$ | $z_4$ | $z_5$ | $z_6$ | $z_7$ |
|---|---|---|---|---|---|---|---|
| | 2 | 2 | 0 | 3 | 1 | 4 | 5 |

| Player 2: | $z_1$ | $z_2$ | $z_3$ | $z_4$ | $z_5$ | $z_6$ | $z_7$ |
|---|---|---|---|---|---|---|---|
| | 3 | 4 | 6 | 2 | 5 | 1 | 0 |

| Player 3: | $z_1$ | $z_2$ | $z_3$ | $z_4$ | $z_5$ | $z_6$ | $z_7$ |
|---|---|---|---|---|---|---|---|
| | 2 | 2 | 0 | 1 | 1 | 0 | 2 |

**(b)** Does the game of Part (a) have a pure-strategy Nash equilibrium?

**(c)** Is the following behavioral-strategy profile a Nash equilibrium of the extensive-form game?

$$\begin{pmatrix} A & B & | & C & D & | & E & F & | & G & H \\ \frac{1}{2} & \frac{1}{2} & | & \frac{2}{3} & \frac{1}{3} & | & \frac{1}{4} & \frac{3}{4} & | & \frac{1}{2} & \frac{1}{2} \end{pmatrix}$$

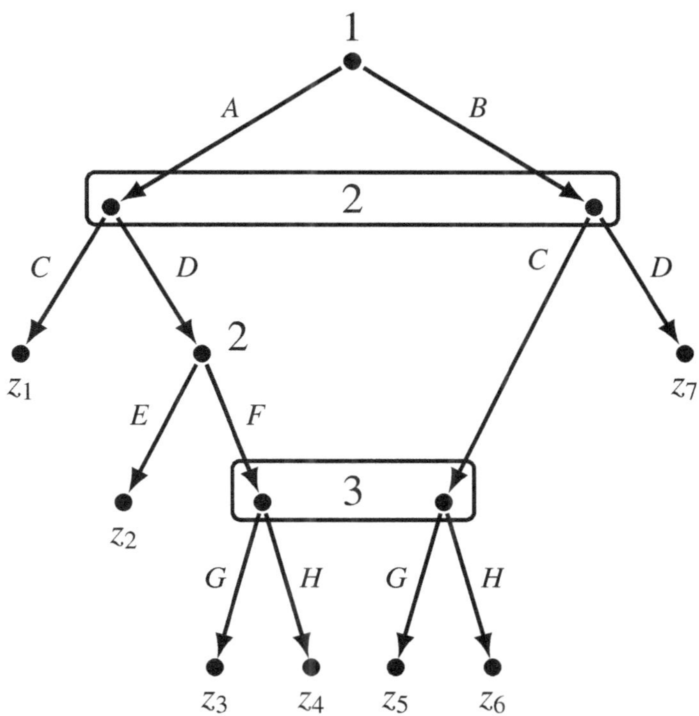


Figure 7.1: The game-frame for Exercise 7.4.

**Solution to Exercise 7.4.**

**(a)** The extensive-form game and the corresponding strategic form are as follows:

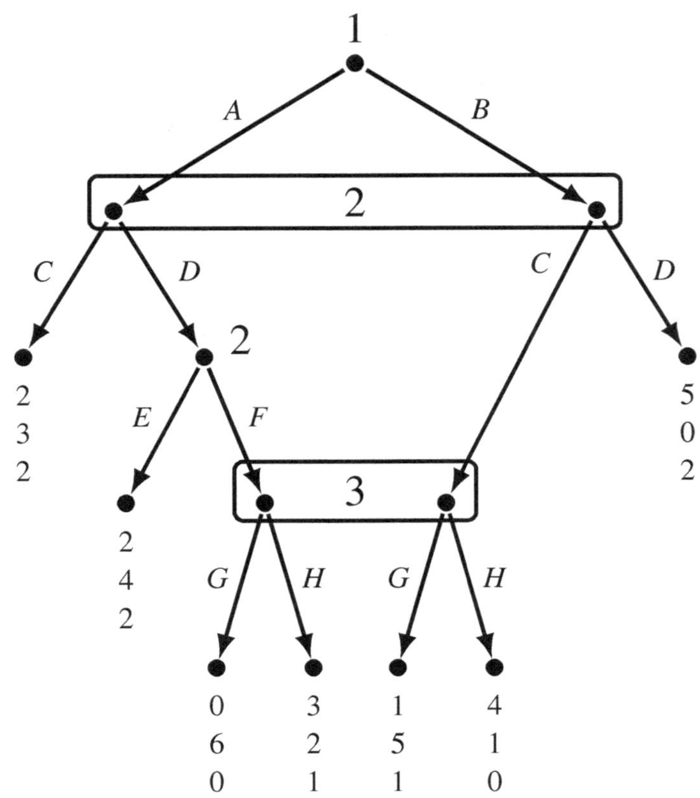


Player 2

| Player 1 | CE | CF | DE | DF |
|---|---|---|---|---|
| A | 2 3 2 | 2 3 2 | 2 4 2 | 0 6 0 |
| B | 1 5 1 | 1 5 1 | 5 0 2 | 5 0 2 |

Player 3: *G*

Player 2

| Player 1 | CE | CF | DE | DF |
|---|---|---|---|---|
| A | 2 3 2 | 2 3 2 | 2 4 2 | 3 2 1 |
| B | 4 1 0 | 4 1 0 | 5 0 2 | 5 0 2 |

Player 3: *H*

**(b)** The game does not have any pure-strategy Nash equilibria.

**(c)** The behavioral strategy $\begin{pmatrix} A & B & \vert & C & D & \vert & E & F & \vert & G & H \\ \frac{1}{2} & \frac{1}{2} & \vert & \frac{2}{3} & \frac{1}{3} & \vert & \frac{1}{4} & \frac{3}{4} & \vert & \frac{1}{2} & \frac{1}{2} \end{pmatrix}$ is not a Nash equilibrium. This can be seen in two ways.

- Method 1: reason directly on the extensive form. For Player 1 it is optimal to randomize between $A$ and $B$ only if – given the behavioral strategies of Players 2 and 3 – Player 1 expects the same payoff from playing $A$ for sure and from playing $B$ for sure. Playing $A$ with probability 1 yields the following expected payoff to Player 1:

$$\frac{2}{3}\times 2+\frac{1}{3}\left[\frac{1}{4}\times 2+\frac{3}{4}\times\left(\frac{1}{2}\times 0+\frac{1}{2}\times 3\right)\right]=\frac{15}{8}=1.875.$$

On the other hand, playing $B$ with probability 1 yields the following expected payoff to Player 1:

$$\frac{1}{3}\times 5+\frac{2}{3}\times\left(\frac{1}{2}\times 1+\frac{1}{2}\times 4\right)=\frac{10}{3}=3.33.$$

Thus Player 1's strategy $\begin{pmatrix} A & B \\ \frac{1}{2} & \frac{1}{2} \end{pmatrix}$ is not a best reply to the behavioral strategies of Players 2 and 3.

- Method 2: note that Player 2's behavioral strategy $\begin{pmatrix} C & D & \vert & E & F \\ \frac{2}{3} & \frac{1}{3} & \vert & \frac{1}{4} & \frac{3}{4} \end{pmatrix}$ is equivalent to any mixed strategy of the form $\begin{pmatrix} CE & CF & DE & DF \\ p & \frac{2}{3}-p & \frac{1}{12} & \frac{1}{4} \end{pmatrix}$ with $0 \le p \le \frac{1}{3}$. Against any such mixed strategy of Player 2, in conjunction with Player 3's mixed strategy $\begin{pmatrix} G & H \\ \frac{1}{2} & \frac{1}{2} \end{pmatrix}$, one can see from the strategic form of the game that, for Player 1 playing $A$ for sure yields an expected payoff of $\frac{1}{2}\times\left(\frac{2}{3}\times 2+\frac{1}{12}\times 2+\frac{1}{4}\times 0\right)+\frac{1}{2}\times\left(\frac{2}{3}\times 2+\frac{1}{12}\times 2+\frac{1}{4}\times 3\right)=\frac{15}{8}$, while playing $B$ for sure yields an expected payoff of $\frac{1}{2}\times\left(\frac{2}{3}\times 1+\frac{1}{12}\times 5+\frac{1}{4}\times 5\right)+\frac{1}{2}\times\left(\frac{2}{3}\times 4+\frac{1}{12}\times 5+\frac{1}{4}\times 5\right)=\frac{10}{3}$. Since these two payoffs are not equal, it is not optimal for Player 1 to play a mixed strategy against the given mixed strategies of Players 2 and 3.

Exercise 7.5 Player 1 is sitting in front of a computer terminal and is about to play a game. Player 2 is sitting in a separate room. Player 1 is told that, when he clicks on START, the computer will randomly pick whether the opponent is another computer (this happens with probability $r$, with $0 < r < 1$) or Player 2 (this happens with probability $1 - r$). Player 1 is **not** informed of what the computer selected. Player 1 then has to choose one of two options: $A$ or $B$. In the case where the opponent is a computer, the computer-opponent receives Player 1's choice as input and is programmed to make the same choice as Player 1. In the case where the opponent is Player 2, Player 2 is informed that she has to play **but she is not told what Player 1 chose** and has to choose between $D$ and $E$. The von Neumann-Morgenstern payoffs are as follows, with $0 < b < c$:

| | Player 1's payoff | Player 2's payoff |
|---|---|---|
| $AD$ | $b$ | $b+c$ |
| $AE$ | $c$ | $b$ |
| $BD$ | $b+c$ | 0 |
| $BE$ | 0 | $c$ |

If Player 1's opponent is a computer then Player 1's payoff is $b$ if he plays $A$ and $c$ if he plays $B$, while Player 2 – who does not get to play – gets a payoff of 0.

**(a)** Draw a two-player extensive-form game to represent the situation described above (the computer is not a player).

**(b)** Write the corresponding strategic-form game.

**(c)** Find the Nash equilibrium of this game for the case where $b = 4$, $c = 6$ and $r = \frac{1}{2}$.

■

**Solution to Exercise 7.5.**

**(a)** The extensive-form game is as follows:

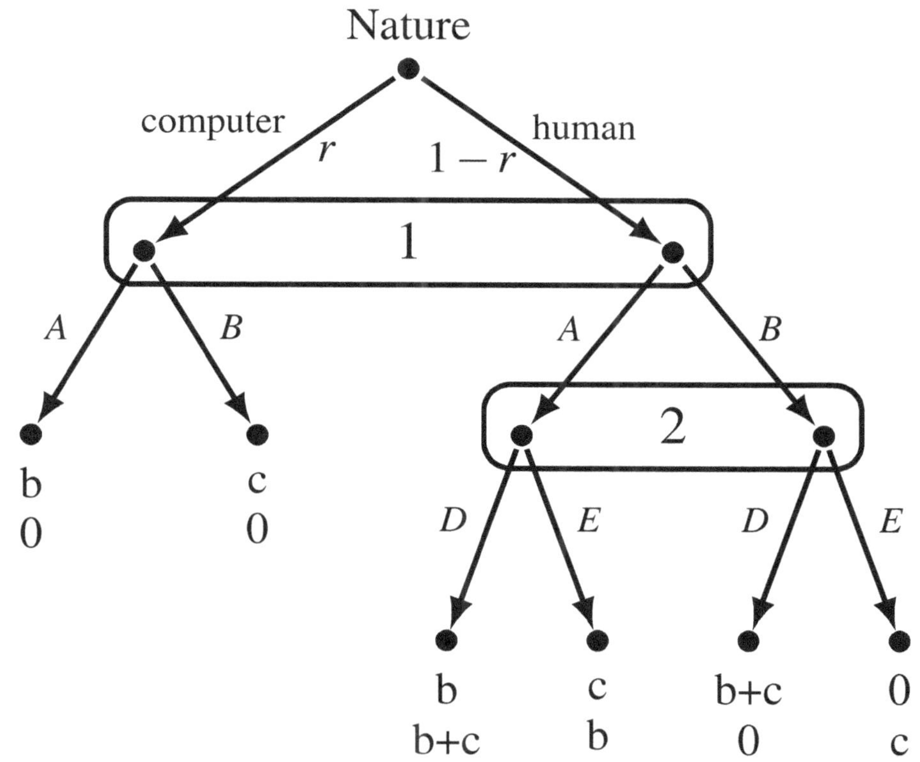


**(b)** The corresponding strategic form is as follows:

| | | Player 2 | | | |
|---|---|---|---|---|---|
| | | *D* | | *E* | |
| Player 1 | *A* | $b$ | $(1-r)(b+c)$ | $rb+(1-r)c$ | $(1-r)b$ |
| | *B* | $c+(1-r)b$ | $0$ | $rc$ | $(1-r)c$ |

**(c)** When $b=4$, $c=6$ and $r=\frac{1}{2}$ the strategic form becomes:

| | | Player 2 | | | |
|---|---|---|---|---|---|
| | | *D* | | *E* | |
| Player 1 | *A* | 4 | 5 | 5 | 2 |
| | *B* | 8 | 0 | 3 | 3 |

This game has no pure-strategy Nash equilibria. To find the mixed-strategy Nash equilibrium, let $p$ be the probability of $A$ and $q$ the probability of $D$; then the following two equations must be satisfied:

$$4q+5(1-q)=8q+3(1-q) \qquad \text{and} \qquad 5p=2p+3(1-p).$$

The solution is $p = \frac{1}{2}$ and $q = \frac{1}{3}$. Thus the Nash equilibrium is

$$\begin{pmatrix} A & B & D & E \\ \frac{1}{2} & \frac{1}{2} & \frac{1}{3} & \frac{2}{3} \end{pmatrix}.$$

**Exercise 7.6** Your name is Sarah. Your best friend calls you and tells you: "I've met the perfect person for you! His name is Alex and I already gave him your phone number. Unfortunately, Alex seems quite shy and I don't know if he will actually call you". You get all excited and start thinking about what to say if Alex calls. Both you and Alex know (in fact it is common knowledge) that of all the blind dates that your friend suggested in the past, $100p\%$ turned out to be people that were a good match. Thus, with probability $p$, Alex will be somebody you like: you will be a good match. In that case, it would be great to go out on a date. However, there is a probability of $(1-p)$ that you will not like Alex (bad match), in which case you would prefer not to go out. If Alex does call then you have to decide between not accepting the invitation, accepting and suggesting to go to the movies or accepting and suggesting to go to dinner. Whatever suggestion you make, he will accept. If Alex is a good match, you prefer to go to dinner and have the chance to know him better, whereas if he is not, then you prefer to go to see a movie. Alex, who is indeed a shy person, would feel terrible if he called you and you rejected the invitation. In that case, he would prefer to act cowardly and not call. If going on a date, he always prefers going to dinner to going to the movies. The von Neumann-Morgenstern utilities are as follows:

| Outcome: | Your utility: | Alex's utility: |
|---|---|---|
| No date (he does not call) | 2 | 3 |
| Rejection (he calls, you say No) | 3 | 0 |
| Good match, movie | 4 | 4 |
| Good match, dinner | 5 | 5 |
| Bad match, movie | 1 | 1 |
| Bad match, dinner | 0 | 2 |

**(a)** Represent this situation as a game in extensive form. [Hint: start with a move by Nature.]
**(b)** Convert the extensive-form game into a strategic-form game.
**(c)** For what values of $p$ is saying 'No' a weakly dominant strategy for you?
**(d)** For what values of $p$ does Alex have a strictly dominated strategy?
**(e)** Are there values of $p$ for which there is a Nash equilibrium where Alex calls and you end up at the movies?
**(f)** Find the pure-strategy Nash equilibria for every value of $p \in (0,1)$. ■

**Solution to Exercise 7.6.**
**(a)** The extensive-form game is as follows:

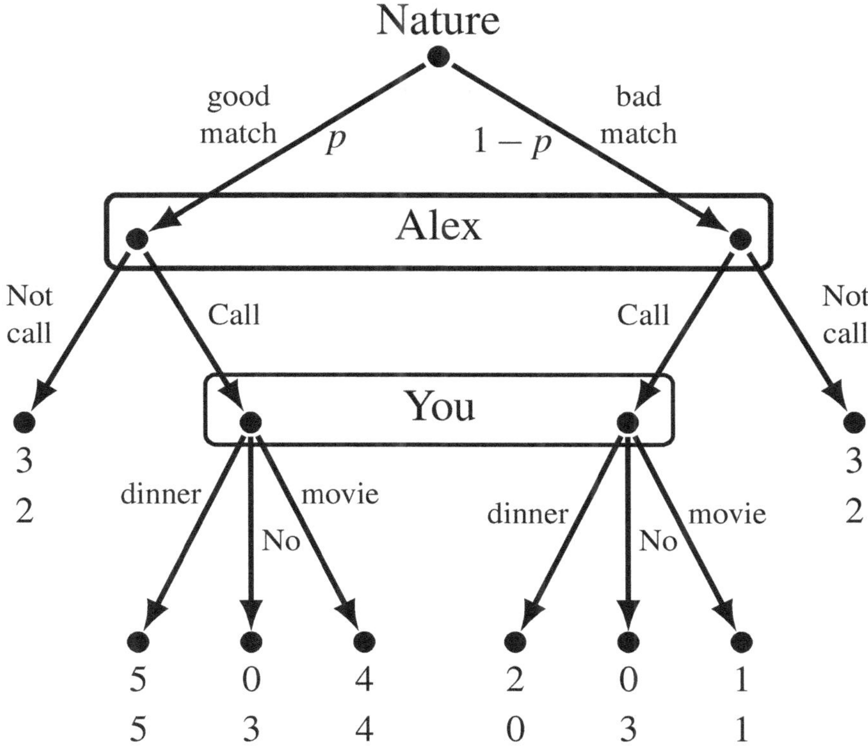


**(b)** The corresponding strategic form is as follows:

| | | Sarah | | | | | |
|---|---|---|---|---|---|---|---|
| | | No | | dinner | | movie | |
| Alex | Not call | 3 | 2 | 3 | 2 | 3 | 2 |
| | Call | 0 | 3 | $2+3p$ | $5p$ | $1+3p$ | $1+3p$ |

**(c)** 'No' is a weakly dominant strategy for you if and only if $3 \geq 5p$ and $3 \geq 1+3p$, that if and only if $p \leq \frac{3}{5}$.

**(d)** 'Call' is a strictly dominated strategy for Alex if and only if $3 > 2+3p$, that is, if and only if $p < \frac{1}{3}$.

**(e)** We will consider all the possibilities.

- For (Call, Movie) to be Nash equilibrium we need $p$ to be such that $1+3p \geq 3$ and $1+3p \geq 5p$. The first inequality requires $p \geq \frac{2}{3}$, while the second requires $p \leq \frac{1}{2}$. These two conditions cannot be simultaneously satisfied. Hence (Call, Movie) is never a Nash equilibrium: if you go out it will for dinner.
- (Not call, No) is always a Nash equilibrium, no matter what the value of $p$.
- If $p \leq \frac{1}{3}$ then (Not call, Dinner) is a Nash equilibrium.
- If $p \leq \frac{2}{3}$ then (Not call, Movie) is a Nash equilibrium
- If $p \geq \frac{3}{5}$ then (Call, Dinner) is a Nash equilibrium.

Thus the pure-strategy Nash equilibria are:

**(1)** If $p \leq \frac{1}{3}$: (Not call, No), (Not call, Dinner) and (Not call, Movie).

**(2)** If $\frac{1}{3} < p < \frac{3}{5}$ (note that $\frac{3}{5} < \frac{2}{3}$): (Not call, No), (Not call, Movie).

**(3)** If $\frac{3}{5} \leq p \leq \frac{2}{3}$: (Not call, No), (Not call, Movie) and (Call, Dinner).

**(4)** If $p > \frac{2}{3}$: (Not call, No) and (Call, Dinner).

Exercise 7.7 Consider the following "truth game." There are two players, 1 and 2, and a game-master. The game-master has a coin that is bent in such a way that, flipped randomly, the coin will come up "heads" 80% of the time. The bias of this coin is known to both players. The game-master flips this coin, and the outcome of the coin flip is shown only to Player 1. Player 1 then makes an announcement to Player 2 about the result of the coin flip; Player 1 is allowed to say either 'heads" or "tails" and nothing else; furthermore, Player 1 can lie if he so chooses. Player 2, having heard what Player 1 said, but not having seen the result of the coin flip, must then guess what the result of the coin flip truly was. That ends the game. The outcomes are as follows. For Player 2 things are quite simple; Player 2 gets \$1 if her guess matches the actual result of the coin flip, and gets \$0 otherwise. For Player 1 things are more complex. He gets \$2 if Player 2's guess is that the coin came up "heads", and \$0 if Player 2's guess is "tails", regardless of how the coin came up. **In addition** to this, Player 1 gets an extra \$1 if what he says to Player 2 is the truth, while he does not get anything extra if his statement to Player 2 is a lie. Each player is selfish (that is, only cares about how much money he/she gets), greedy (prefers more money to less) and risk neutral.

**(a)** Draw an extensive-form representation of this game.

**(b)** Convert the extensive form game into a strategic-form game.

**(c)** Find the pure-strategy Nash equilibria.

**(d)** What is the output of the cardinal version of the iterated deletion of strictly dominated strategies (IDSDS)?

■

**Solution to Exercise 7.7.**

**(a)** Denote an outcome as $(\$x, \$y)$, where $\$x$ is the amount of money given to Player 1 and $\$y$ is the amount of money given to Player 2. Then the stated preferences can be captured by the following utility functions: $U_1(\$x, \$y) = x$ and $U_2(\$x, \$y) = y$. The extensive-form game is as follows:

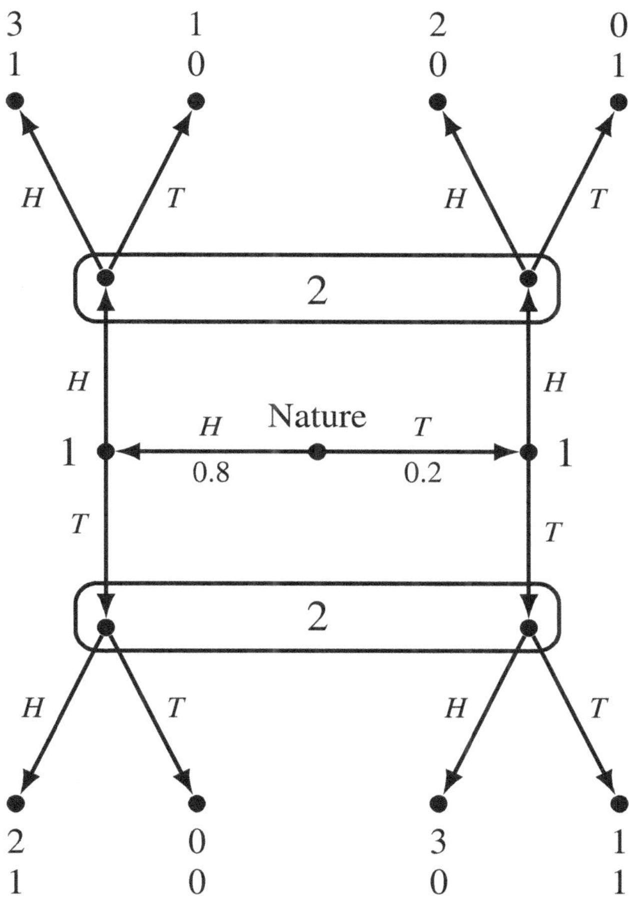


**(b)** The corresponding strategic-form game is as follows. Strategy $xy$ of Player 1 is to be interpreted as "if the result of the coin toss was $H$ then say $x$ and if the result was $T$ then say $y$"; thus, for example, $HT$ is the strategy to always tell the truth, while $TH$ is the strategy to always lie. Similarly, strategy $xy$ of Player 2 is to be interpreted as "if Player 1 says $H$ then guess $x$ and if Player 1 says $T$ then guess $y$"; thus, for example, $HT$ is the strategy to match the guess to Player 1's statement, while $TH$ is the strategy to guess the opposite of what Player 1 says.

Player 2

| | | *HH* | *HT* | *TH* | *TT* |
|---|---|---|---|---|---|
| | *HH* | 2.8 0.8 | 2.8 0.8 | 0.8 0.2 | 0.8 0.2 |
| Player | *HT* | 3 0.8 | 2.6 1 | 1.4 0 | 1 0.2 |
| 1 | *TH* | 2 0.8 | 0.4 0 | 1.6 1 | 0 0.2 |
| | *TT* | 2.2 0.8 | 0.2 0.2 | 2.2 0.8 | 0.2 0.2 |

**(c)** The pure-strategy Nash equilibria are: $(HH,HT)$ with payoffs (2.8, 0.8) and $(TT,TH)$ with payoffs (2.2, 0.8).

**(d)** For Player 1, $TH$ is strictly dominated by the mixed strategy $\begin{pmatrix} HT & TT \\ \frac{1}{2} & \frac{1}{2} \end{pmatrix}$. After deleting Player 1's strategy $TH$, for Player 2 $TT$ becomes strictly dominated by the mixed strategy $\begin{pmatrix} HT & TT \\ \frac{1}{2} & \frac{1}{2} \end{pmatrix}$. The procedure stops here. Thus, the output of the IDSDS if the following set of strategy profiles:

$$\begin{gathered}\{(HH,HH),(HH,HT),(HH,TH),\\ (HT,HH),(HT,HT),(HT,TH),\\ (TT,HH),(TT,HT),(TT,TH)\}\end{gathered}$$

**Exercise 7.8** Players 1 and 2 are given \$9 each. Then, with equal probability, Player 1 is dealt card H (high) or card L (low). Player 2 is not dealt a card, and never gets to look at Player 1's card until the end of the game. After looking at his card, Player 1 decides whether to play or fold. If he folds, he pays \$1 to Player2 and the game ends (so that Player 1 ends up with \$8 and Player 2 with \$10). If Player 1 plays, the Player 2 is informed and must decide whether to fold or see. If she decides to fold, she pays \$1 to Player 1 and the game ends (so that Player 2 ends up with \$10 and Player 2 with \$8). If Player 2 sees, the card given to Player 1 is shown and if the card is H, then Player 1 wins \$4 from Player 2; if it is L, Player 2 wins \$A from Player 1, with $1 \leq A \leq 9$. Assume that both players are selfish and greedy (i.e. each player only cares about how much money he/she ends up with) and is risk neutral.

**(a)** Draw an extensive-form game to represent this situation.

**(b)** Write the corresponding strategic-form game.

**(c)** Find the Nash equilibria of this game as a function of the parameter $A$.

■

**Solution to Exercise 7.8.**

**(a)** Denote an outcome as $(\$x,\$y)$, where $\$x$ is the amount of money Player 1 ends up with and $\$y$ is the amount of money Player 2 ends up with. Then the stated preferences can be captured by the following utility functions: $U_1(\$x,\$y) = x$ and $U_2(\$x,\$y) = y$. The extensive-form game is as follows:

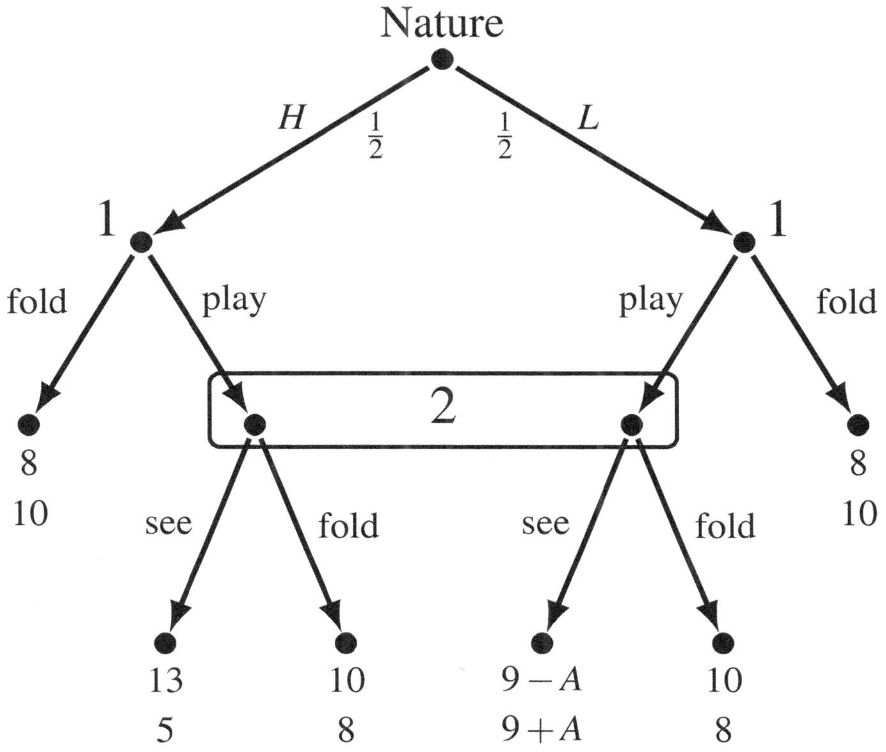


**(b)** The corresponding strategic form is as follows:

| | | Player 2: see | | Player 2: fold | |
|---|---|---|---|---|---|
| Player 1 | always play | $11-\frac{A}{2}$ | $7+\frac{A}{2}$ | 10 | 8 |
| | always fold | 8 | 10 | 8 | 10 |
| | play if $H$ fold if $L$ | 10.5 | 7.5 | 9 | 9 |
| | play if $L$ fold if $H$ | $8.5-\frac{A}{2}$ | $9.5+\frac{A}{2}$ | 9 | 9 |

First note that, for Player 1,

- "always fold" is strictly dominated by "play if H, fold if L"
- "play if L, fold if H" is strictly dominated by "always play".

After deleting the two strictly dominated strategies of Player 1, the game reduces to:

| | | Player 2 | | | |
|---|---|---|---|---|---|
| | | see | | fold | |
| Player 1 | always play | $11-\frac{A}{2}$ | $7+\frac{A}{2}$ | 10 | 8 |
| | play if $H$ fold if $L$ | 10.5 | 7.5 | 9 | 9 |

Thus,

- If $A < 2$ then, for Player 2 "fold" strictly dominates "play" and thus the unique Nash equilibrium of the game is the pure-strategy profile (always play, fold).
- If $A = 2$ then, for every $q \leq \frac{2}{3}$ the following is a Nash equilibrium:

$$\left(\begin{array}{c|cc} \text{always play} & \text{see} & \text{fold} \\ 1 & q & 1-q \end{array}\right).$$

  The constraint $q \leq \frac{2}{3}$ reflects the requirement that, for Player 1, "always play" should yield at least as high a payoff as "play if $H$, fold if $L$" against the mixed strategy of Player 2 that assigns probability $q$ to "see": $10q + 10(1-q) \geq 10.5q + 9(1-q)$.

- If $A > 2$ then there is no pure-strategy Nash equilibrium. The mixed-strategy Nash equilibrium is given by

$$\left(\begin{array}{cc|cc} \text{always play} & \text{play if H, fold if L} & \text{see} & \text{fold} \\ \frac{3}{A+1} & \frac{A-2}{A+1} & \frac{2}{A+1} & \frac{A-1}{A+1} \end{array}\right).$$

  The probabilities are obtained by solving the following equations (where $p$ is the probability of "always see" and $q$ the probability of "see"):

$$\left(11 - \tfrac{A}{2}\right) q + 10(1-q) = 10.5q + 9(1-q)$$

$$\left(7 + \tfrac{A}{2}\right) p + 7.5(1-p) = 8p + 9(1-p).$$

**Exercise 7.9** Players 1 and 2 are adversaries in a paintball shooting event. Player 2, who is unarmed, is approaching Player 1's hideout. Player 1 has a paintball rifle with only one bullet in it. He can either shoot when Player 2 is still far away or wait; if he shoots when Player 2 is far away, he only has a probability $p$ of hitting Player 2, with $0 < p < 1$. If Player 2 is hit, then Player 1 wins. Player 2 is deaf, so he doesn't know whether Player 1 has fired or not, unless of course he is hit. Player 2 – if not hit (whether because Player 1 did not fire or because Player 1 fired and missed) – has a choice between running away, in which case Player 1 wins the game, or proceeding. In the latter case, if Player 1 did not fire before, his rifle will now fire automatically (detecting Player 2's presence) and Player 2 will be hit for sure and the winner will be Player 1; if, on the other hand, Player 1 fired previously (and missed), then Player 2 can claim victory. The von Neumann-Morgenstern payoffs are as follows:

- For Player 1: utility of winning without firing a shot is 9, utility of winning by firing a shot is 8 (bullets are very expensive!), utility of losing is 0.
- For Player 2: utility of winning is 9, utility of running away is 6 and utility of losing is 0.

**(a)** Draw an extensive-form game that represents this interaction.

**(b)** Write the corresponding strategic-form game.

**(c)** Find the Nash equilibrium for every $p \in (0,1)$.

**Solution to Exercise 7.9.**

**(a)** The extensive-form game is as follows:

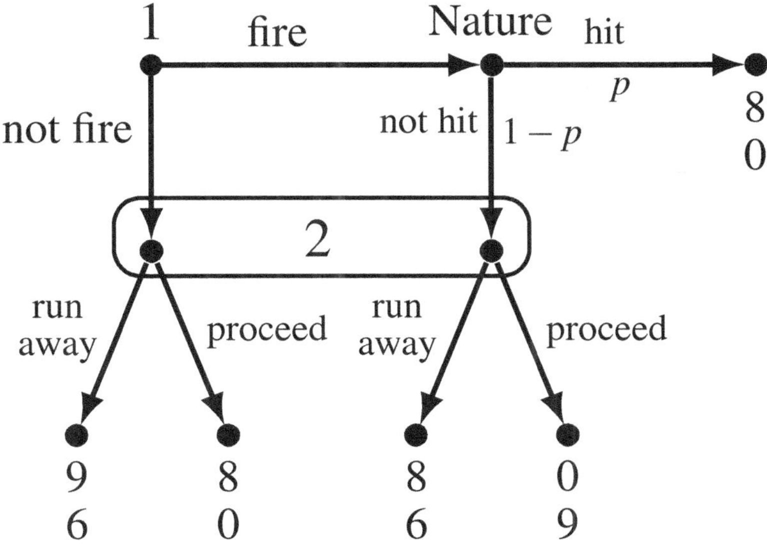


**(b)** The corresponding strategic-form game is as follows:

| | | Player 2 | |
|---|---|---|---|
| | | proceed | run away |
| Player 1 | fire | $8p$ $\quad 9(1-p)$ | $8$ $\quad 6(1-p)$ |
| | not fire | $8$ $\quad 0$ | $9$ $\quad 6$ |

**(c)** Since $p < 1$, for Player 1 "not fire" strictly dominates "fire" and thus, for every $p \in (0,1)$, there is a unique Nash equilibrium, namely (not fire, run away).

Exercise 7.10 Consider the following game. Player 1 moves first and chooses between $A$, $B$ and $C$. If Player 1 chooses $A$ the game ends and the outcome is the lottery

$$\begin{pmatrix} z_1 & z_2 & z_4 \\ \frac{2}{3} & \frac{1}{5} & \frac{2}{15} \end{pmatrix}.$$

If Player 1 chooses $B$ or $C$ then it is Player 2's turn to move. Player 2 is only told that Player 1 chose either $B$ or $C$ (but not which of $B$ or $C$) and she has to choose between $D$ and $E$ and then the game ends. The outcome that follows from $B$ and $D$ is $z_2$, the outcome that follows from $B$ and $E$ is $z_3$, the outcome that follows from $C$ and $D$ is $z_4$ and the outcome that follows from $C$ and $E$ is $z_5$. The two players have preferences over lotteries over the set $Z = \{z_1, z_2, z_3, z_4, z_5\}$ that satisfy the axioms of Expected Utility Theory.

Player 1's ranking of the basic outcomes is $z_5 \succ z_2 \succ z_3 \succ z_1 \succ z_4$.

Player 2's ranking of the basic outcomes is $z_3 \succ z_4 \succ z_1 \sim z_5 \succ z_2$.

Let $L_1$, $L_2$, $L_3$, $L_4$ and $L_5$ be the following lotteries:

$$L_1 = \begin{pmatrix} z_4 & z_5 \\ \frac{1}{6} & \frac{5}{6} \end{pmatrix} \quad L_2 = \begin{pmatrix} z_4 & z_5 \\ \frac{1}{3} & \frac{2}{3} \end{pmatrix} \quad L_3 = \begin{pmatrix} z_4 & z_5 \\ \frac{1}{2} & \frac{1}{2} \end{pmatrix}$$

$$L_4 = \begin{pmatrix} z_2 & z_3 \\ \frac{1}{3} & \frac{2}{3} \end{pmatrix} \quad L_5 = \begin{pmatrix} z_2 & z_3 \\ \frac{2}{3} & \frac{1}{3} \end{pmatrix}$$

Player 1 is indifferent between $L_1$ and $z_2$ for sure, is indifferent between $L_2$ and $z_3$ for sure and is indifferent between $L_3$ and $z_1$ for sure, while Player 2 is indifferent between $L_4$ and $z_4$ for sure and is indifferent between $L_5$ and $z_1$ for sure.

**(a)** Construct a von Neumann-Morgenstern utility function for each player with 0 as the lowest value and 6 as the largest value.

**(b)** Draw an extensive-form game that represents the interaction described above.

**(c)** Write the corresponding strategic-form game.

**(d)** Find Nash equilibrium.

■

**Solution to Exercise 7.10.**

**(a)** The von Neumann-Morgenstern utility functions are as follows:

- For Player 1: $\begin{matrix} z_1 & z_2 & z_3 & z_4 & z_5 \\ 3 & 5 & 4 & 0 & 6 \end{matrix}$.
- For Player 2: $\begin{matrix} z_1 & z_2 & z_3 & z_4 & z_5 \\ 2 & 0 & 6 & 4 & 2 \end{matrix}$.

For Player 1 we start by setting $U_1(z_5) = 6$ and $U_1(z_4) = 0$. Then from the first indifference we get that $U_1(z_2) = \frac{1}{6} \times 0 + \frac{5}{6} \times 6 = 5$, from the second indifference we get that $U_1(z_3) = \frac{2}{3} \times 6 = 4$ and from the third indifference we get that $U_1(z_1) = \frac{1}{2} \times 6 = 3$. The reasoning for Player 2 is similar.

**(b)** The extensive-form game-frame is as follows:

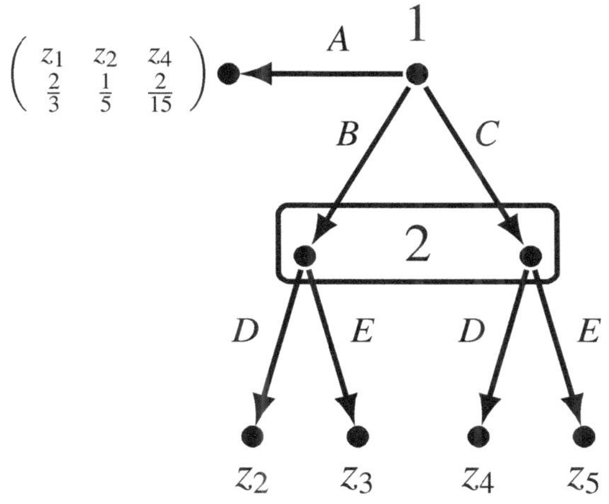


The expected utility of lottery $\begin{pmatrix} z_1 & z_2 & z_4 \\ \frac{2}{3} & \frac{1}{5} & \frac{2}{15} \end{pmatrix}$ is 3 for Player 1 and $\frac{28}{15}$ for Player 2.

Thus the extensive-form game is as follows:

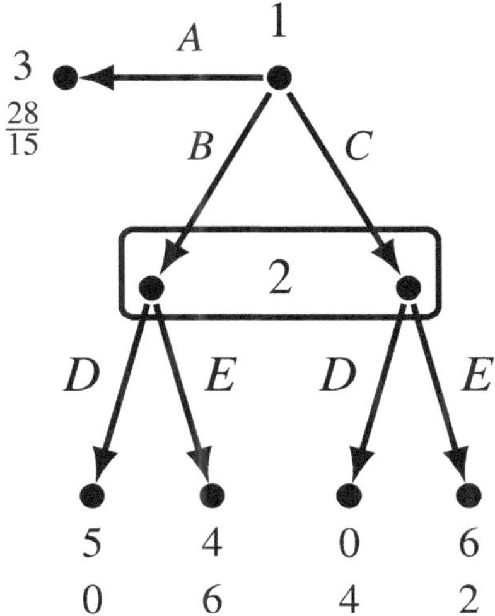

**(c)** The corresponding strategic form is as follows:

| | | Player 2 | |
|---|---|---|---|
| | | $D$ | $E$ |
| Player 1 | $A$ | $3 \quad \frac{28}{15}$ | $3 \quad \frac{28}{15}$ |
| | $B$ | $5 \quad 0$ | $4 \quad 6$ |
| | $C$ | $0 \quad 4$ | $6 \quad 2$ |

This game has no pure-strategy Nash equilibria. To find the mixed-strategy Nash equilibrium first note that $A$ is strictly dominated by $B$ for Player 1; thus $A$ must be played with zero probability. Let $p$ be the probability with which Player 1 plays $B$ and $q$ the probability with which Player 2 plays $D$. Then $p$ and $q$ must satisfy the following equations: $5q+4(1-q)=6(1-q)$ and $4(1-p)=6p+2(1-p)$, whose solutions are $q=\frac{2}{7}$ and $p=\frac{1}{4}$. Thus the Nash equilibrium is given by $\begin{pmatrix} A & B & C & | & D & E \\ 0 & \frac{1}{4} & \frac{3}{4} & | & \frac{2}{7} & \frac{5}{7} \end{pmatrix}$.

## 7.2 Subgame-perfect equilibrium revisited

The exercises in this section deal with the notion of subgame-perfect equilibrium in extensive-form games with cardinal payoffs (Volume 1, Chapter 7, Section 2).

Exercise 7.11 Consider a game in which Player 1 first chooses between *In* and *Out*. If Player 1 selects *Out* then the game ends with payoff $x$ for Player 1, with $0<x<3$, and payoff 1 for Player 2. If Player 1 selects *In* then this is is revealed to Player 2 and then the two players play a game where they simultaneously and independently choose between $A$ and $B$. If they both choose $A$ then Player 1 gets a payoff of 3 and Player 2 a payoff of 1. If they both choose $B$ then Player 1 gets a payoff of 1 and Player 2 a payoff of 3. If they make different choices then they both get a payoff of 0. All of the above payoffs are von Neumann-Morgenstern payoffs.

**(a)** Draw and extensive-form game that represents the interaction described above.

**(b)** Write the corresponding strategic form.

**(c)** Find the pure-strategy Nash equilibria and note how they depend on $x$.

**(d)** Find the pure-strategy subgame-perfect equilibria and note how they depend on $x$.

**(e)** For the case where $1<x<3$, find a pure-strategy Nash equilibrium which is not subgame perfect.

**(f)** What are the mixed-strategy subgame-perfect equilibria of this game when $x=\frac{1}{2}$? You can interpret 'strategy' as 'behavioral strategy'.

■

**Solution to Exercise 7.11.**

**(a)** The extensive-form game is as follows (alternatively, one could switch the order in which the players choose in the simultaneous game after *In*):

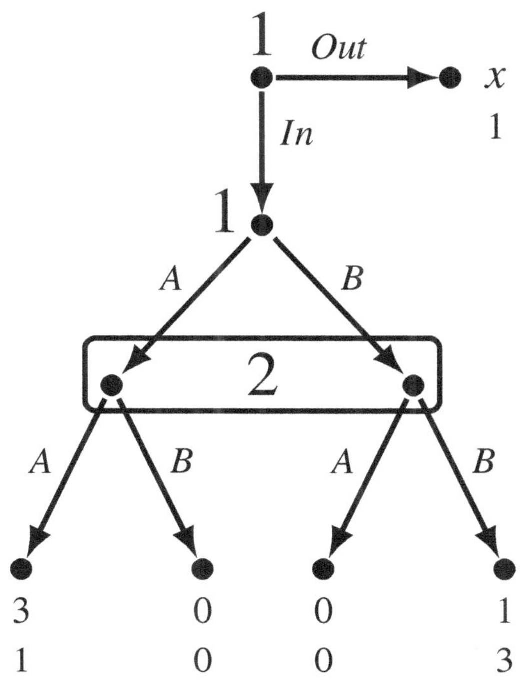


**(b)** The corresponding strategic-form game is as follows:

| Player 1 \ Player 2 | A | B |
|---|---|---|
| *Out, A* | $x$   1 | $x$   1 |
| *Out, B* | $x$   1 | $x$   1 |
| *In, A* | 3   1 | 0   0 |
| *In, B* | 0   0 | 1   3 |

**(c)** The pure-strategy Nash equilibria are:

$$\begin{cases} (InA,A) \text{ and } (InB,B) & \text{if } x<1 \\ (InA,A),\ (InB,B),\ (OutA,B) \text{ and } (OutB,B) & \text{if } x=1 \\ (InA,A),\ (OutA,B) \text{ and } (OutB,B) & \text{if } 1<x<3 \end{cases}$$

**(d)** To find the subgame-perfect equilibria first we solve the subgame, whose strategic form is as follows:

| | | Player 2 | | | |
|---|---|---|---|---|---|
| | | $A$ | | $B$ | |
| Player 1 | $A$ | 3 | 1 | 0 | 0 |
| | $B$ | 0 | 0 | 1 | 3 |

The pure-strategy Nash equilibria of this game are $(A,A)$ and $(B,B)$. Thus the pure-strategy subgame-perfect equilibria of the original game are:

$$\begin{cases} (InA,A) \text{ and } (InB,B) & \text{if } x < 1 \\ (InA,A),\ (InB,B), \text{ and } (OutB,B) & \text{if } x = 1 \\ (InA,A) \text{ and } (OutB,B) & \text{if } 1 < x < 3 \end{cases}$$

**(e)** When $1 < x < 3$, a Nash equilibrium which is not subgame-perfect is $(OutA,B)$ because the restriction of this strategy profile to the subgame gives $(A,B)$, which is not a Nash equilibrium of the subgame.

**(f)** Let $x = \frac{1}{2}$. First we need to find the mixed-strategy equilibrium of the subgame. Let $p$ be the probability with which Player 1 selects $A$ and $q$ the probability with which Player 2 selects $A$. Then it must be that $p = 3(1-p)$ and $3q = 1-q$, that is, $p = \frac{3}{4}$ and $q = \frac{1}{4}$. Thus the mixed-strategy Nash equilibrium of the subgame is:

$$\left(\begin{array}{cc|cc} A & B & A & B \\ \frac{3}{4} & \frac{1}{4} & \frac{1}{4} & \frac{3}{4} \end{array}\right) \quad \text{with a payoffs of } \frac{3}{4} \text{ for both Player 1 and Player 2.}$$

Thus, when $x = \frac{1}{2}$, the subgame-perfect equilibria of the original game are:

$$(InA,A), \quad (InB,B) \quad \text{and} \quad \left(\begin{array}{cc|cc|cc} Out & In & A & B & A & B \\ 0 & 1 & \frac{3}{4} & \frac{1}{4} & \frac{1}{4} & \frac{3}{4} \end{array}\right).$$

Exercise 7.12 Consider the extensive-form game shown in Figure 7.2, where the payoffs are von Neumann-Morgenstern payoffs. Note that Player 3 makes the first move but the payoffs are are given in the following order: the top number is Player 1's payoff, the middle number is Player 2's payoff and the bottom number is Player 3's payoff.

**(a)** Write all the strategies of Player 3.

**(b)** Find three subgame-perfect equilibria and specify the payoffs at those equilibria.

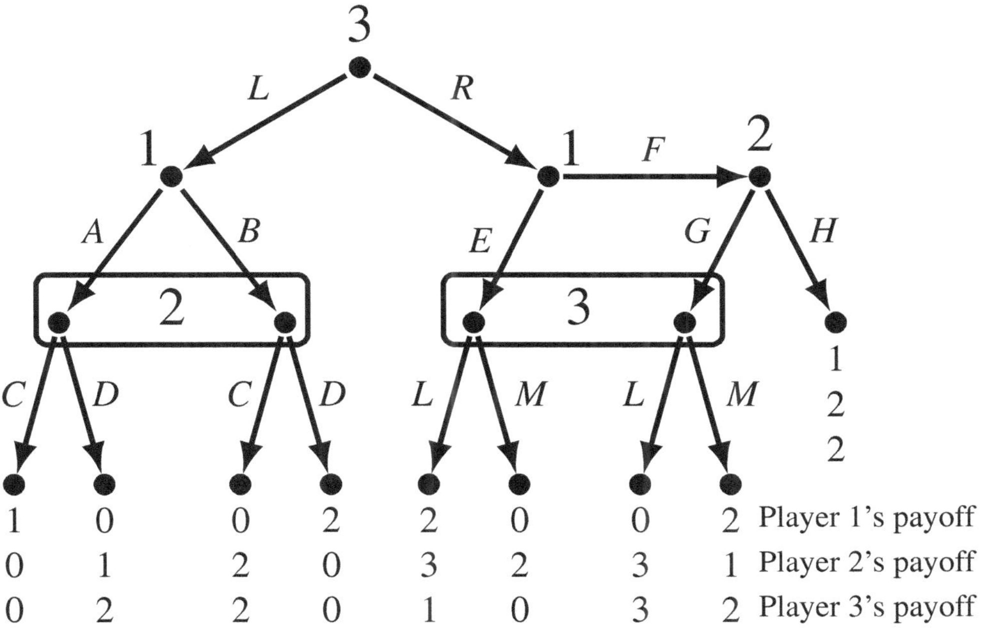


Figure 7.2: The game for Exercise 7.12.

**Solution to Exercise 7.12.**

**(a)** Player 3 has four strategies: $(L,L)$, $(L,M)$, $(R,L)$ and $(R,M)$, where the first element of the pair denotes Player 3's choice at the root of the tree and the second element her choice at her information set on the right.

**(b)** First we solve the proper subgame on the left, whose strategic form is as follows:

Player 2

| | | $C$ | $D$ |
|---|---|---|---|
| Player 1 | $A$ | 1 0 | 0 1 |
| | $B$ | 0 2 | 2 0 |

This game has no pure-strategy Nash equilibria. To find the mixed-strategy equilibrium, let $p$ be the probability of $A$ and $q$ the probability of $C$. Then it must be that $q = 2(1-q)$ and $2(1-p) = p$, that is, $p = q = \frac{2}{3}$. The expected payoff at the equilibrium is $\frac{2}{3}$ for both Player 1 and Player 2 and it is $\frac{4}{9} \times 0 + \frac{2}{9} \times 2 + \frac{2}{9} \times 2 + \frac{1}{9} \times 0 = \frac{8}{9}$ for Player 3.

Next we solve the proper subgame on the right. Its strategic form is:

Player 3: $L$

| | | Player 2: $G$ | | | Player 2: $H$ | | |
|---|---|---|---|---|---|---|---|
| Player 1 | $E$ | 2 | 3 | 1 | 2 | 3 | 1 |
| | $F$ | 0 | 3 | 3 | 1 | 2 | 2 |

Player 3: $M$

| | | Player 2: $G$ | | | Player 2: $H$ | | |
|---|---|---|---|---|---|---|---|
| Player 1 | $E$ | 0 | 2 | 0 | 0 | 2 | 0 |
| | $F$ | 2 | 1 | 2 | 1 | 2 | 2 |

This game has three pure-strategy Nash equilibria, namely $(E,G,L)$, $(E,H,L)$ and $(F,H,M)$.

Thus, for the original game, the following are subgame-perfect equilibria:

- $\left(\begin{array}{cc|cc|cc|cc|cc|cc} A & B & C & D & E & F & G & H & L & R & L & M \\ \frac{2}{3} & \frac{1}{3} & \frac{2}{3} & \frac{1}{3} & 1 & 0 & 1 & 0 & 0 & 1 & 1 & 0 \end{array}\right)$ with payoffs $(2,3,1)$.

- $\left(\begin{array}{cc|cc|cc|cc|cc|cc} A & B & C & D & E & F & G & H & L & R & L & M \\ \frac{2}{3} & \frac{1}{3} & \frac{2}{3} & \frac{1}{3} & 1 & 0 & 0 & 1 & 0 & 1 & 1 & 0 \end{array}\right)$ with payoffs $(2,3,1)$.

- $\left(\begin{array}{cc|cc|cc|cc|cc|cc} A & B & C & D & E & F & G & H & L & R & L & M \\ \frac{2}{3} & \frac{1}{3} & \frac{2}{3} & \frac{1}{3} & 0 & 1 & 0 & 1 & 0 & 1 & 0 & 1 \end{array}\right)$ with payoffs $(1,2,2)$.

**Exercise 7.13** Consider the extensive-form game with cardinal payoffs shown in Figure 7.3, where $0 < p < 1$. Find the subgame-perfect equilibrium for every value of $p \in (0,1)$.

■

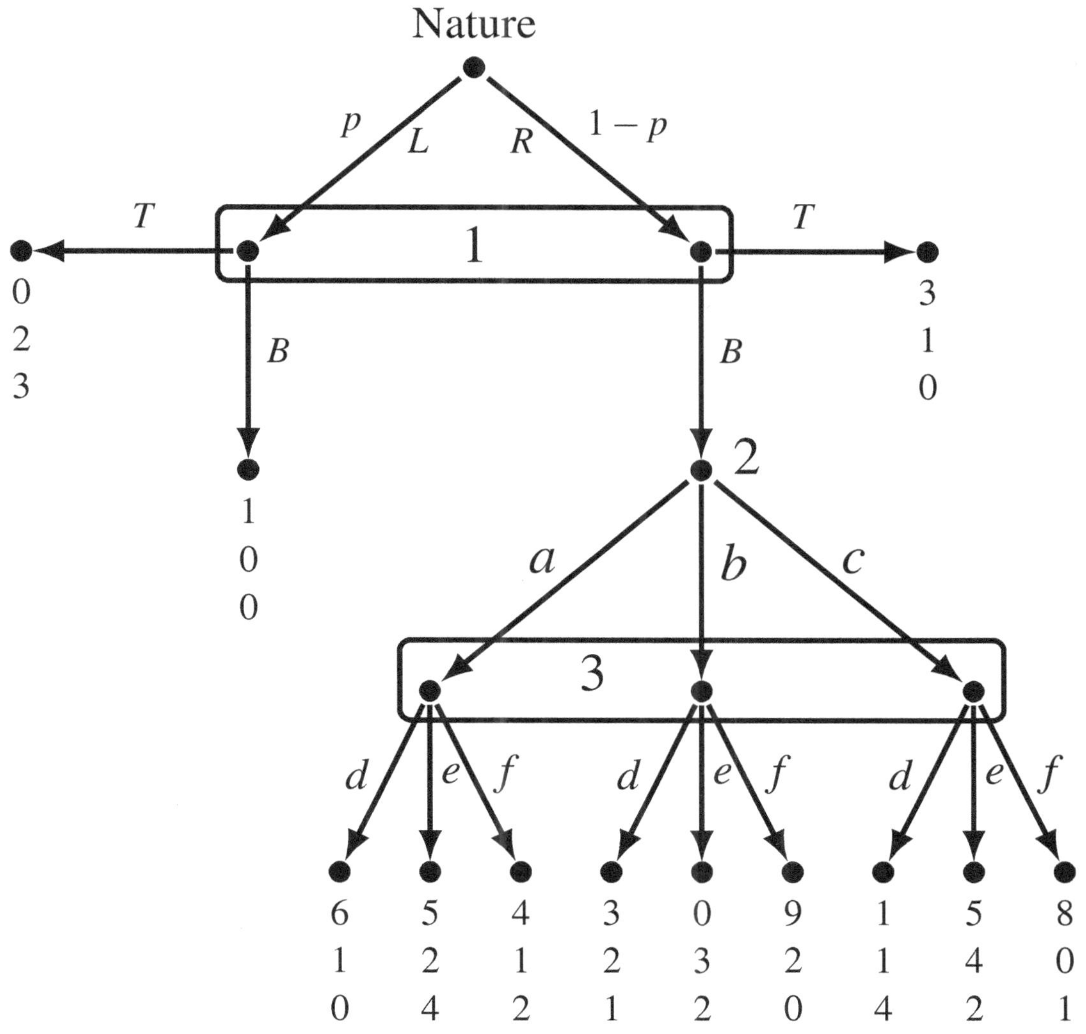


Figure 7.3: The game for Exercise 7.13.

**Solution to Exercise 7.13.** First we solve the proper subgame that starts at the singleton information set of Player 2. Its strategic form is as follows:

Player 3

| Player 2 | *d* | | *e* | | *f* | |
|---|---|---|---|---|---|---|
| *a* | 1 | 0 | 2 | 4 | 1 | 2 |
| *b* | 2 | 1 | 3 | 2 | 2 | 0 |
| *c* | 1 | 4 | 4 | 2 | 0 | 1 |

This game does not have any pure-strategy Nash equilibria. At a mixed-strategy Nash equilibrium Player 1 must play *a* with zero probability (because *a* is strictly dominated by *b*) and Player 2 must play *f* with zero probability (because *f* is strictly dominated by *e*). Let $q$ be the probability of $b$ and $r$ the probability of $d$. The $q$ and $r$ must satisfy the following equations:

$$2r+3(1-r)=r+4(1-r) \quad \text{and} \quad q+4(1-q)=2.$$

The solution is $q=\frac{3}{3}$ and $r=\frac{1}{2}$. Thus the Nash equilibrium of the subgame is

$$\left(\begin{array}{ccc|ccc} a & b & c & d & e & f \\ 0 & \frac{2}{3} & \frac{1}{3} & \frac{1}{2} & \frac{1}{2} & 0 \end{array}\right)$$

The corresponding payoffs are as follows:

- For Player 1: $\frac{2}{6}\times 3+\frac{2}{6}\times 0+\frac{1}{6}\times 1+\frac{1}{6}\times 5=2.$
- For Player 2: $\frac{2}{6}\times 2+\frac{2}{6}\times 3+\frac{1}{6}\times 1+\frac{1}{6}\times 4=2.5.$
- For Player 3: $\frac{2}{6}\times 1+\frac{2}{6}\times 2+\frac{1}{6}\times 4+\frac{1}{6}\times 2=2.$

Thus the game can be simplified as follows:

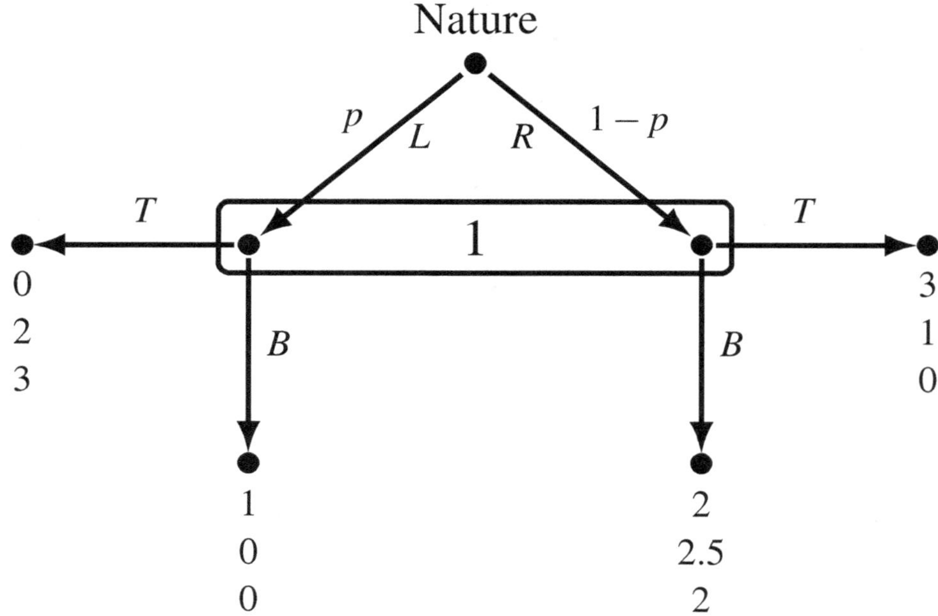


In this simplified game, for Player 1 $T$ gives an expected payoff of $3(1-p)=3-3p$ while $B$ gives an expected payoff of $p+2(1-p)=2-p$. Hence Player 1 will play $T$ if $3-3p>2-p$, that is, if $p<\frac{1}{2}$, will play $B$ if $p>\frac{1}{2}$ and will be indifferent between playing $B$ and playing $T$ if $p=\frac{1}{2}$. Thus the subgame-perfect equilibria of the game of Figure 7.3 are:

**(1)** If $p<\frac{1}{2}$: $\left(\begin{array}{cc|ccc|ccc} B & T & a & b & c & d & e & f \\ 0 & 1 & 0 & \frac{2}{3} & \frac{1}{3} & \frac{1}{2} & \frac{1}{2} & 0 \end{array}\right)$

**(2)** If $p > \frac{1}{2}$: $\left(\begin{array}{cc|ccc|ccc} B & T & a & b & c & d & e & f \\ 1 & 0 & 0 & \frac{2}{3} & \frac{1}{3} & \frac{1}{2} & \frac{1}{2} & 0 \end{array}\right)$

**(3)** If $p = \frac{1}{2}$, for every $q \in (0,1)$,

$$\left(\begin{array}{cc|ccc|ccc} B & T & a & b & c & d & e & f \\ q & 1-q & 0 & \frac{2}{3} & \frac{1}{3} & \frac{1}{2} & \frac{1}{2} & 0 \end{array}\right).$$

Exercise 7.14 Consider first the extensive-form game shown in Figure 7.4.

**(a)** Find the subgame-perfect equilibrium.

**(b)** Write the corresponding strategic-form game.

**(c)** Find all the pure-strategy Nash equilibria.

**(d)** Are there any pure-strategy Nash equilibria where the strategy of at least one player is weakly dominated by another pure strategy?

Now modify the game as follows, turning it into a game with imperfect information. If Player 1 starts with move $i$, then Nature moves and, with probability $1 - \varepsilon$ (with $0 < \varepsilon < 1$), the game continues as above, while, with probability $\varepsilon$, Player 2 is replaced by a robot that is programmed to always choose $f$. Player 1 does not know if he is facing Player 1 or the robot (that is, does not know what Nature chose). The corresponding payoffs to Players 1 and 2 are still as given in Figure 7.4 (the robot is not treated as a player).

**(e)** Draw the imperfect-information game described above.

**(f)** Write the corresponding strategic-form game.

**(g)** Find all the pure-strategy Nash equilibria.

**(h)** As $\varepsilon \to 0$ the imperfect-information game of Part (e) tends to the perfect-information game given in Figure 7.4. Does the set of pure-strategy Nash equilibria of the imperfect-information game tend to the set of Nash equilibria of the perfect information game? In particular, as $\varepsilon$ goes to zero, is the backward induction solution of the perfect information game of Part (a) a Nash equilibrium of the imperfect-information game of Part (e)?

■

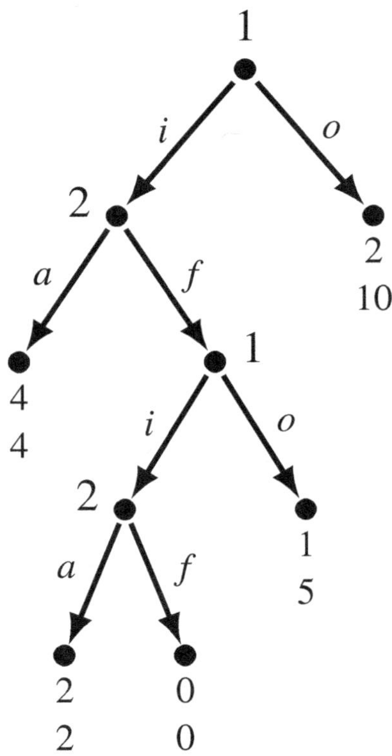


Figure 7.4: The game for Exercise 7.14.

**Solution to Exercise 7.14.**

**(a)** The subgame-perfect equilibrium (i.e. backward-induction solution) is $(ii, aa)$.

**(b)** The strategic-form game is as follows:

| | | Player 2 | | | |
|---|---|---|---|---|---|
| | | $aa$ | $af$ | $fa$ | $ff$ |
| | $ii$ | 4 4 | 4 4 | 2 2 | 0 0 |
| Player | $io$ | 4 4 | 4 4 | 1 5 | 1 5 |
| 1 | $oi$ | 2 10 | 2 10 | 2 10 | 2 10 |
| | $oo$ | 2 10 | 2 10 | 2 10 | 2 10 |

**(c)** There are six pure-strategy Nash equilibria:

$$(ii, aa), (ii, af), (oi, fa), (oi, ff), (oo, fa), (oo, ff).$$

**(d)** Since, for Player 2, $ff$ is weakly dominated by $fa$, the Nash equilibria $(oi, ff)$ and $(oo, ff)$ involve a weakly dominated strategy for Player 2.

**(e)** The new extensive-form game is as follows:

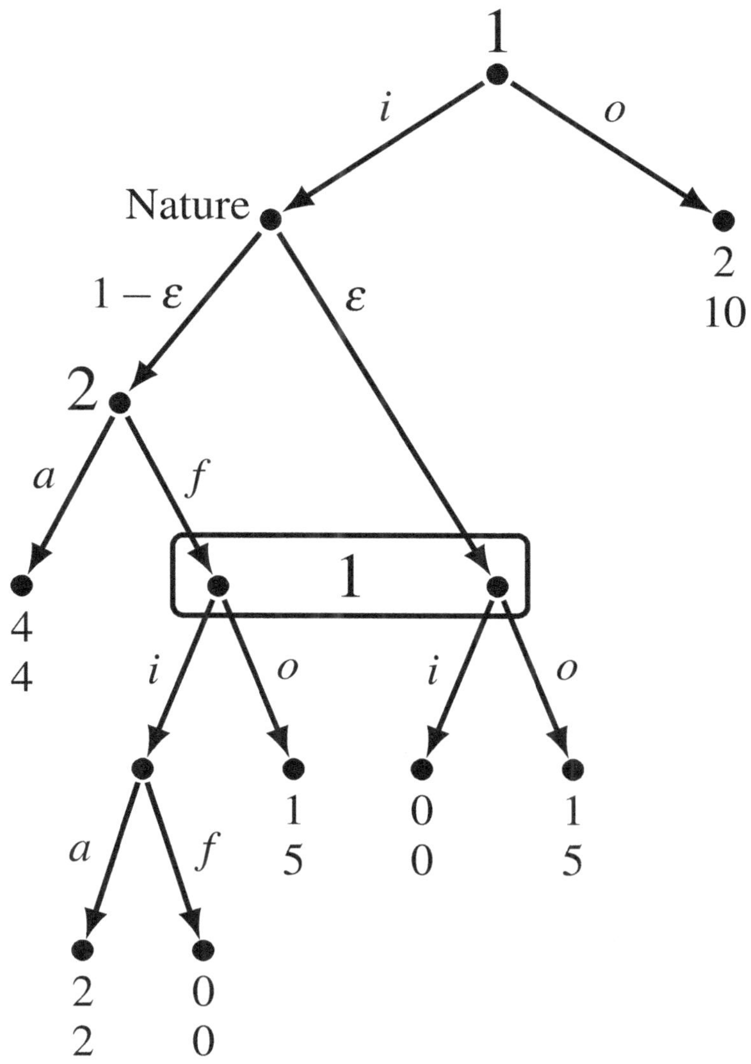


**(f)** The corresponding strategic-form is as follows:

| | | Player 2 | | | |
|---|---|---|---|---|---|
| | | $aa$ | $af$ | $fa$ | $ff$ |
| | $ii$ | $4(1-\varepsilon)$  $4(1-\varepsilon)$ | $4(1-\varepsilon)$  $4(1-\varepsilon)$ | $2(1-\varepsilon)$  $2(1-\varepsilon)$ | 0  0 |
| Player | $io$ | $4-3\varepsilon$  $4+\varepsilon$ | $4-3\varepsilon$  $4+\varepsilon$ | 1  5 | 1  5 |
| 1 | $oi$ | 2  10 | 2  10 | 2  10 | 2  10 |
| | $oo$ | 2  10 | 2  10 | 2  10 | 2  10 |

**(g)** The following are Nash equilibria, for every $\varepsilon \in (0,1)$:

$$(oi, fa), \quad (oi, ff), \quad (oo, fa), \quad (oo, ff).$$

The following are Nash equilibria if and only if $\varepsilon \geq \frac{2}{3}$:

$$(oi, aa), \quad (oi, af), \quad (oo, aa), \quad (oo, af).$$

There are no other pure-strategy Nash equilibria.

**(h)** As $\varepsilon \to 0$, only the first set of equilibria survives and this set does **not** include $(ii, aa)$, which is the backward-induction solution of the perfect-information game of Part (a).

Exercise 7.15 Consider the two-player game shown in Figure7.5, where the payoffs are von Neumann-Morgenstern payoffs (as usual, the first number is Player 1's payoff and the second number is Player 2's payoff).

**(a)** Find the Nash equilibrium of this game.

**(b)** Assume now that the game is played sequentially, with Player 1 moving first and Player 2 moving second, after learning Player 1's choice. Draw the corresponding extensive-form game and find the backward-induction solution. Does Player 1 gain from being the first mover, relative to the simultaneous game of Part (a)?

**(c)** Assume now that the game is still played sequentially. Player 1 moves first and communicates his choice to an artificial intelligence agent (AIA), which then sends a message to Player 2 saying either "Player 1 chose $S$" or "Player 1 chose $C$"; the AIA is programmed to send a truthful message to Player 2 with probability $(1-\varepsilon)$ and a deceitful message with probability $\varepsilon$, where $0 < \varepsilon < 1$; this fact is common knowledge between Players 1 and 2. After receiving AIA's message, Player 2 makes her choice between $S$ and $C$. The payoffs are still as shown in Figure7.5 (the AIA is not a player, so it is still a two-player game).

1. Draw the extensive-form game that represents this new situation. [Hint: use a move by Nature to represent AIA's choice of messages.]
2. Write the corresponding strategic-form game.
3. Find the pure-strategy Nash equilibrium.
4. Consider the case where $\varepsilon \to 0$. Is there anything that you find surprising in this game?

■

| | | Player 2 | |
|---|---|---|---|
| | | $S$ | $C$ |
| Player 1 | $S$ | 5  2 | 3  1 |
| | $C$ | 6  3 | 4  4 |

Figure 7.5: The game for Exercise 7.15.

**Solution to Exercise 7.15.**

**(a)** Since $C$ is a strictly dominant strategy for Player 1, $(C,C)$ is the unique Nash equilibrium.

**(b)** The perfect-information game is as follows:

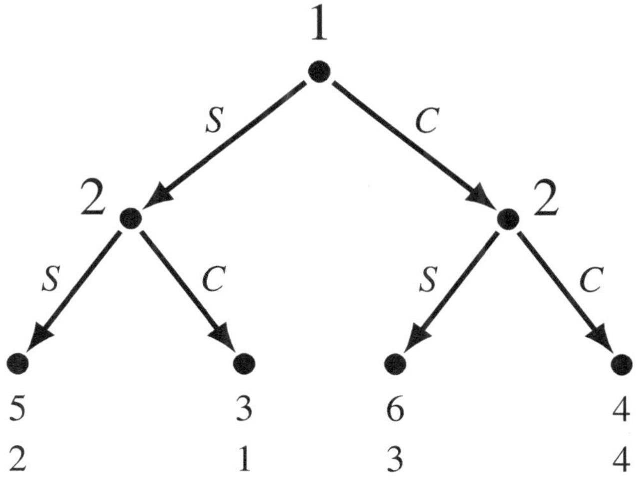


**(c)** The backward induction solution is: Player 1 plays $S$, Player 2's strategy is "play $S$ if Player 1 plays $S$ and play $C$ if Player 1 plays $C$". The payoffs are $(5,2)$ whereas in the simultaneous game of Part (a) they are $(4,4)$. Thus Player 1 indeed gains from being the first mover.

**(d)** **1.** The imperfect-information game is as follows:

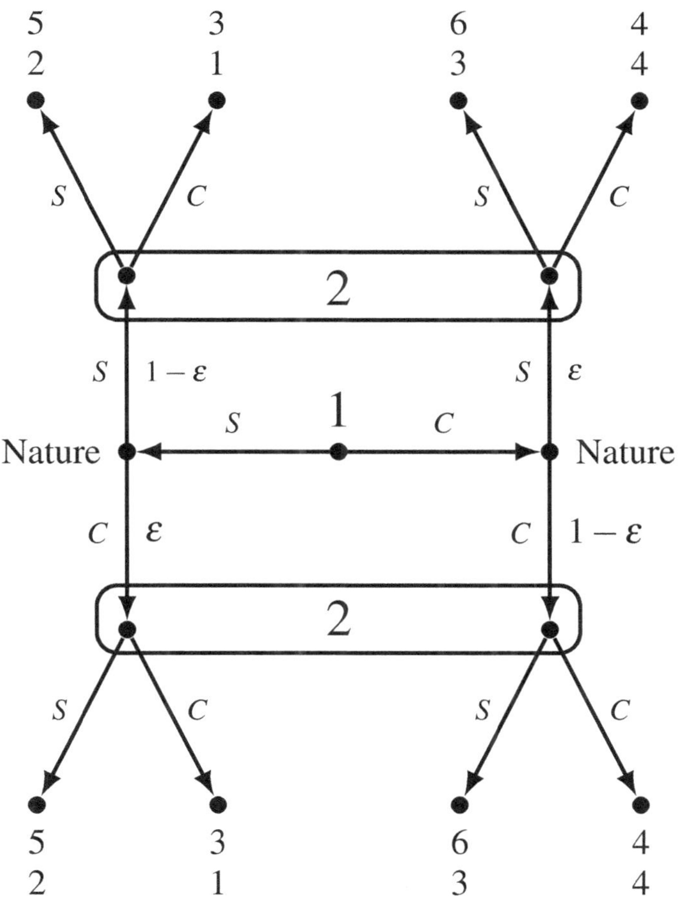

**2.** The corresponding strategic form is as follows. A strategy $(x,y)$ for Player 2 is to be interpreted as "if the message is '$S$' then play $x$ and if the message is '$C$' then play $y$".

| | | Player 2 | | | | | | | |
|---|---|---|---|---|---|---|---|---|---|
| | | $SS$ | | $SC$ | | $CS$ | | $CC$ | |
| Player 1 | $S$ | 5 | 2 | $5-2\varepsilon$ | $2-\varepsilon$ | $3+2\varepsilon$ | $1+\varepsilon$ | 3 | 1 |
| | $C$ | 6 | 3 | $4+2\varepsilon$ | $4-\varepsilon$ | $6-2\varepsilon$ | $3+\varepsilon$ | 4 | 4 |

**3.** No matter the value of $\varepsilon$, the only pure-strategy Nash equilibrium is $(C, (C,C))$ with payoffs $(4,4)$.

**4.** What is surprising is that, if $\varepsilon$ is very small, then we are very close to the game of Part (b) where Player 1 gains from being the first mover. Yet, with even a tiny probability that his first move is not perfectly observed by Player 2, we are back to the outcome of the simultaneous game.

Exercise 7.16 Find the subgame-perfect equilibria (including the mixed strategy ones) of the extensive-form game with cardinal payoffs shown in Figure 7.6 and calculate the payoff of every player at each subgame-perfect equilibrium. ■

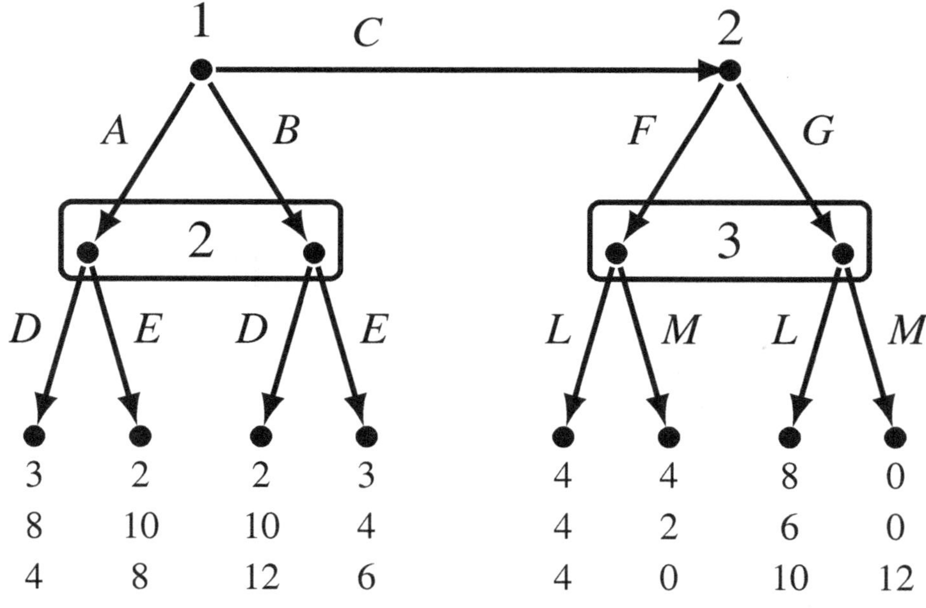


Figure 7.6: The game for Exercise 7.16.

**Solution to Exercise 7.16.** There is only one proper subgame, starting from the node of Player 2 after choice C of Player 1. The strategic form of that game is

| | | Player 3 | |
|---|---|---|---|
| | | $L$ | $M$ |
| Player 2 | $F$ | 4 4 | 2 0 |
| | $G$ | 6 10 | 0 12 |

This game does not have any pure-strategy Nash equilibria. To find the mixed-strategy Nash equilibrium, let $p$ be the probability of $F$ and $q$ the probability of $L$. Then $q$ must satisfy the equation $4q+2(1-q)=6q$. Similarly, $p$ must satisfy the equation $4p+10(1-p)=12(1-p)$. Thus $p=\frac{1}{3}$ and $q=\frac{1}{2}$. Hence the Nash equilibrium of the subgame is

$$\left(\begin{array}{cc|cc} F & G & L & M \\ \frac{1}{3} & \frac{2}{3} & \frac{1}{2} & \frac{1}{2} \end{array}\right)$$

with corresponding expected payoffs of 3 for Player 2 and 8 for Player 3. The expected payoff of Player 1 in the subgame is $4\left(\frac{1}{6}\right)+4\left(\frac{1}{6}\right)+8\left(\frac{2}{6}\right)+0\left(\frac{2}{6}\right)=4$. Thus the game can be simplified as follows:

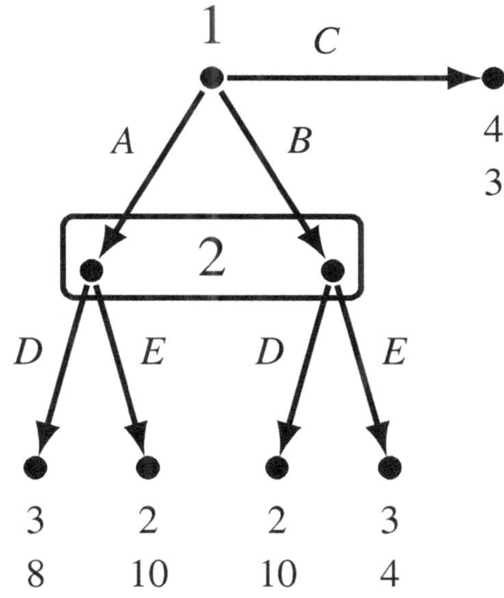


The strategic form of this reduced game is as follows:

| | | Player 2 | |
|---|---|---|---|
| | | $D$ | $E$ |
| Player 1 | $A$ | 3 8 | 2 10 |
| | $B$ | 2 10 | 3 4 |
| | $C$ | 4 3 | 4 3 |

In this game, for Player 1, both $A$ and $B$ are strictly dominated by $C$. Thus, at a Nash equilibrium $C$ must be played with probability 1 and, for Player 2, any mixture of $D$ and $E$ is a best reply. Hence, for every $p \in [0,1]$, the following is a Nash equilibrium of this game:

$$\begin{pmatrix} A & B & C & D & E \\ 0 & 0 & 1 & p & 1-p \end{pmatrix}.$$

Thus, in conclusion, the subgame perfect equilibria of the game of Figure 7.6 are as follows: for every $0 \le p \le 1$,

$$\begin{pmatrix} A & B & C & D & E & F & G & L & M \\ 0 & 0 & 1 & p & 1-p & \frac{1}{3} & \frac{2}{3} & \frac{1}{2} & \frac{1}{2} \end{pmatrix}.$$

The payoffs are the same for all the equilibria, namely $(4,3,8)$.

**Exercise 7.17** Consider the extensive-form game-frame shown in Figure 7.7. The preferences of all three players over lotteries over the set $Z = \{z_1, z_2, \ldots, z_8\}$ satisfy the axioms of Expected Utility Theory. Player 1's ranking of the basic outcomes is

$$z_1 \sim_1 z_3 \sim_1 z_5 \sim_1 z_7 \succ_1 z_2 \sim_1 z_4 \sim_1 z_6 \sim_1 z_8.$$

Player 2's ranking is:

$$z_3 \sim_2 z_8 \succ_2 z_5 \sim_2 z_6 \succ_2 z_4 \succ_2 z_1 \sim_2 z_2 \sim_2 z_7.$$

Furthermore, she is indifferent between $z_6$ and the lottery $\begin{pmatrix} z_1 & z_3 \\ 0.2 & 0.8 \end{pmatrix}$ and is also indifferent between $z_4$ and the lottery $\begin{pmatrix} z_3 & z_4 & z_7 \\ 0.2 & 0.5 & 0.3 \end{pmatrix}$.

Player 3's ranking is:

$$z_2 \sim_3 z_6 \succ_3 z_1 \sim_3 z_7 \succ_3 z_3 \sim_3 z_5 \succ_3 z_4 \sim_3 z_8.$$

Furthermore, he is indifferent between $z_7$ and the lottery $\begin{pmatrix} z_2 & z_6 & z_8 \\ 0.4 & 0.4 & 0.2 \end{pmatrix}$ and is also indifferent between $z_5$ and the lottery $\begin{pmatrix} z_6 & z_8 \\ 0.4 & 0.6 \end{pmatrix}$.

**(a)** Construct a von Neumann-Morgenstern utility function $U_1$ that represents Player 1's preferences, such that $U_1$ assigns utility 10 to the best outcome(s) and 0 to the worst outcome(s).

**(b)** Construct a von Neumann-Morgenstern utility function $U_2$ that represents Player 2's preferences, such that $U_2$ assigns utility 10 to the best outcome(s) and 0 to the worst outcome(s).

**(c)** Construct a von Neumann-Morgenstern utility function $U_3$ that represents Player 3's preferences, such that $U_3$ assigns utility 10 to the best outcome(s) and 0 to the worst outcome(s).

**(d)** Find the subgame perfect equilibrium of the corresponding extensive-form game. ■

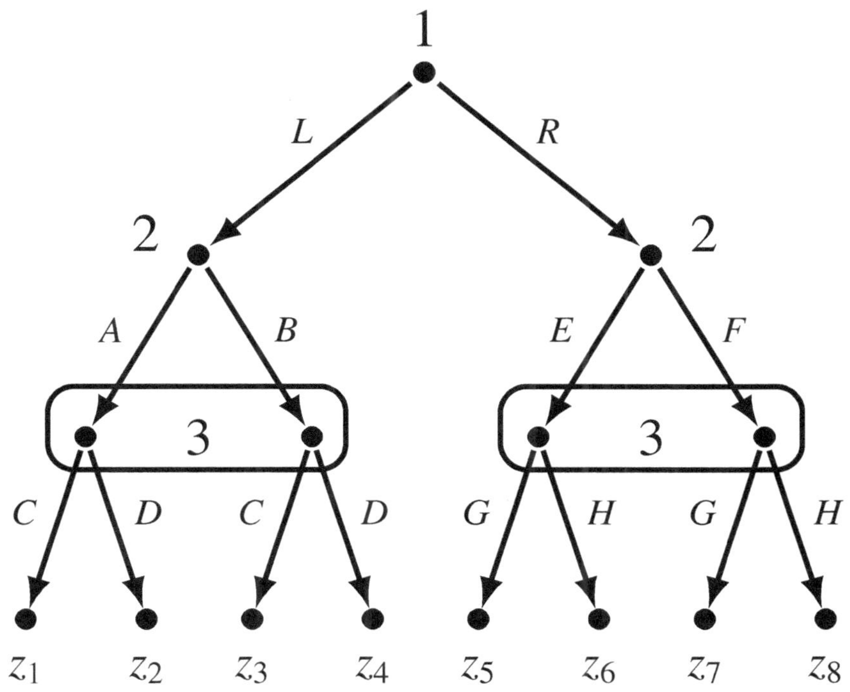

1
L R
2 2
A B E F
3 3
C D C D G H G H
$z_1$ $z_2$ $z_3$ $z_4$ $z_5$ $z_6$ $z_7$ $z_8$

Figure 7.7: The game for Exercise 7.17.

**Solution to Exercise 7.17.**

**(a)** $U_1(z_1) = U_1(z_3) = U_1(z_5) = U_1(z_7) = 10$, $U_1(z_2) = U_1(z_4) = U_1(z_6) = U_1(z_8) = 0$.

**(b)** First of all, $U_2(z_3) = U_2(z_8) = 10$ and $U_2(z_1) = U_2(z_2) = U_2(z_7) = 0$. From the first indifference we get that $U_2(z_5) = U_2(z_6) = 8$ and from the second indifference we get that $U_2(z_4) = 4$.

**(c)** First of all, $U_3(z_2) = U_3(z_6) = 10$ and $U_3(z_4) = U_3(z_8) = 0$. From the first indifference we get that $U_3(z_1) = U_3(z_7) = 8$ and from the second indifference we get that $U_3(z_3) = U_3(z_5) = 4$.

Thus, replacing the basic outcomes with the corresponding utilities we obtain the following game:

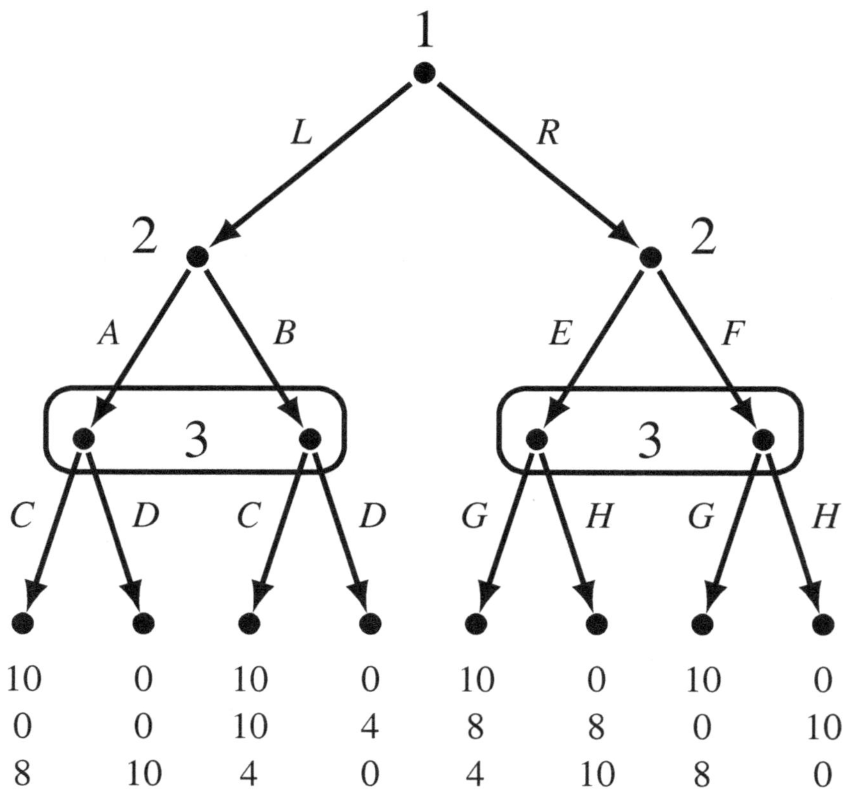


**(d)** To find the subgame-perfect equilibrium, let us first solve the proper subgame on the left, whose strategic form is as follows:

| | | Player 3: $C$ | Player 3: $D$ |
|---|---|---|---|
| Player 2 | $A$ | 0   8 | 0   10 |
| | $B$ | 10   4 | 4   0 |

The unique Nash equilibrium of this game is $(B,C)$.

Next we solve the proper subgame on the right, whose strategic form is as follows:

| | | Player 3: $G$ | Player 3: $H$ |
|---|---|---|---|
| Player 2 | $E$ | 8   4 | 8   10 |
| | $F$ | 0   8 | 10   0 |

This game has no pure-strategy Nash equilibria. To find the mixed-strategy equilibrium, let $p$ be the probability of $E$ and $q$ the probability of $G$. Then it must be that $8 = 10(1-q)$, that is, $q = \frac{1}{5}$ and $4p + 8(1-p) = 10p$, that is, $p = \frac{4}{7}$. Thus the mixed-strategy equilibrium is

$$\left(\begin{array}{cc|cc} E & F & G & H \\ \frac{4}{7} & \frac{3}{7} & \frac{1}{5} & \frac{4}{5} \end{array}\right).$$

The expected payoffs of the players in this subgame are: 2 for Player 1, 8 for Player 2 and $\frac{40}{7}$ for Player 3. Thus the original game reduces to:

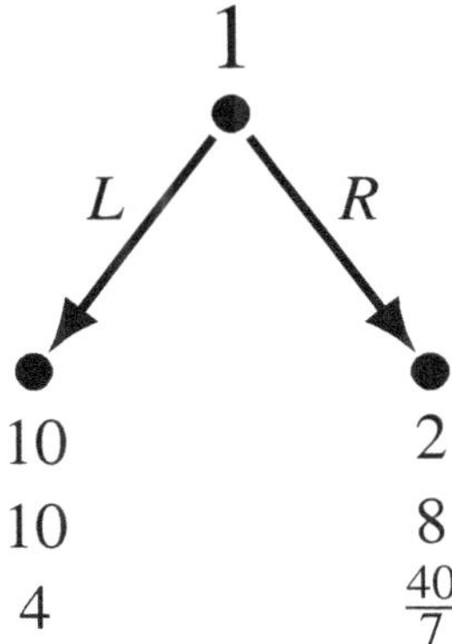


where $L$ is the optimal choice for Player 1.

Thus the subgame-perfect equilibrium of the original game is

$$\left(\begin{array}{cc|cc|cc|cc|cc} L & R & A & B & E & F & C & D & G & H \\ 1 & 0 & 0 & 1 & \frac{4}{7} & \frac{3}{7} & 1 & 0 & \frac{1}{5} & \frac{4}{5} \end{array}\right).$$

Exercise 7.18 Consider first the strategic-form game shown in Figure 7.8, where the payoffs are von Neumann-Morgenstern payoffs.

**(a)** Find the Nash equilibrium.

Now suppose that Player 3 is first given the choice between "play" and "not play". If she chooses "not play" then she gets a payoff of 0.7 while the other two players get a payoff of 0; if she chooses "play" then the three players play the simultaneous game of Figure 7.8.

**(b)** Draw an extensive-form game that represents this interaction.

**(c)** Find the subgame-perfect equilibrium.

■

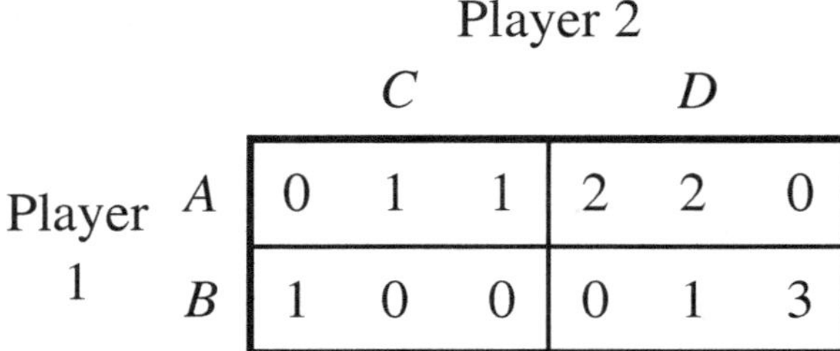

Player 2

| Player 1 | C | D |
|---|---|---|
| A | 0 1 1 | 2 2 0 |
| B | 1 0 0 | 0 1 3 |

Player 3: $E$

Player 2

| Player 1 | C | D |
|---|---|---|
| A | 2 0 0 | 0 1 1 |
| B | 1 2 1 | 4 3 0 |

Player 3: $F$

Figure 7.8: The game for Exercise 7.18.

**Solution to Exercise 7.18.**

**(a)** For Player 2, $C$ is strictly dominated by $D$. Thus, at a Nash equilibrium, Player 2 must play $D$ with probability 1. Hence the game can be reduced as follows:

Player 3

| Player 1 | E | F |
|---|---|---|
| A | 2 0 | 0 1 |
| B | 0 3 | 4 0 |

This game has no pure-strategy Nash equilibria. To find the mixed-strategy Nash equilibrium, let $p$ be the probability of $A$ and $q$ the probability of E. Then $p$ and $q$ must be the solutions to: $2q = 4(1-q)$ and $3(1-p) = p$. Thus $p = \frac{3}{4}$ and $q = \frac{2}{3}$. Hence the Nash equilibrium of the original game is

$$\left(\begin{array}{cc|cc|cc} A & B & C & D & E & F \\ \frac{3}{4} & \frac{1}{4} & 0 & 1 & \frac{2}{3} & \frac{1}{3} \end{array}\right).$$

Player 1's expected payoff is $\frac{4}{3}$, Player 2's expected payoff is $\frac{5}{3}$ and Player 3's expected payoff is $\frac{3}{4} = 0.75$.

**(b)** The extensive-form game is as follows, where, at each terminal node, the top number is Player 1's payoff, the middle number Player 2's payoff and the bottom number Player 3's payoff:

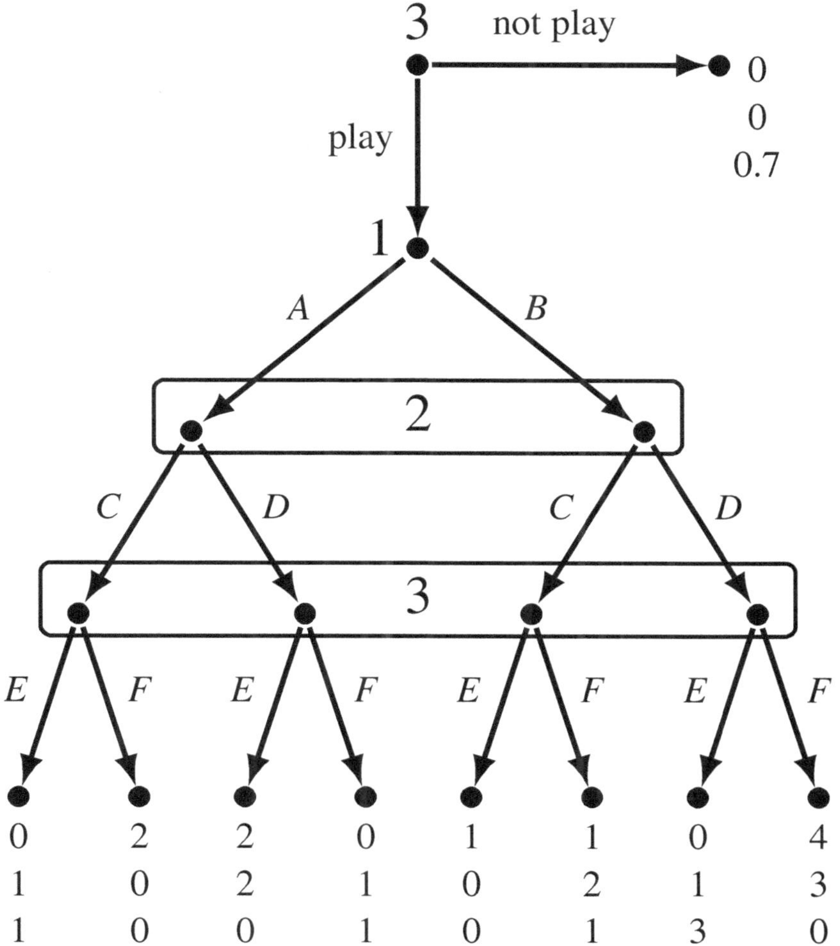


The Nash equilibrium of the subgame was calculated in Part (a); since Player 3's payoff in the subgame is 0.75, she will choose "play". Thus the subgame-perfect equilibrium is given by

$$\left(\begin{array}{cc|cc|cc|cc} A & B & C & D & \text{play} & \text{not play} & E & F \\ \frac{3}{4} & \frac{1}{4} & 0 & 1 & 1 & 0 & \frac{2}{3} & \frac{1}{3} \end{array}\right).$$

Exercise 7.19 Consider the extensive-form game of Figure 7.9.

**(a)** How many proper subgames are there?

**(b)** List all the pure strategies of Player 1 and all the pure strategies of Player 2.

**(c)** Find the subgame-perfect equilibrium.

■

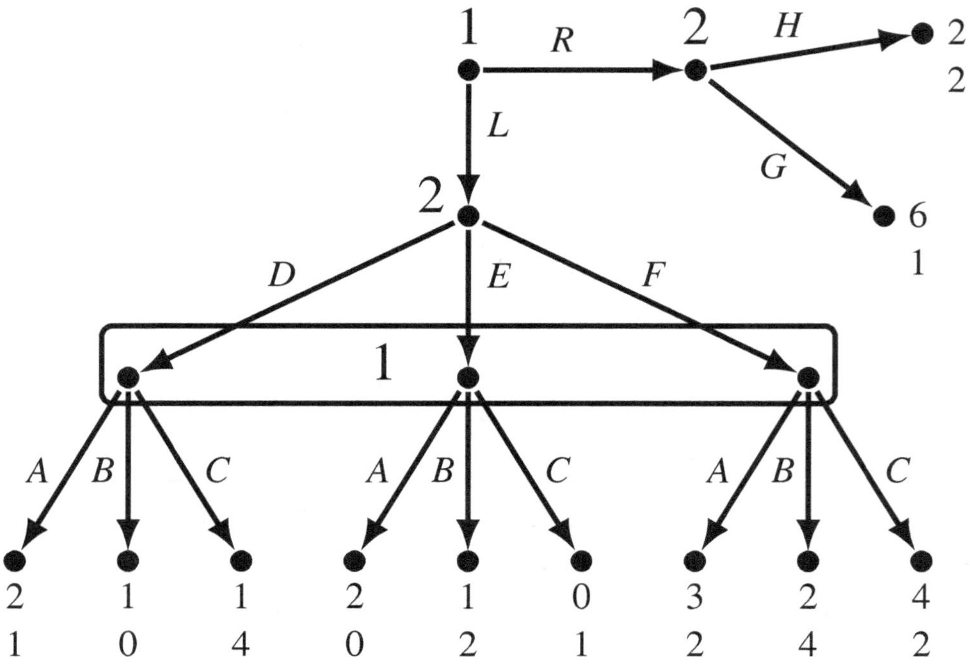


Figure 7.9: The game for Exercise 7.19.

**Solution to Exercise 7.19.**

**(a)** Two: one starting at the left node of Player 2 and the other at the right node of Player 2.

**(b)** Player 1's strategies: $LA, LB, LC, RA, RB, RC$.

Plater 2's strategies: $GD, HD, GE, HE, GF, HF$.

**(c)** First we solve the subgames. The subgame on the right is a trivial game, whose solution is for Player 2 to choose $H$. The strategic form of the subgame on the left is as follows:

| | | Player 2 | | |
|---|---|---|---|---|
| | | $D$ | $E$ | $F$ |
| Player 1 | $A$ | 2 1 | 2 0 | 3 2 |
| | $B$ | 1 0 | 1 2 | 2 4 |
| | $C$ | 1 4 | 0 1 | 4 2 |

This game does not have any pure-strategy Nash equilibria. Thus we need to find a mixed-strategy equilibrium. Note that a strictly dominated strategy can never be played with positive probability at a mixed-strategy Nash equilibrium. Since, for Player 1, $B$ is strictly dominated (by $A$), Player 1 must assign zero probability to $B$. Similarly, $E$ is strictly dominated for Player 2 (by $F$), thus Player 2 must assign zero probability to $E$. Thus a mixed-strategy equilibrium must be of the form

$$\left(\begin{array}{ccc|ccc} A & B & C & D & E & F \\ p & 0 & 1-p & q & 0 & 1-q \end{array}\right).$$

To find the equilibrium we need to solve the following two equations: $2q+3(1-q)=q+4(1-q)$ and $p+4(1-p)=2p+2(1-p)$ . The solution is: $p=\frac{2}{3}$, and $q=\frac{1}{2}$. Thus the mixed-strategy Nash equilibrium of the subgame is given by:

$$\left(\begin{array}{ccc|ccc} A & B & C & D & E & F \\ \frac{2}{3} & 0 & \frac{1}{3} & \frac{1}{2} & 0 & \frac{1}{2} \end{array}\right)$$

Player 1's expected payoff is 2.5 and Player 2's expected payoff is 2. Thus Player 1 will want to choose $L$. Hence the subgame-perfect equilibrium of the entire game, given in terms of behavioral strategies, is

$$\left(\begin{array}{ccccc|ccccc} L & R & A & B & C & D & E & F & G & H \\ 1 & 0 & \frac{2}{3} & 0 & \frac{1}{3} & \frac{1}{2} & 0 & \frac{1}{2} & 0 & 1 \end{array}\right)$$

with expected payoffs of 2.5 for Player 1 and 2 for Player 2. In terms of mixed strategies the equilibrium is given by

$$\left(\begin{array}{cccccccccccc} LA & LB & LC & RA & RB & RC & GD & GE & GF & HD & HE & HF \\ \frac{2}{3} & 0 & \frac{1}{3} & 0 & 0 & 0 & 0 & 0 & 0 & \frac{1}{2} & 0 & \frac{1}{2} \end{array}\right).$$

## 7.3 More difficult exercises

The exercises in this section are more difficult and challenging than the exercises in the previous sections.

**Exercise 7.20** Consider the following situation. There are three people: Amy, Bart and Chris. Amy and Bart have hats in their hands. These three people are arranged in a room so that only the following is true: Bart can see Amy and Chris can see Bart; thus, Chris cannot see Amy, Amy can see neither Bart nor Chris and Bart cannot see Chris.

The game is as follows. First, Amy chooses either to put her hat on her head (abbreviated by H) or on the floor (F).

After observing Amy's move, Bart chooses between putting his hat on his head or on the floor.

If Bart puts his hat on his head, then everybody learns this, the game ends and everyone gets a payoff of 1. If Bart puts his hat on the floor then Chris must guess whether Amy's hat is on her head or on the floor by saying either "head" or "floor". This ends the game. If Chris guesses correctly then he gets a payoff of 2 and Amy gets a payoff of 0. If Chris guesses incorrectly, then these payoffs are reversed. Bart's payoff is the same as Amy's. All of these payoffs are von Neumann-Morgenstern payoffs.

**(a)** Represent this situation as an extensive-form game.

**(b)** Write the corresponding strategic form.

**(c)** Find the pure-strategy Nash equilibria, and at least three Nash equilibria that are not in pure strategies (that is, where at least one player does no play a pure strategy).

**(d)** Are all the Nash equilibria also subgame-perfect?

■

**Solution to Exercise 7.20.**

**(a)** The extensive-form game is as follows:

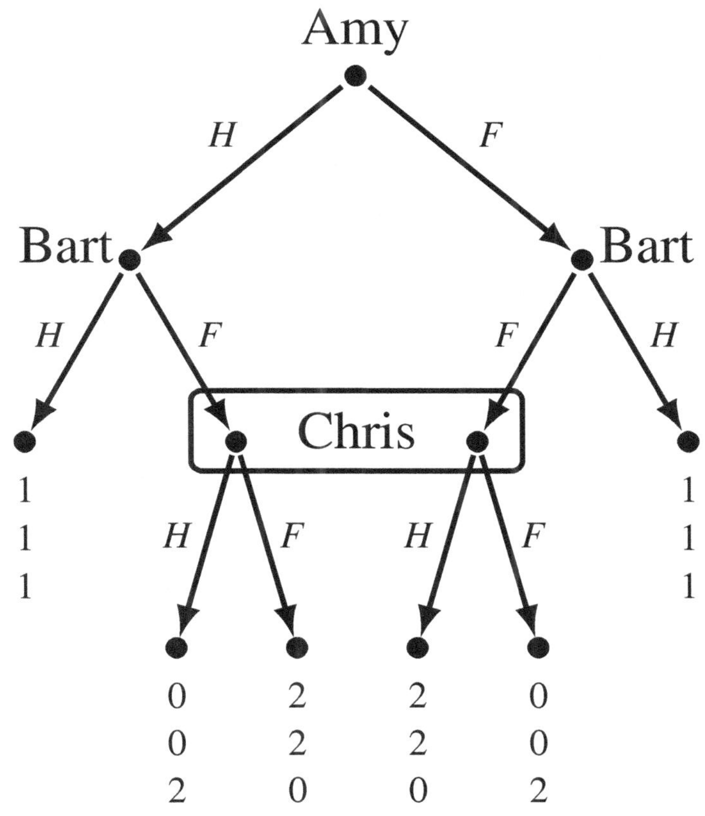


**(b)** The strategic form is as follows:

BART

| AMY | $HH$ | $HF$ | $FH$ | $FF$ |
|---|---|---|---|---|
| $H$ | 1 1 1 | 1 1 1 | 0 0 2 | 0 0 2 |
| $F$ | 1 1 1 | 2 2 0 | 1 1 1 | 2 2 0 |

CHRIS: $H$

BART

| AMY | $HH$ | $HF$ | $FH$ | $FF$ |
|---|---|---|---|---|
| $H$ | 1 1 1 | 1 1 1 | 2 2 0 | 2 2 0 |
| $F$ | 1 1 1 | 0 0 2 | 1 1 1 | 0 0 2 |

CHRIS: $F$

**(c)** The pure-strategy Nash equilibria are $(H, HH, H)$ and $(F, HH, F)$.

To find the mixed-strategy equilibria, first think intuitively. If Chris chooses $H$ and $F$ with equal probability, then Bart gets an expected payoff of 1 no matter what Amy does and no matter what he himself does (at his two decision nodes). Since Amy has the same payoff as Bart, Amy also gets an expected payoff of 1 no matter what she herself does and no matter what Bart does. When will Chris want to choose $H$ and $F$ with equal probability? When (either his information set is not reached or) when the two nodes in his information set have the same probability.

Let us see this by reasoning in terms of behavioral strategies. Let $p$ be the probability of $H$ for Amy, $q$ the probability of $H$ for Bart at the left node, $s$ the probability of $H$ for Bart at the right node and $r$ the probability of $H$ for Chris at his information set. Then the probability of Chris's left-hand node is $p(1-q)$ and the probability of the right-hand node is $(1-p)(1-s)$. Then $r, p, q, s$, such that $r = \frac{1}{2}$ and $p(1-q) = (1-p)(1-s)$ gives a Nash equilibrium. For example $p = q = r = s = \frac{1}{2}$ is a Nash equilibrium, that is, the behavioral-strategy profile where, at each information set, each player plays both actions with equal probability is a Nash equilibrium.

There are also other mixed-behavioral-strategy equilibria with $r \neq \frac{1}{2}$. For example,

- $(p,q,s,r)$ with $p = q = s = 1$ and $1/2 < r < 1$.
- $(p,q,s,r)$ with $p = 0,\ q = s = 1$ and $0 < r < 1/2$.

One can also see this in the strategic form in terms of mixed strategies. For example the first behavioral strategy profile described above is equivalent to the following mixed-strategy profile:

$$\left(\begin{array}{cc|cccc|cc} H & F & HH & HF & FH & FF & H & F \\ \frac{1}{2} & \frac{1}{2} & \frac{1}{4} & \frac{1}{4} & \frac{1}{4} & \frac{1}{4} & \frac{1}{2} & \frac{1}{2} \end{array}\right).$$

Given the mixed strategies of the opponents,

- Amy's expected payoff from $H$ is 1 and so is her expected payoff from $L$,
- Bart's expected payoff from $HH$ is 1 and so is his expected payoff from $HF$ and from $FH$ and from $FF$,
- Chris's expected payoff from $H$ is 1 and so his expected payoff from $F$.

**Exercise 7.21** The CEO of a company is worried because there has been an ongoing conflict between the two vice-presidents: VP1 and VP2. It is not clear who is at fault and, of course, each VP blames the other. The CEO puts pressure on the two to reach a peaceful settlement. She asks VP1, who is the more senior of the two, to make a serious "peace offering" to VP2. Of course, VP1 can just pretend to oblige. If VP1 makes a serious proposal for peace, VP2 can either accept, in which case peace is achieved and the game ends, or reject the proposal. If peace is not achieved then the CEO cannot tell whether it was because VP1 did not make a good-faith effort or because VP2 rejected the serious proposal of VP1 (each would claim it was the other party's lack of cooperation). If peace is not achieved, the CEO – not knowing the true cause – has to decide whether to demote VP1 or demote VP2: she makes a public announcement that, in the absence of peace, one of the two VPs will be demoted, without specifying in advance whether it will be VP1 or VP2. Denote the possible outcomes be as follows:

$z_1$ : Peace is achieved.

$z_2$ : Peace is not achieved because VP2 rejected the serious proposal of VP1 and this is followed by the CEO demoting VP1.

$z_3$ : Peace is not achieved because VP2 rejected the serious proposal of VP1 and this is followed by the CEO demoting VP2.

$z_4$ : Peace is not achieved because VP1 did not make a good-faith proposal to VP2 and this is followed by the CEO demoting VP1.

$z_5$ : Peace is not achieved because VP1 did not make a good-faith proposal to VP2 and this is followed by the CEO demoting VP2.

**(a)** Draw a three-player extensive-form game-frame that represents this situation. Note: if VP1 does not make a serious proposal, then there is no need to have VP2 react to it.

**(b)** Write the corresponding strategic-form game-frame.

Now add the following von Neumann-Morgenstern preferences over lotteries over the set $Z = \{z_1, z_2, z_3, z_4, z_5\}$. First of all, the players rank the basic outcomes as follows:

$$\begin{array}{rl} \text{VP1:} & z_3 \succ z_5 \succ z_1 \succ z_4 \succ z_2 \\ \text{VP2:} & z_2 \sim z_4 \succ z_1 \succ z_3 \sim z_5 \\ \text{CEO:} & z_1 \succ z_3 \sim z_4 \succ z_2 \sim z_5. \end{array}$$

Furthermore,

- VP1 is indifferent between the following two lotteries:

$$\begin{pmatrix} z_2 & z_3 \\ \frac{1}{3} & \frac{2}{3} \end{pmatrix} \quad \text{and} \quad \begin{pmatrix} z_5 \\ 1 \end{pmatrix}$$

and is also indifferent between the following two lotteries:

$$\begin{pmatrix} z_2 & z_5 \\ \frac{1}{3} & \frac{2}{3} \end{pmatrix} \quad \text{and} \quad \begin{pmatrix} z_1 \\ 1 \end{pmatrix}$$

and is also indifferent between the following two lotteries:

$$\begin{pmatrix} z_1 & z_2 \\ \frac{3}{4} & \frac{1}{4} \end{pmatrix} \quad \text{and} \quad \begin{pmatrix} z_4 \\ 1 \end{pmatrix}.$$

- VP2 is indifferent between the following two lotteries:

$$\begin{pmatrix} z_2 & z_3 & z_4 & z_5 \\ \frac{1}{3} & \frac{1}{9} & \frac{1}{9} & \frac{4}{9} \end{pmatrix} \quad \text{and} \quad \begin{pmatrix} z_1 \\ 1 \end{pmatrix}.$$

- The CEO is indifferent between the following two lotteries:

$$\begin{pmatrix} z_1 & z_2 & z_3 & z_5 \\ \frac{1}{8} & \frac{1}{4} & \frac{1}{2} & \frac{1}{8} \end{pmatrix} \quad \text{and} \quad \begin{pmatrix} z_4 \\ 1 \end{pmatrix}.$$

**(c)** Representing the above preferences with von Neumann-Morgenstern utility functions with 0 as the lowest utility and 36 as the largest utility, convert the strategic-form game-frame of Part (b) into a game.

**(d)** Find all the pure-strategy Nash equilibria of the game of Part (c).

**(e)** Find a mixed-strategy Nash equilibrium at which VP2's strategy is to reject the serious proposal with probability 1, while the other two players do not employ a pure strategy.

**(f)** What are the players' expected payoffs at the mixed-strategy Nash equilibrium of Part (e)?

**(g)** Write a system of equations whose solution gives the completely mixed-strategy Nash equilibrium (that is, the Nash equilibrium where every strategy is played with positive probability).

**Solution to Exercise 7.21.**

**(a)** The extensive-form game-frame is as follows, where 'P' means "propose seriously", 'N' means "not serious proposal", 'A' means "accept", 'R' means "reject", 'D1' means "demote VP1" and 'D2' means "demote VP2".

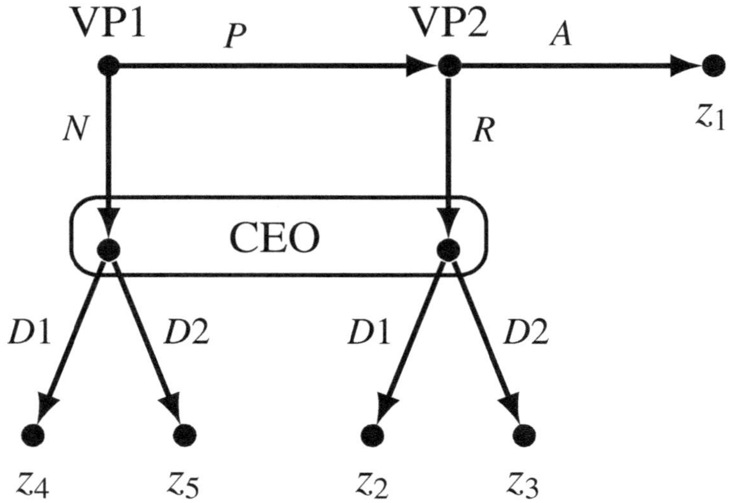

**(b)** The corresponding strategic-form game-frame is as follows:

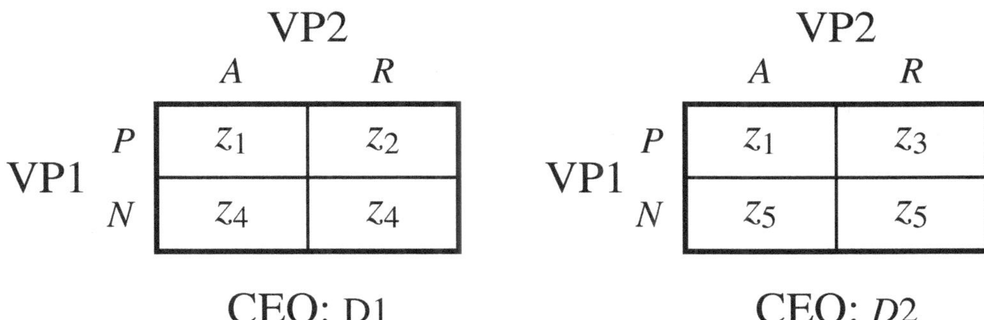

| VP1 \ VP2 | $A$ | $R$ |
|---|---|---|
| $P$ | $z_1$ | $z_2$ |
| $N$ | $z_4$ | $z_4$ |

CEO: D1

| VP1 \ VP2 | $A$ | $R$ |
|---|---|---|
| $P$ | $z_1$ | $z_3$ |
| $N$ | $z_5$ | $z_5$ |

CEO: $D2$

**(c)** Let us begin with VP1: $U_{VP1}(z_3) = 36$ and $U_{VP1}(z_2) = 0$. From the first indifference we get that $U_{VP1}(z_5) = \frac{1}{3} \times 0 + \frac{2}{3} \times 36 = 24$. From the second indifference we get that $U(z_1) = \frac{2}{3} \times 24 = 16$. From the last indifference we get that $U(z_4) = \frac{3}{4} \times 16 = 12$.

Next VP2. $U_{VP2}(z_2) = U_{VP2}(z_4) = 36$ and $U_{VP2}(z_3) = U_{VP2}(z_5) = 0$. From the indifference we get that $U_{VP2}(z_1) = \frac{1}{3} \times 36 + \frac{1}{9} \times 36 = 16$.

Finally the CEO. $U_{CEO}(z_1) = 36$ and $U_{CEO}(z_2) = U_{CEO}(z_5) = 0$. From the indifference we have that $\frac{1}{8} \times 36 + \frac{1}{2} \times U_{CEO}(z_3) = U_{CEO}(z_4)$, so that (since $U_{CEO}(z_3) = U_{CEO}(z_4)$) $U_{CEO}(z_3) = U_{CEO}(z_4) = 9$.

Thus the strategic-form game is as follows:

| VP1 \ VP2 | $A$ | $R$ |
|---|---|---|
| $P$ | 16 16 36 | 0 36 0 |
| $N$ | 12 36 9 | 12 36 9 |

CEO: D1

| VP1 \ VP2 | $A$ | $R$ |
|---|---|---|
| $P$ | 16 16 36 | 36 0 9 |
| $n$ | 24 0 0 | 24 0 0 |

CEO: $D2$

**(d)** There is only one pure-strategy equilibrium: (N, R, D1), with payoffs (12, 36, 9).

**(e)** Suppose that VP2 chooses $R$ with probability 1. Let $p \in (0,1)$ be the probability that VP1 chooses $P$ and let $q \in (0,1)$ be the probability that the CEO chooses $D1$. Then the following must be true:

- VP1 must be indifferent between $P$ and $N$: $(1-q)36 = q12 + (1-q)24$, which gives $q = \frac{1}{2}$.
- The CEO must be indifferent between $D1$ and $D2$: $(1-p)9 = p9$, which gives $p = \frac{1}{2}$.

Given these probabilities, VP2 indeed strictly prefers $R$ (which yields an expected payoff of 18) to $A$ (which yields an expected payoff of 17). Thus the mixed-strategy Nash equilibrium is $\begin{pmatrix} P & N & \vline & A & R & \vline & D1 & D2 \\ \frac{1}{2} & \frac{1}{2} & \vline & 0 & 1 & \vline & \frac{1}{2} & \frac{1}{2} \end{pmatrix}$.

**(f)** The expected payoffs at the Nash equilibrium are as follows: VP1: 18, VP2: 18, CEO: $\frac{9}{2} = 4.5$.

**(g)** Let $p$ be the probability of $P$, $q$ the probability of $A$ and $r$ the probability of $D1$. Then $p$, $q$ and $r$ must be the solution to the following equations:

- $16q + 36(1-q)(1-r) = 12r + 24(1-r)$,
- $16p + 36(1-p)r = 36r$,
- $36pq + 9(1-p) = 36pq + 9p(1-q)$, that is, $1 - p = p(1-q)$.

(The solution is $p = \frac{3}{5}$, $q = \frac{1}{3}$, $r = \frac{4}{9}$.)

Exercise 7.22 Suppose that a player in an extensive-form game has 4 information sets. At one he has 3 choices, at another he has 2 choices, at the third he has 5 choices and at the fourth he has 4 choices.

**(a)** How many pure strategies does this player have?

**(b)** What is the dimension of her set of mixed strategies?

**(c)** What is the dimension of her set of behavior strategies?

**(d)** Now answer (a)-(c) for the general case where the player has $m$ information sets and that at his $k^{th}$ information set she can choose among $n_k$ actions.

**Solution to Exercise 7.22.**

**(a)** $3 \times 2 \times 5 \times 4 = 120$.

**(b)** The set of mixed strategies is the simplex in $\mathbb{R}^{120}$ which has dimension 119. (If $A = \{a_1, \ldots, a_n\}$ is a finite set with $n$ elements, then the set of probability distributions over $A$, denoted by $\Delta(A)$, is the simplex in $\mathbb{R}^n$, which has dimension $(n-1)$: once you specify the probabilities $p(a_1)$ through $p(a_{n-1})$, $p(a_n)$ is determined by the constraint that the sum of the probabilities must be equal to 1.)

**(c)** It is the sum of the dimension of the simplex in $\mathbb{R}^3$ plus the dimension of the simplex in $\mathbb{R}^2$, etc. Thus it is 2 + 1 + 4 + 3 = 10. A much smaller space that the space of mixed strategies!
(The dimension of the Cartesian product of a collection of subsets of Euclidean spaces is the sum of the dimensions of the sets.)

**(d)** Denote by $B_k$ the set of choices at this player's $k^{th}$ information set ($B_k$ is a finite set with $n_k$ elements). Then a behavior strategy at the $k^{th}$ information set is an element of $\Delta(B_k)$, whose dimension is $(n_k - 1)$.

Thus, (a) $\prod_{k=1}^{m} n_k$. (b) $\left(\prod_{k=1}^{m} n_k\right) - 1$. (c) $\sum_{k=1}^{n} (n_k - 1)$.

Exercise 7.23 Consider the extensive-form game-frame shown in Figure 7.10. The preferences of all the players satisfy the axioms of Expected Utility Theory. The players rank the outcomes as indicated below (if basic outcome $w$ is above basic outcome $y$ then $w$ is strictly preferred to $y$, and if $w$ and $y$ are written next to each other then the player is indifferent between the two):

$$Player\,1: \begin{pmatrix} z_7, z_9 \\ z_1, z_2, z_4, z_5 \\ z_{10} \\ z_3, z_6, z_8 \end{pmatrix} \quad Player\,2: \begin{pmatrix} z_1, z_3, z_{10} \\ z_4, z_5 \\ z_2, z_7, z_8 \\ z_6 \\ z_9 \end{pmatrix} \quad Player\,3: \begin{pmatrix} z_2, z_7, z_{10} \\ z_8 \\ z_1, z_4, z_9 \\ z_3, z_5, z_6 \end{pmatrix}.$$

Furthermore, Player 2 is indifferent between $z_4$ and the lottery $\begin{pmatrix} z_1 & z_2 \\ \frac{1}{2} & \frac{1}{2} \end{pmatrix}$ and Player 3 is indifferent between $z_1$ and the lottery $\begin{pmatrix} z_2 & z_5 \\ \frac{1}{2} & \frac{1}{2} \end{pmatrix}$.

Find the subgame-perfect equilibrium of the corresponding game.

[Hint: you do not have enough information to construct the von Neumann-Morgenstern utility functions of the players over the set $Z = \{z_1, z_2, \ldots, z_{10}\}$, but you do have enough information to find the subgame-perfect equilibrium.]

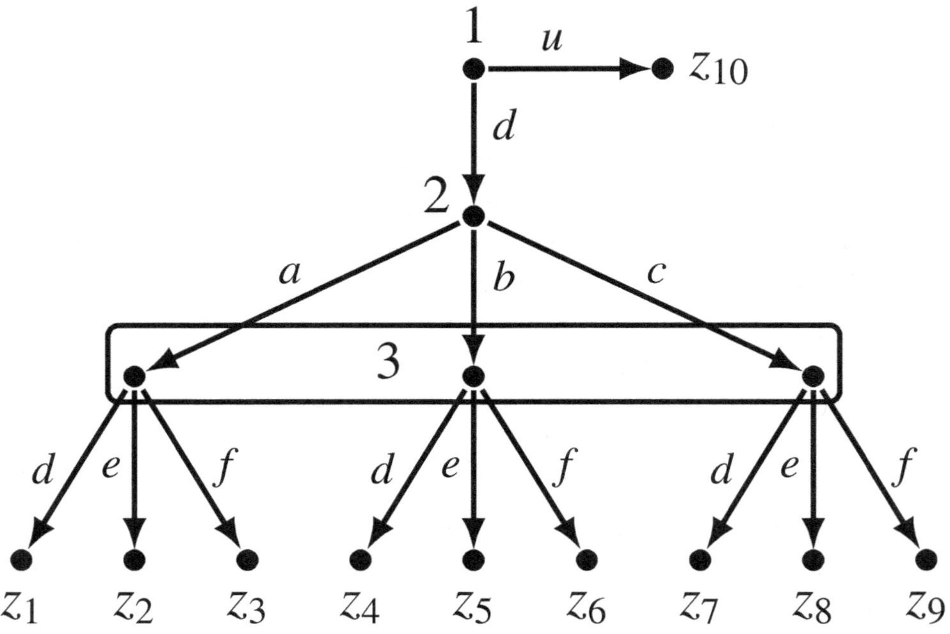


Figure 7.10: The game for Exercise 7.23.

**Solution to Exercise 7.23.** The strategic-form game-frame corresponding to the proper subgame is as follows:

| | | Player 3 | | |
|---|---|---|---|---|
| | | $d$ | $e$ | $f$ |
| Player 2 | $a$ | $z_1$ | $z_2$ | $z_3$ |
| | $b$ | $z_4$ | $z_5$ | $z_6$ |
| | $c$ | $z_7$ | $z_8$ | $z_9$ |

For Player 2 strategy $c$ is strictly dominated by strategy $b$, because $z_4 \succ_2 z_7$, $z_5 \succ_2 z_8$ and $z_6 \succ_2 z_9$. For Player 3 strategy $f$ is strictly dominated by strategy $d$, because $z_1 \succ_3 z_3$, $z_4 \succ_3 z_6$ and $z_7 \succ_3 z_9$. Thus, at a Nash equilibrium, $c$ and $f$ must be played with zero probability. Hence the game can be reduced as follows:

| | | Player 3 | |
|---|---|---|---|
| | | $d$ | $e$ |
| Player 1 | $a$ | $z_1$ | $z_2$ |
| | $b$ | $z_4$ | $z_5$ |

Restricted to these outcomes the payers' rankings are:

$$Player 2: \begin{pmatrix} z_1 \\ z_4, z_5 \\ z_2 \end{pmatrix} \qquad Player 3: \begin{pmatrix} z_2 \\ z_1, z_4 \\ z_5 \end{pmatrix}.$$

Let $U_2$ be Player 2's von Neumann-Morgenstern normalized utility function over the set $\{z_1, z_2, z_4, z_5\}$. Then $U_2(z_1) = 1$ and $U_2(z_2) = 0$. Furthermore, since Player 2 is indifferent between $z_4$ and $z_5$ and also between $z_4$ and the lottery $\begin{pmatrix} z_1 & z_2 \\ \frac{1}{2} & \frac{1}{2} \end{pmatrix}$, $U_2(z_4) = U_2(z_5) = \frac{1}{2}$.
Let $U_3$ be Player 3's von Neumann-Morgenstern normalized utility function over the set $\{z_1, z_2, z_4, z_5\}$. Then $U_3(z_2) = 1$ and $U_3(z_5) = 0$. Furthermore, since Player 3 is indifferent between $z_1$ and $z_4$ and also between $z_1$ and the lottery $\begin{pmatrix} z_2 & z_5 \\ \frac{1}{2} & \frac{1}{2} \end{pmatrix}$, $U_3(z_1) = U_3(z_4) = \frac{1}{2}$.
Hence the game is as follows (for convenience we have changed from the normalized utility function to a new function obtained by multiplying all the payoffs by 2):

| | | Player 3 | |
|---|---|---|---|
| | | $d$ | $e$ |
| Player 1 | $a$ | 2 1 | 0 2 |
| | $b$ | 1 1 | 1 0 |

This game has no pure-strategy Nash equilibria. Let $p$ be the probability of $a$ and $q$ the probability of $d$. Then for a Nash equilibrium we need $2q = 1$ and $1 = 2p$. Hence in the subgame the outcome will be $\begin{pmatrix} z_1 & z_2 & z_4 & z_5 \\ \frac{1}{4} & \frac{1}{4} & \frac{1}{4} & \frac{1}{4} \end{pmatrix}$. Since, for Player 1, all of these outcomes are better than $z_{10}$, Player 1 will play $d$. Thus the subgame-perfect equilibrium is

$$\left(\begin{array}{cc|ccc|ccc} d & u & a & b & c & d & e & f \\ 1 & 0 & \frac{1}{2} & \frac{1}{2} & 0 & \frac{1}{2} & \frac{1}{2} & 0 \end{array}\right).$$

Exercise 7.24 Consider the following situation. You go to a car dealership with the intention of buying a second-hand car. The dealer shows you a car. He knows what the quality of the car is. Let $\theta$ denote the quality of the car (measured in expected number of days of trouble-free running). You think that there is a $\frac{1}{3}$ probability that $\theta = 5,000$ and a $\frac{2}{3}$ probability that $\theta = 3,000$; the dealer knows that these are your beliefs. The dealer can either claim that the car is of quality $\theta = 5,000$ or that it is of quality $\theta = 3,000$; the dealer could be telling the truth or could be lying. After you have heard the dealer's claim, you decide whether to offer to pay $6,000 or $4,000. The dealer can either accept your offer or reject it. If he rejects it, you go back home without the car. Your von Neumann-Morgenstern utility function $U$ is as follows:

- if you buy a car of quality $\theta$ at price $p$: $U = \frac{\theta - p}{1,000} + 2$,
- if you don't buy the car: $U = 0$.

The dealer's von Neumann-Morgenstern utility function $V$ is:

- if he sells a car of quality $\theta$ to you at price $p$: $V = \frac{p-\theta}{1,000}$,
- if he does not sell the car to you: $V = 0$.

**(a)** Draw an extensive-form game to represent this situation.

**(b)** Find all the pure-strategy subgame-perfect equilibria.

**(c)** Find a mixed-strategy subgame-perfect equilibrium where You play a pure strategy while the Dealer does not play a pure strategy.

**(d)** Suppose that the true quality of the car is 5,000. Is there a pure-strategy subgame-perfect equilibrium at which you buy the car?

**(e)** Suppose that the true quality of the car is 3,000. Is there a pure-strategy subgame-perfect equilibrium at which you buy the car?

**Solution to Exercise 7.24.**

**(a)** The extensive-form game is as follows ('D' stands for "Dealer", 'A' for "Accept" and 'R' for "Reject"; at each terminal node, the top number is the Dealer's payoff and the bottom number is your payoff):

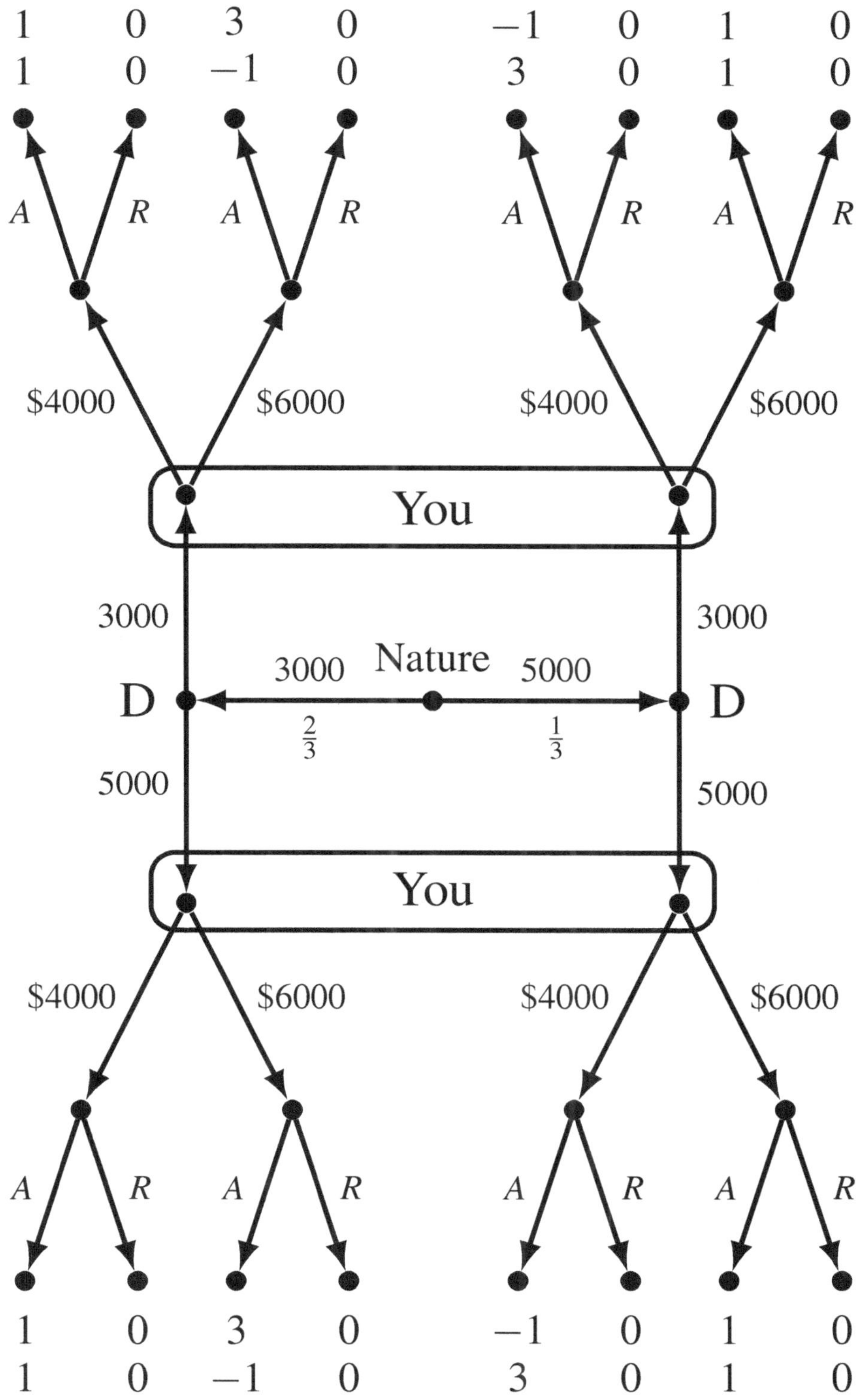


**(b)** At a subgame-perfect equilibrium the dealer whose car is of quality 5000 will Reject if and only if you offer to pay \$4000, while if his car is of quality 3000 then he will Accept any offer. Thus the game simplifies to the following:

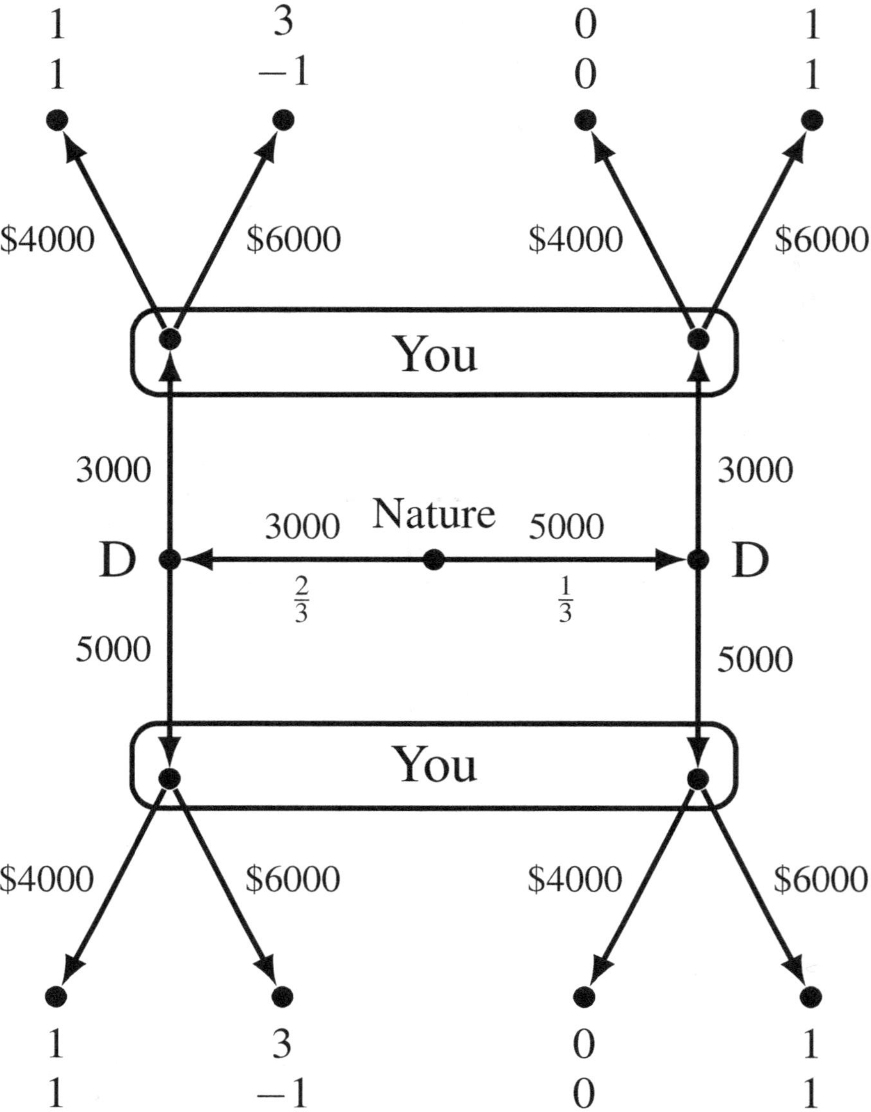


The strategic form corresponding to this reduced game is shown on the next page. This game has only two pure-strategy Nash equilibria:

- First equilibrium: the dealer always claims that the quality is 5000 and you always offer to pay $4000.
- Second equilibrium: the dealer always claims that the quality is 3000 and you always offer to pay $4000.

The pure-strategy subgame-perfect equilibria of the original game are obtained by augmenting the above strategies of the Dealer with the optimal Accept/Reject decisions explained above.

| Dealer \ You | \$6000 always | \$6000 if 5000<br>\$4000 if 3000 | \$4000 if 5000<br>\$6000 if 3000 | \$4000 always |
|---|---|---|---|---|
| 5000 always | $\frac{7}{3}$   $-\frac{1}{3}$ | $\frac{7}{3}$   $-\frac{1}{3}$ | $\frac{2}{3}$   $\frac{2}{3}$ | $\frac{2}{3}$   $\frac{2}{3}$ |
| 5000 if $\theta = 5000$<br>3000 if $\theta = 3000$ | $\frac{7}{3}$   $-\frac{1}{3}$ | 1   1 | 2   $-\frac{2}{3}$ | $\frac{2}{3}$   $\frac{2}{3}$ |
| 3000 if $\theta = 5000$<br>5000 if $\theta = 3000$ | $\frac{7}{3}$   $-\frac{1}{3}$ | 2   $-\frac{2}{3}$ | 1   1 | $\frac{2}{3}$   $\frac{2}{3}$ |
| 3000 always | $\frac{7}{3}$   $-\frac{1}{3}$ | $\frac{2}{3}$   $\frac{2}{3}$ | $\frac{7}{3}$   $-\frac{1}{3}$ | $\frac{2}{3}$   $\frac{2}{3}$ |

**(c)** For every $p$ such that $\frac{1}{4} \leq p \leq 1$, the following is a mixed strategy equilibrium: the dealer plays "5000 always" with probability "p" and "always tell the truth" (that is, "5000 if 5000 and 3000 if 3000") with probability $(1-p)$, and you choose "4,000 always" with probability 1 (of course, for this to be a subgame-perfect equilibrium of the original game, the dealer's strategy has to be augmented with the optimal Accept/Reject decisions explained above).

**(d)** and **(e)** Since, at a pure-strategy subgame-perfect equilibrium, the dealer will not accept your offer of \$4,000 if the true quality of the car is 5000, it follows that in equilibrium if you buy a car it must be one whose quality is 3000. Thus the answer to (d) is No and the answer to (e) is Yes.

www.ingramcontent.com/pod-product-compliance
Lightning Source LLC
Chambersburg PA
CBHW082202220526
45470CB00010B/3020

* 9 7 9 8 3 2 6 2 9 6 1 4 6 *